D1064140

Handbook of Electronic Test Procedures

TK 7870 .L4573
Lenk, John D.
Handbook of electronic test
procedures /

Handbook of Electronic Test Procedures

JOHN D. LENK

Consulting Technical Writer

PRENTICE-HALL, INC., *Englewood Cliffs, N.J. 07632*

Library of Congress Cataloging in Publication Data

Lenk, John D.
Handbook of electronic test procedures.

Includes index.
1. Electronic apparatus and appliances—Testing—Handbooks, manuals, etc. I. Title.
TK7870.L4573 621.381'043 81-5218
ISBN 0-13-377457-0 AACR2

Editorial/production supervision and
Interior design: Ellen DeFilippis Denning
Manufacturing buyer: Gordon Osbourne
Jacket design: Dawn L. Stanley

© 1982 by Prentice-Hall, Inc., Englewood Cliffs, N.J. 07623

All rights reserved. No part of this book may be reproduced in any form or by any means without permission in writing from the publisher.

Printed in the United States of America

10 9 8 7 6 5 4 3 2 1

PRENTICE-HALL INTERNATIONAL, INC., *London*
PRENTICE-HALL OF AUSTRALIA PTY. LIMITED, *Sydney*
PRENTICE-HALL OF CANADA, LTD., *Toronto*
PRENTICE-HALL OF INDIA PRIVATE LIMITED, *New Delhi*
PRENTICE-HALL OF JAPAN, INC., *Tokyo*
PRENTICE-HALL OF SOUTEAST ASIA PTE. LTD., *Singapore*
WHITEHALL BOOKS LIMITED, WELLINGTON, *New Zealand*

*To Irene, the "Sandpiper Lady,"
and Mr. Lamb, the "Magic Lamb."
Thank you for making me a best-selling
technical author.*

Contents

PREFACE xiii

CHAPTER 1. TWO-JUNCTION TRANSISTOR TESTS 1

1-1. Basic two-junction transistor tests, 1
1-2. Two-junction transistor leakage tests, 2
1-3. Two-junction transistor breakdown tests, 5
1-4. Two-junction transistor gain tests, 6
1-5. Two-junction transistor switching tests, 12
1-6. Testing transistors in circuit, 16
1-7. Two-junction transistor tests using a curve tracer, 18
1-8. Transistor current gain tests using a curve tracer, 21
1-9. Transistor breakdown voltage tests using a curve tracer, 26
1-10. Transistor leakage tests using a curve tracer, 27
1-11. Transistor saturation voltage tests using a curve tracer, 28

1-12. Transistor output admittance and impedance tests using a curve tracer, 29
1-13. Testing effects of temperature on transistors using a curve tracer, 31

CHAPTER 2. FIELD-EFFECT TRANSISTOR TESTS 34

2-1. FET operating modes, 34
2-2. Handling MOSFETs, 36
2-3. MOSFET protection circuits, 37
2-4. FET control voltage tests, 39
2-5. FET operating voltage tests, 40
2-6. FET operating current tests, 41
2-7. FET breakdown voltage tests, 42
2-8. FET gate leakage tests, 45
2-9. Dual-gate FET tests, 46
2-10. FET dynamic characteristics, 48
2-11. FET forward transadmittance (transconductance) tests, 49
2-12. FET reverse transadmittance tests, 52
2-13. FET output admittance tests, 52
2-14. FET input admittance tests, 53
2-15. FET amplification factor, 54
2-16. FET input capacitance tests, 54
2-17. FET output capacitance tests, 54
2-18. FET reverse transfer capacitance tests, 55
2-19. FET element capacitance tests, 58
2-20. FET channel resistance tests, 58
2-21. FET switching-time tests, 59
2-22. FET gain tests, 60
2-23. FET noise-figure tests, 63
2-24. FET cross-modulation tests, 65
2-25. FET intermodulation tests, 67
2-26. FET tests using a curve tracer, 67

CHAPTER 3. UNIJUNCTION TRANSISTOR TESTS 74

3-1. Basic UJT and PUT functions, 74
3-2. UJT characteristics, 77
3-3. PUT characteristics, 84
3-4. Testing UJTs with circuits, 88
3-5. UJT tests using a curve tracer, 98

CHAPTER 4. SOLID-STATE DIODE TESTS 101

4-1. Basic diode tests, 101
4-2. Diode continuity tests, 102
4-3. Diode reverse leakage tests, 104
4-4. Diode forward voltage drop tests, 105
4-5. Diode dynamic tests, 105
4-6. Diode switching tests, 109
4-7. Zener diode tests, 111
4-8. Tunnel diode tests, 115
4-9. Signal and power diode tests using a curve tracer, 119
4-10. Zener diode test using a curve tracer, 122
4-11. Tunnel diode test using a curve tracer, 122

CHAPTER 5. THYRISTOR AND CONTROL RECTIFIER (SCR) TESTS 125

5-1. Thyristor and control rectifier basics, 125
5-2. Control rectifier and thyristor test parameters, 133
5-3. Basic control rectifier and thyristor tests, 138
5-4. Blocking voltage and leakage current tests, 145
5-5. Gate trigger voltage and current tests, 147
5-6. Latching and holding current tests, 151
5-7. Average forward voltage test, 153
5-8. Control rectifier and thyristor tests using a curve tracer, 154

CHAPTER 6. AUDIO-CIRCUIT TESTS 159

6-1. Frequency response of audio circuits, 159
6-2. Basic frequency-response tests, 160
6-3. Voltage-gain tests, 163
6-4. Power output and gain tests, 163
6-5. Power-bandwidth tests, 163
6-6. Load-sensitivity tests, 164
6-7. Dynamic output impedance tests, 165
6-8. Dynamic input impedance tests, 165
6-9. Audio-circuit signal-tracing tests, 166
6-10. Checking distortion by sine-wave analysis tests, 166

6-11. Checking distortion by square-wave analysis tests, 166
6-12. Harmonic distortion tests, 169
6-13. Intermodulation distortion tests, 170
6-14. Background noise tests, 172
6-15. Feedback audio-amplifier tests, 173
6-16. Analyzing experimental and design problems with test results, 179

CHAPTER 7. POWER-SUPPLY-CIRCUIT TESTS 189

7-1. Power-supply-output tests, 189
7-2. Power-supply-regulation tests, 191
7-3. Power-supply-internal-resistance test, 191
7-4. Power-supply-ripple tests, 192
7-5. Measuring transformer characteristics, 194

CHAPTER 8. RADIO-FREQUENCY-CIRCUIT TESTS 200

8-1. Basic RF voltage measurement, 201
8-2. Measuring the resonant frequency of *LC* circuits, 203
8-3. Measuring the inductance of a coil, 206
8-4. Measuring the self-resonance and distributed capacitance of a coil, 208
8-5. Measuring the *Q* of resonant circuits, 209
8-6. Measuring the impedance of resonant circuits, 212
8-7. Testing transmitter RF circuits, 214
8-8. Testing receiver RF circuits with a meter and signal generator, 215
8-9. Testing receiver RF circuits with a sweep generator/oscilloscope, 220

CHAPTER 9. COMMUNICATIONS EQUIPMENT TESTS 229

9-1. Using oscilloscopes in communications equipment tests, 229
9-2. Using probes in communications equipment tests, 239

9-3. Using frequency meters and counters in communications equipment tests, 245
9-4. Using a dummy load in communications equipment tests, 255
9-5. Using RF wattmeters in communications equipment tests, 257
9-6. Using field strength meters in communications equipment tests, 258
9-7. Standing-wave-ratio (SWR) measurement, 259
9-8. Dip meters, 262
9-9. Antenna and transmission line measurements, 265
9-10. Oscillator tests, 277
9-11. Using spectrum analyzers in communications equipment tests, 282
9-12. Special test sets for communications equipment, 289

INDEX 295

Preface

As the title implies, this is a handbook of electronic test procedures. The entire book is devoted to step-by-step procedures for the test of electronic devices, components, and circuits. The book not only tells you how to perform the test, but describes what is being tested and why the test is required. This combination of theory-plus-application makes the handbook suitable as a reference text for student technicians, hobbyists, and experimenters, and as a guidebook for experienced, working technicians.

Where practical throughout the book, three sets of procedures are given for the tests. The first set of procedures describes tests that can be performed with elementary test equipment such as meters. These procedures are especially useful for the home experimenter and hobbyist. The second set of procedures describes the same tests using more advanced equipment such as oscilloscopes. These procedures are primarily for the advanced student and working technician. The third set of procedures cover the same ground using even more specialized and sophisticated test equipment, such as curve tracers and special test sets. These last procedures, although presented in a simple, readily understandable fashion, are slanted for the laboratory technician.

Although the emphasis is on test, the book goes much further than the usual collection of test procedures. For example, this book tells you what kind of oscilloscope displays or meter readouts to expect from tests for *both good and bad* components and circuits. In many cases, the book goes on to tell what is *probably wrong* if the expected test results are not obtained. Thus, the procedures described in this book form the basis for troubleshooting, and the starting point for analysis of experimental or design circuits.

Chapter 1 is devoted entirely to test procedures for two-junction or bipolar transistors. The first sections of this chapter describe transistor characteristics and test procedures from the practical standpoint. The information in these sections permits you to test all the important two-junction transistor characteristics using basic shop equipment. The sections also help you understand the basis for such tests. The remaining sections of the chapter describe how the same tests, and additional tests, are performed using more sophisticated equipment. Chapters 2 through 5 provide similar coverage for field-effect transistors (FETs), unijunction transistors (UJTs), solid-state diodes, and thyristors (SCRs, triacs, diacs, etc.).

Chapter 6 is devoted entirely to test procedures for audio circuits. These procedures can be applied to complete audio equipment (such as a stereo system), or to specific circuits (such as the audio circuits of a radio transmitter or receiver). Also, the procedures can be applied to audio circuits at any time during design or experimentation. The test procedures include a series of notes regarding the effect of changes in component values on test results. This information is summarized at the end of the chapter, and is of particular interest to hobbyists and experimenters. Chapters 7 through 9 provide similar coverage for power-supply circuits, radio-frequency circuits, and communications equipment.

Many professionals have contributed their talent and knowledge to the preparation of this handbook. The author acknowledges that the tremendous effort to make this book such a comprehensive work is impossible for one person, and he wishes to thank all who have contributed directly and indirectly. The author wishes to give special thanks to the following: B&K Precision Dynascan Corporation, General Electric Semiconductor Products Department, Heath Company, Hewlett-Packard, Motorola Semiconductor Products, Inc., Pace Communications, Radio Shack, Solid State Division of RCA Corporation, Tektronix Inc., and Texas Instruments.

The author extends his gratitude to Dave Boelio, Hank Kennedy, John Davis, Jerry Slawney, Art Rittenberg, and Don Schaefer of Prentice-Hall. Their faith in the author has given him encouragement, and their editorial/marketing expertise has made many of the author's books best-sellers. The author also wishes to thank Mr. Joseph A. Labok of Los Angeles Valley College for his help and encouragement.

JOHN D. LENK

Handbook of Electronic Test Procedures

1

Two-Junction Transistor Tests

This chapter is devoted entirely to test procedures for two-junction or bipolar transistors. The first sections of this chapter describe transistor characteristics and test procedures from the practical standpoint. The information in these sections permits you to test all of the important two-junction transistor characteristics using basic shop equipment. The sections also help you understand the basis for such tests. The remaining sections of the chapter describe how the same tests, and additional tests, are performed using more sophisticated equipment, such as the oscilloscope curve tracer.

1-1 BASIC TWO-JUNCTION TRANSISTOR TESTS

Transistors are subjected to a variety of tests during manufacture. It is neither practical nor necessary to duplicate all these tests in the field. There are only four basic tests required in most practical applications: gain, leakage, breakdown, and switching time. Unless a transistor is used for pulse or digital work, the switching characteristics are not of great importance.

In the final analysis, the only true test of a transistor is in the circuit with which the transistor is to be used. Except for special circumstances, however, a transistor will operate properly in-circuit provided that:

1. The transistor shows the proper gain.
2. The transistor does not break down under the maximum operating voltages.
3. Leakage, if any, is within tolerance.
4. In the case of pulse circuits, the switching characteristics (such as delay time and storage time) are within tolerance.

There are two exceptions to this rule. Transistor characteristics change with variations in operating frequency and temperature. For example, a transistor may be tested at 1 megahertz (MHz) and show more than enough gain to meet circuit requirements. At 10 MHz, however, the gain of the same transistor may be zero. This can be due to a number of factors. Any transistor has some capacitance at the input and output. As frequency increases, the capacitive reactance changes and, at some frequency, the transistor becomes unsuitable for the circuit (insufficient gain, unstable oscillation, etc.).

In the case of temperature, the current flow in any junction of the transistor increases with increases in temperature. A transistor may be tested for leakage at a normal ambient temperature and show a leakage well within tolerance. When the same transistor is used "in-circuit," the temperature increases, increasing the leakage to an unsuitable level.

It is usually not practical to test transistors over the entire range of operating frequencies and temperature with which the transistor will be used. Instead, the transistor is generally tested under the conditions specified in the datasheet. Then the transistor characteristics are predicted at other frequencies and temperatures using datasheet graphs. In the case of some high-frequency radio-frequency (RF) applications, the transistor must be tested at the intended operating frequency in a special test circuit that is similar (if not identical) to the operating circuit.

1-2 TWO-JUNCTION TRANSISTOR LEAKAGE TESTS

For test purposes, both NPN and PNP transistors can be considered as two diodes connected back to back. Thus, the procedures for transistor leakage tests are similar to those of diodes, as described in Chapter 4. In theory, there should be no current flow across a diode junction when the junction is reverse-biased. Any current flow under these conditions is the result of leakage. In the case of a transistor, the collector–base junction is reverse-biased and should show no current flow. However, in most practical applications, there is some collector–base current flow, particularly as the collector voltage is operated near its limits and as the operating temperature is increased.

1-2.1 Collector Leakage

Collector leakage current is designated as I_{cbo} or I_{bo} on most datasheets. Collector leakage may be termed "collector cutoff current" on other datasheets, where a nominal and/or maximum current is specified for a given collector-base voltage and ambient temperature. Collector-base leakage is normally measured with the emitter open but can also be measured with the emitter shorted to the base or connected to the base through a resistance.

Figure 1-1 shows the basic circuits for the collector-base leakage test. Any of the circuits can be used, but those of Fig. 1-1(a) and (d) are the most popular. The procedure is the same for all the circuits shown in Fig. 1-1. The voltage source is adjusted for a given value (thus providing a given reverse bias), and the current (if any) is read on the meter. This current must be below a given maximum for a given reverse bias.

Temperature is often a critical factor in leakage measurements. Using a 2N332 transistor as an example, the maximum collector leakage current at 25°C is 2 microamperes (μA) [with 30 volts (V) applied between collector and base]. When the collector-base voltage is lowered to 5 V, and the temperature is raised to 150°C, the maximum collector leakage is 50 μA.

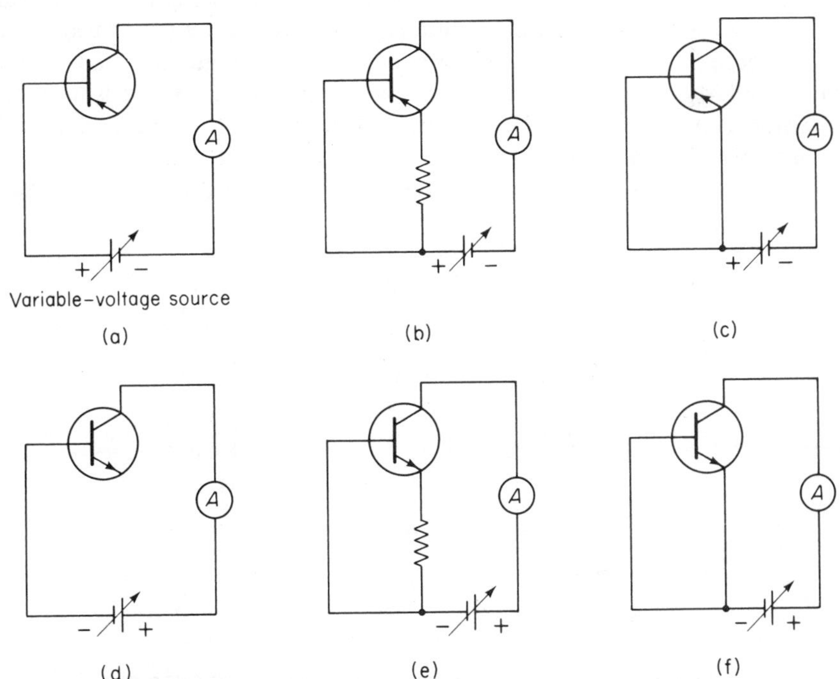

FIGURE 1-1 Collector-base leakage test circuits.

1-2.2 Emitter Leakage

Some datasheets also specify emitter–base current leakage. This is not usually the case, however, since the emitter–base junction is usually forward-biased in most circuits. Should it become necessary to test the emitter–base current leakage (I_{eo} or I_{ebo}), the circuits of Fig. 1–1 can be used, except that the collector and emitter connections are interchanged. The emitter–base junction is reverse-biased, the collector is left open, and the meter is placed in the emitter–base circuit. The procedures for testing emitter–base leakage are identical to those for collector–base voltage.

1-2.3 Testing Two-Junction Transistor Leakage with an Ohmmeter

It is possible to make a quick check of transistor leakage with an ohmmeter. For the purpose of this test, a transistor is considered as two diodes connected back to back. Each diode should show low forward resistance and high reverse resistance. These resistances can be measured with an ohmmeter as shown in Fig. 1–2. The same ohmmeter range is used for each pair of measurements (base to emitter, base to collector, and collector to emitter). On low-power transistors, there may be a few ohms indicated from collector to emitter. Avoid using the $R \times 1$ range of an ohmmeter with a high internal-battery voltage. Either of these conditions can damage a low-power transistor.

If both forward and reverse readings are very high, the transistor is open. Similarly, if any of the readings show a short or very low resistance, the transistor is shorted or leaking badly. Also, if the forward and reverse readings are the same (or nearly equal), the transistor is defective.

A typical forward resistance is 300 to 700 ohms (Ω). Typical reverse

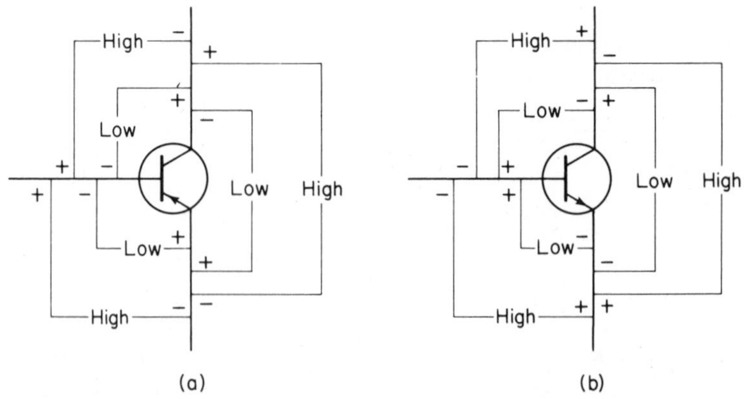

FIGURE 1-2 Transistor leakage tests with an ohmmeter: (a) PNP; (b) NPN.

resistances are 10 to 60 kilohms (kΩ). Actual resistance values depend on ohmmeter range and battery voltage rather than on the transistor characteristics. Thus, the *ratio of forward to reverse resistance* is the best indicator. Almost any transistor will show a 30:1 ratio. Many transistors show ratios of 100:1 or greater.

CAUTION: Transistors should not be tested for any characteristic unless all the characteristics are known. The transistor can be damaged if this rule is not followed. Even if no damage occurs, the test results can be inaccurate. Never test a transistor with voltages, or currents, higher than the rated values. The *maximum current rating* is often overlooked. For example, if a transistor is designed to operate with a maximum of 45 V at the collector, it could be assumed that a 9-V battery is safe for all measurements involving the collector. However, assume that the internal (emitter-to-collector) resistance of the transistor is 90 Ω and the maximum rated emitter-collector current is 25 milliamperes (mA). With 9 V connected directly between the emitter and collector, the emitter-collector current is 100 mA, four times the maximum rated 25 mA. This can cause the junctions to overheat and damage the transistor.

1-3 TWO-JUNCTION TRANSISTOR BREAKDOWN TESTS

The circuits and procedures for transistor breakdown tests are similar to those for leakage tests. The most important breakdown test is to determine the collector-base breakdown voltage. In this test, the collector and base are reverse-biased, with the emitter open, and the voltage source is adjusted to a *given value of leakage current*. The voltage is then compared with the minimum collector breakdown voltage specified for the transistor.

For example, the minimum collector breakdown voltage specified for a 2N332 transistor is 45 V (with 50 μA flowing and an ambient temperature of 25°C). If 50 μA flows with less than 45 V, the transistor collector-base junction is breaking down.

Another breakdown test specified on some transistor datasheets is the collector-emitter breakdown voltage. In this test, the collector and emitter are reverse-biased, with the base open. The voltage source is then adjusted for a given value of leakage current through both the emitter-base and collector-base junctions. The collector-emitter breakdown voltage test determines the condition of both junctions simultaneously.

Breakdown voltage is designated as BV_{cbo} (collector to base, emitter open), BV_{ces} (emitter shorted to base), or BV_{ceo}(collector to emitter, base open) on most datasheets. Breakdown is normally measured with the emitter (or base) open but can also be measured with the emitter shorted to the base or connected to the base through a resistor or with the emitter and base reverse-biased.

Figure 1–3 shows the basic circuits for breakdown tests. The circuits shown are for PNP transistors. The same circuit can be used for NPN transistors when the voltage polarity is reversed. In all cases, the voltage source is adjusted for a given leakage current flow. Then the voltage is compared with a minimum specified voltage.

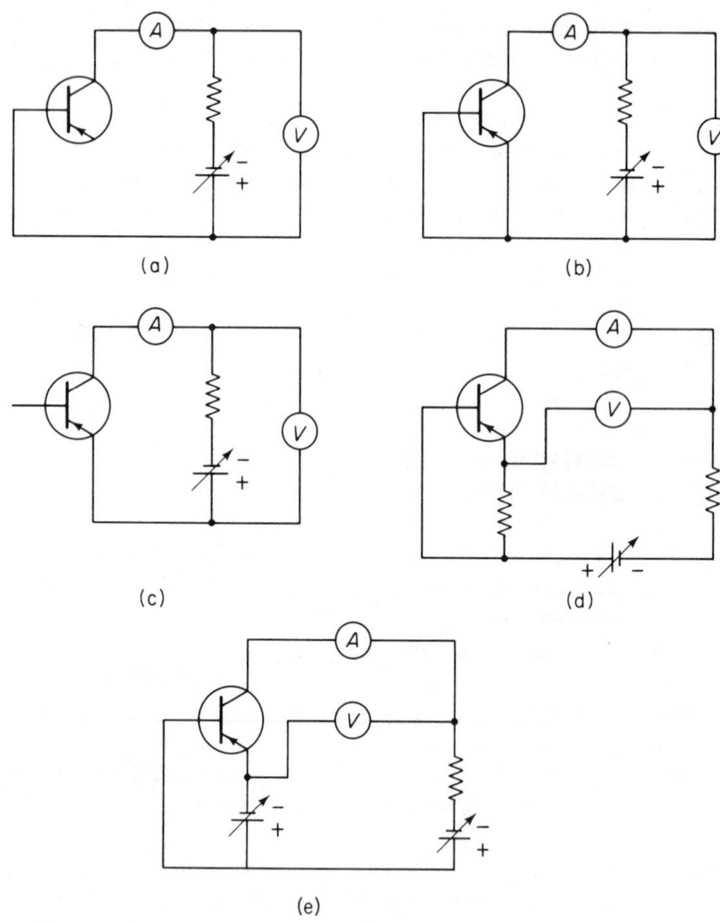

FIGURE 1-3 Breakdown voltage test circuits: (a) BV_{CBO}; (b) BV_{CES}; (c) BV_{CEO}; (d) BV_{CER}; (e) BV_{CEX}.

1-4 TWO-JUNCTION TRANSISTOR GAIN TESTS

The dynamic gain of a transistor is determined by the amount of change in output for a given change in input. Two-junction transistors are tested for *current gain*. The change in output current for a given change in input current is measured without changing the output voltage.

When a transistor is connected in a common-base circuit, the collector forms the output circuit, and the emitter forms the input circuit. Common-base current gain is known as alpha, indicated by the lowercase Greek letter (α).

Most datasheets specify gain with the transistor connected in a common-emitter circuit rather than a common-base circuit. In the common-emitter circuit, the base is the input and the collector is the output. Current gain for a common-emitter circuit is known as beta, indicated by the lowercase Greek letter (β).

In addition to alpha and beta, datasheets use several other terms to specify gain. The term "forward current transfer ratio" and the letters h_{fe} are the most popular means of indicating current gain for two-junction transistors, even though some manufacturers use "collector-to-base current gain."

Hybrid System. The *h* in the letters h_{fe} refers to the hybrid of transistor-equivalent-model circuits. Transistor test circuits are often structured along the same lines. In the hybrid system, the transistor and the test or operating circuits are considered as a "black box" with an input and an output rather than individual components.

When lowercase letters h_{fe} (or sometimes improperly H_{fe}) are used in transistor specifications, this indicates that the current gain is measured by noting the change in collector alternating current for a given change in base alternating current. This is also known as "a-c beta" or "dynamic beta."

When capital letters H_{FE} or h_{FE} are used in transistor specifications, the current gain is measured by noting the collector direct current for a given base direct current. This is generally known as "d-c beta."

Direct versus Alternating Gain Measurements. Direct-current gain measurements apply under a wider range of conditions and are easier to make. Alternating-current gain measurements require more elaborate test circuits, and the test results will vary with the frequency of the alternating current used for the test. A-c measurements are more realistic, however, since transistors are normally used with a-c signals.

There are a number of circuits for both a-c and d-c gain tests and a number of test procedures. Similarly, there are many commercial transistor testers as well as adapters that permit transistors to be tested with oscilloscopes. Some testers permit transistors to be tested while still connected in the circuit. It is impractical to cover the use and operation of all such testers and oscilloscope adapters in this book. Also, detailed instructions are provided with the testers. These instructions must be followed in all cases. Instead of attempting to duplicate the operating instructions, the following paragraphs describe the operating principles of the tests.

1-4.1 Basic Transistor Gain Tests

Alpha Tests. Figure 1-4 shows the basic circuits for alpha measurement of PNP and NPN transistors. Both emitter current I_E and collector current I_C are measured under static conditions. Then the emitter current I_E is changed a given amount by varying the resistance of R_1 or by changing the emitter-base source voltage. The collector voltage must remain the same.

The difference in collector current I_C is noted, and the value of alpha is calculated using the equation shown. For example, assume that the emitter current I_C is changed 4 mA and that this results in a change of 3 mA in collector current I_C. This means a current gain of 0.75.

Beta Tests. Figure 1-5 shows the basic circuits for beta measurement of PNP and NPN transistors. Both base current I_B and the collector current I_C are measured under static conditions. Then, without changing the collector voltage, the base current I_B is changed by a given amount and the difference in collector current I_C is noted.

For example, assume that when the circuit is first connected, the base current I_B is 7 mA and the collector current I_C is 43 mA. When the base current I_B is increased to 10 mA (a 3-mA increase), the collector current I_C is increased to 70 (a 27-mA increase). This represents a 27-mA increase in collector current I_C for a 3-mA increase in base current I_B, or a current gain of 9.

Precautions in Transistor Gain Tests. Certain precautions should be observed if transistors are to be tested using *noncommercial* test circuits. The most important of these are as follows:

1. The collector and emitter (or base) load resistances (represented by R_1 and R_2 in Figs. 1-4 and 1-5) should be of such value that the max-

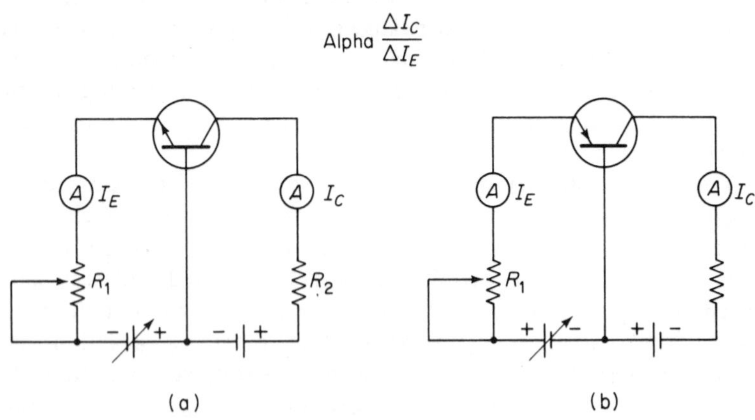

FIGURE 1-4 Basic dc alpha test circuits.

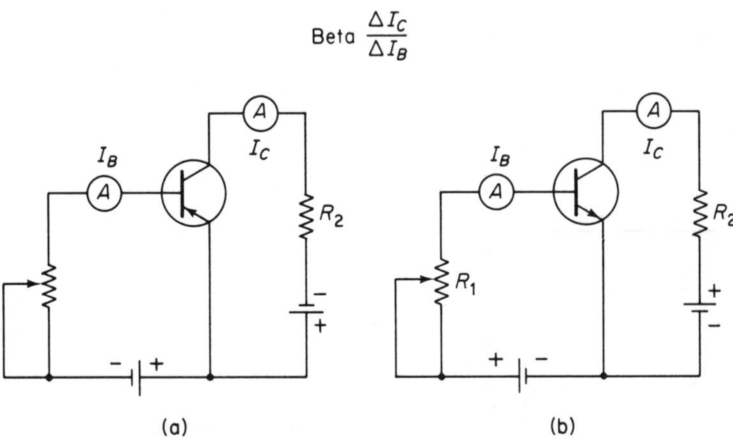

FIGURE 1-5 Basic dc beta test circuits.

imum current limitations of the transistor are not exceeded. In the case of power transistors, the wattage rating of the load resistance should be large enough to dissipate the heat.

2. Where there is a large collector leakage current, this must be accounted for in test conditions. Measure leakage using the same voltages and currents as for gain tests. Then subtract any significant leakage currents from the gain test currents.
3. The effect of meters used in the test circuits must also be taken into account.

1-4.2 Basic Two-Junction Transistor a-c Gain Tests

There are several types of circuits used for the a-c or dynamic test of transistors. Some of the commercial transistor testers use the same basic circuits shown in Figs. 1-4 and 1-5, except that an a-c signal is introduced into the input and gain is measured at the output. Usually, these testers provide a 60- or 1000-Hz signal for injection into the input. Where it is desirable to test transistors at higher frequencies, some tester circuits permit an external high-frequency signal to be injected.

Feedback-Gain Tester. One method used in shop-type (nonlaboratory) transistor testers for a-c gain measurement is the feedback circuit. A typical feedback-gain test circuit is shown in Fig. 1-6. Here, the transistor under test is inserted into an audio-oscillator circuit. The amount of feedback is adjusted by means of a calibrated control until the circuit begins to oscillate. Oscillation is indicated by a tone on the loudspeaker. The transistor current gain at the oscillation starting point can be read directly from the dial calibration.

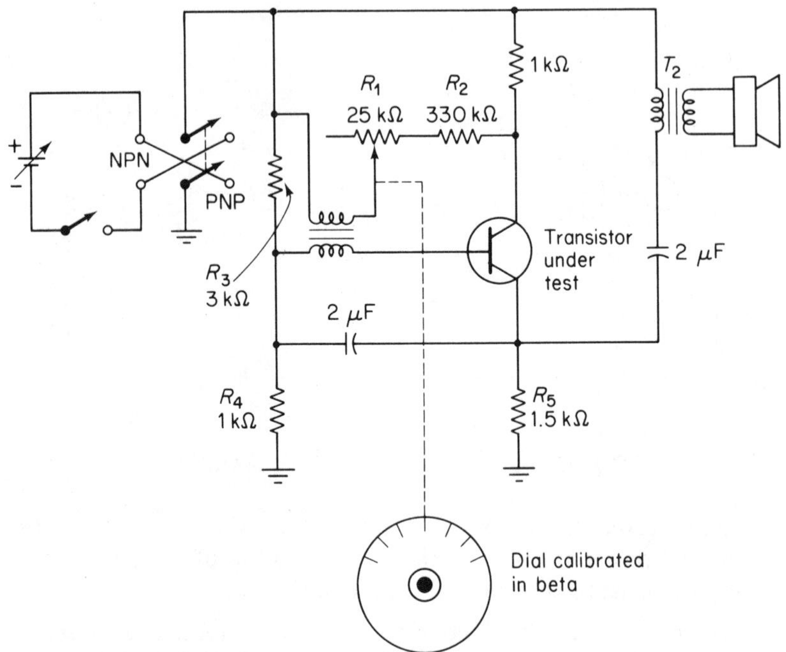

FIGURE 1-6 Feedback-type gain (beta) test circuit.

The setting of the feedback control and the transistor gain determine the point at which oscillation starts. For example, if a large feedback is required to produce oscillation, the transistor gain is low. If the circuit oscillates with very little feedback, the transistor gain is high. Thus, the dial connected to variable feedback resistor R_1 can be calibrated directly in terms of gain.

1-4.3 Testing Two-Junction Transistor Gain with an Ohmmeter

It is possible to make a quick check of transistor gain with an ohmmeter. The basic circuit is shown in Fig. 1–7. Normally, there is little or no current flow between emitter and collector until the base-emitter junction is forward-biased. This fact can be used to provide a basic gain test of two-junction transistors.

In this test, the R_1 range of the ohmmeter should be used. Any internal-battery voltage can be used provided that it does not exceed the maximum collector–emitter breakdown voltage. In position *A* of switch S_1, there is no voltage applied to the base and the base-emitter junction is not forward-biased. Thus, the ohmmeter reads a high resistance. When switch S_1 is set to *B,* the base–emitter circuit is forward-biased (by the voltage across R_1 and R_2) and current flows in the emitter–collector circuit. This is indicated by a lower

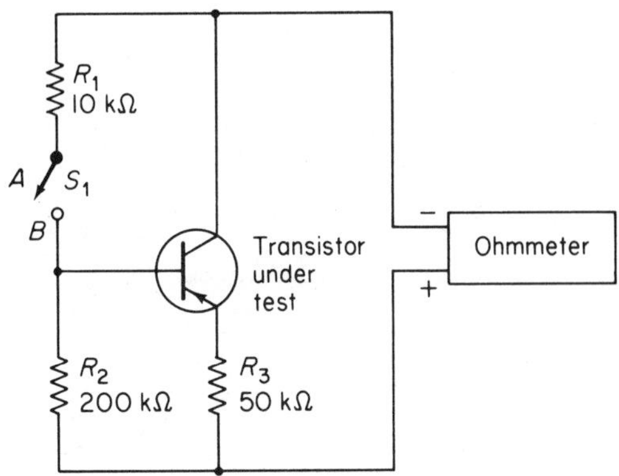

FIGURE 1-7 Transistor gain test using an ohmmeter.

resistance reading on the ohmmeter. A 10:1 (or better) resistance ratio is typical for an audio-frequency transistor.

1-4.4 Testing Two-Junction Transistor RF Gain

The tests described thus far can be used to establish the gain of a two-junction transistor operating at low frequencies. However, the tests do not establish the gain in a high-frequency (RF) circuit. The only true test of circuit gain (either voltage gain or power gain) is to operate the device in a working circuit and to measure actual gain. The most practical method is to operate the device in the circuit with which the device is to be used. However, it may be convenient to have a "standard" or "universal" circuit for gain test.

Very often, the datasheets for transistors to be used in RF work show the circuits used to test gain. Such a circuit is shown in Fig. 1-8. The circuit shown is for RF gain tests of a 2N5160 transistor operating at 175 MHz. The circuit can be modified to test other transistors, at other frequencies, by changing the circuit values. If you are not familiar with the design of such RF circuits, your attention is invited to the author's best-selling *Handbook of Simplified Solid-State Circuit Design* (Prentice-Hall, Inc., Englewood Cliffs, N.J., 1978), or the *Handbook of Electronic Circuit Designs* (Prentice-Hall, Inc., Englewood Cliffs, N.J., 1976). These books give step-by-step details for design of RF circuits.

With the transistor operating in a circuit similar to that of Fig. 1-8, you simply introduce an RF signal (at the frequency of interest) to the input, and measure both the input and output voltage with an RF voltmeter (or meter with an RF probe). The procedures for RF voltage measurement are discussed fully in Chapters 8 and 9.

If power-gain measurement is desired, use input and output load resistances (noninductive or composition resistors). Measure the RF voltages

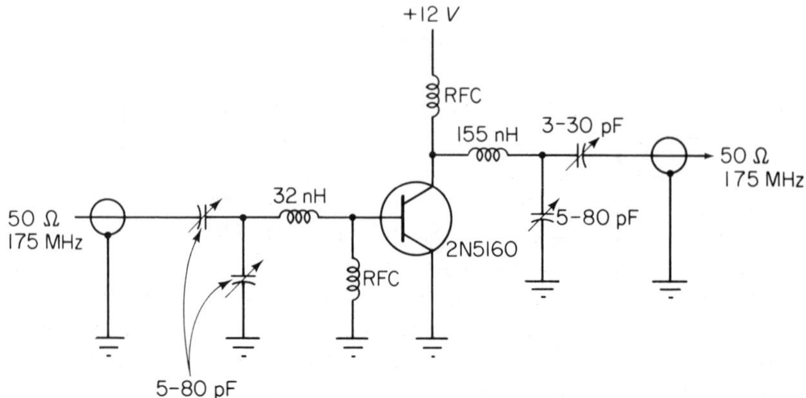

FIGURE 1-8 Typical RF test amplifier circuit for two-junction transistors.

across the input and output resistances and then calculate the power gain using $P = E^2/R$. The ratio of output power divided by input power is the power gain.

1-5 TWO-JUNCTION TRANSISTOR SWITCHING TESTS

Transistors to be used in pulse or digital applications must be tested for switching characteristics. For example, when a pulse is applied to the input of a transistor, there is a measurable time delay before the pulse starts to appear at the output. Similarly, after the pulse is removed, there is additional time delay before the transistor output returns to its normal level. These "switching times" or "turn-on" and "turn-off" times are usually on the order of a few microseconds (or possibly nanoseconds) for high-speed pulse transistors.

The switching characteristics of transistors designed for computer or digital work are listed on the datasheets. Each manufacturer lists its own set of specifications. However, there are four terms (rise time, fall time, delay time, and storage time) common to most datasheets for transistors used in pulse work. These switching characteristics are of particular importance where the pulse durations are short. For example, assume that the turn-on time of a transistor is 10 nanoseconds (ns) and that a 5-ns pulse is applied to the transistor input. There will be no pulse output, or the pulse will be drastically distorted.

1-5.1 Pulse and Square-Wave Definitions

The following terms are commonly used in describing transistor switching characteristics. The terms are illustrated in Fig. 1-9. The input pulse represents an ideal input waveform for comparison purposes. The other

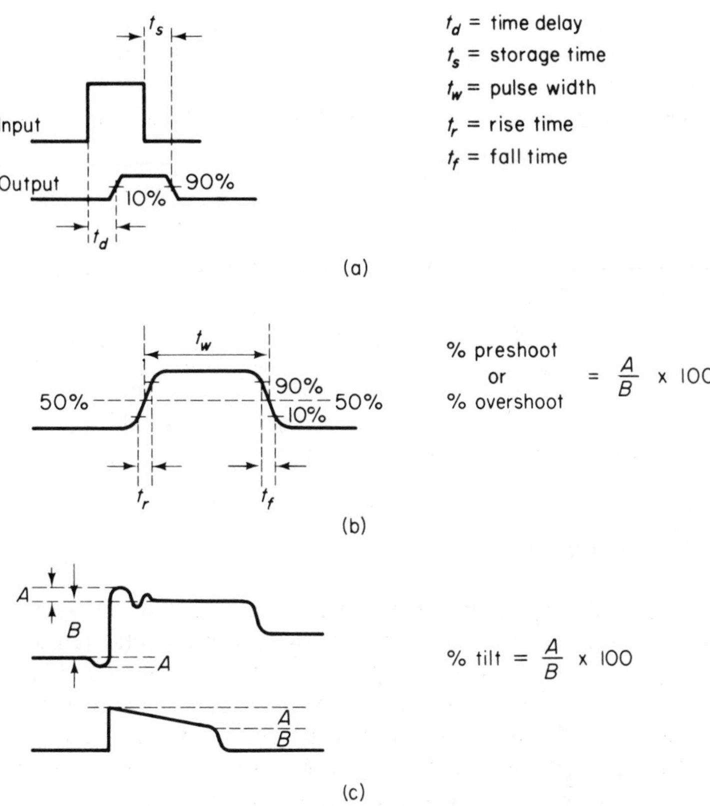

FIGURE 1-9 Commonly used pulse definitions.

waveforms in Fig. 1-9 represent typical output waveforms in order to show the relationship. The terms are defined as follows:

Rise time t_r: the time interval during which the amplitude of the output voltage changes from 10% to 90% of the rising portion of the pulse.

Fall time t_f: the time interval during which the amplitude of the output voltage changes from 90% to 10% of the falling portion of the pulse.

Time delay t_d (or delay time): the time interval between the beginning of the input pulse (time zero) and the time when the rising portion of the output pulse attains an arbitrary amplitude of 10% above the baseline.

Storage time t_s: the time interval between the end of the input pulse (trailing edge) and the time when the falling portion of the output pulse drops to an arbitrary amplitude of 90% from the baseline.

Pulse width (or pulse duration) t_w: the time duration of the output pulse measured between two 50% amplitude levels of the rising and falling portions of the waveform.

Tilt (also known as sag or droop): a measure of the full amplitude (flat top) portion of a pulse. The tilt measurement is usually expressed as a percentage of the amplitude of the rising portion of the pulse.

Overshoot: a measure of the overshoot generally occurring above the 100% amplitude level. This measurement is also expressed as a percentage of the pulse rise.

Preshoot: a measure of the preshoot generally occurring below the baseline (or zero line). This measurement is usually expressed as a percentage of the 100% amplitude level.

These definitions are for guide purposes only. When pulses are very irregular (such as excessive tilt, droop, sag, preshoot, overshoot, ringing, and oscillation), the definitions may become ambiguous.

1-5.2 Testing Two-Junction Transistors for Switching Time

Figure 1-10 shows two circuits for testing the switching characteristics of transistors used in pulse or computer work. Figure 1-10(a) is used when a dual-trace oscilloscope is available. Figure 1-10(b) requires a conventional, single-trace oscilloscope. In either case, the oscilloscope must have wide frequency response and good transient characteristics.

In either circuit of Fig. 1-10, the oscilloscope vertical channel is voltage-calibrated in the normal manner, whereas the horizontal channel must be time-calibrated (rather than sweep-frequency-calibrated). The transistor is tested by applying a pulse to the base of the transistor under test, with a specific bias applied to the transistor base. The same pulse is applied to one of the oscilloscope vertical inputs [Fig. 1-10(a)] or to the pulse monitor oscilloscope [Fig. 1-10(b)]. The transistor collector output (inverted 180° by the common-emitter circuit) is applied to the other oscilloscope vertical input [Fig. 1-10(a)] or to the output monitor oscilloscope [Fig. 1-10(b)]. In the Fig. 1-10(b) circuit, the same oscilloscope can be moved between the two monitor points. The two pulses (input and output) are then compared as to rise time, fall time, delay time, storage time, and so on. The transistor output pulse characteristics can then be compared with transistor specifications.

1-5.3 Rule of Thumb for Switching Tests

Since rise-time and fall-time measurements are of special importance in switching tests, the relationship between oscilloscope rise time and the rise time of the transistor must be taken into account. Obviously, the accuracy of rise-time measurements can be no greater than the rise time of the oscilloscope. Also, if the transistor is tested by means of an external pulse from a

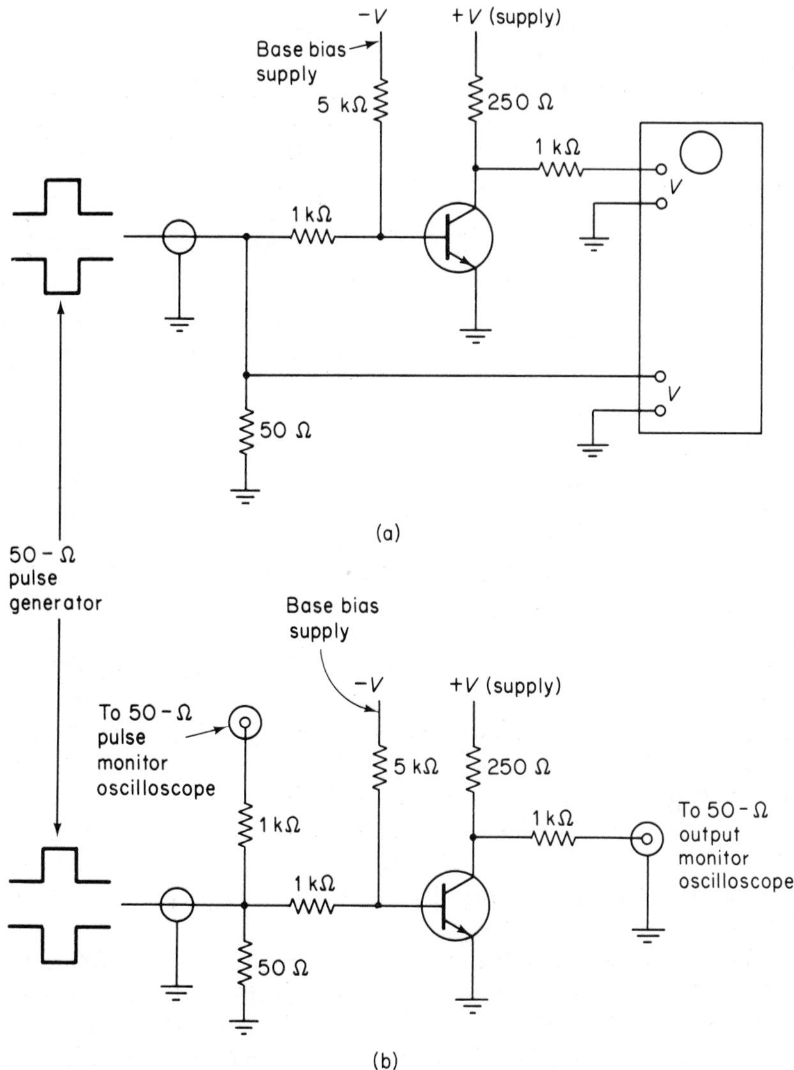

FIGURE 1-10 Two-junction transistor switching test circuits.

pulse generator, the rise time of the pulse generator must also be taken into account.

For example, if an oscilloscope with a 20-ns rise time is used to measure the rise time of a 15-ns pulse, the measurement will be hopelessly inaccurate. Similarly, if a 20-ns pulse generator and a 15-nA oscilloscope are used to measure the rise time, the fastest rise time for accurate measurement is something greater than 20 ns.

There are two basic rules of thumb that can be applied to time measurements (rise time, fall time, etc.).

The first method (known as the *root of the sum of the squares*) involves finding the square of all the rise times associated with the test, adding these squares together, and finding the square root of this sum. For example, using the 20-ns pulse generator and 15-ns oscilloscope, the calculation is:

$$20 \times 20 = 400$$

$$15 \times 15 = 225$$

$$400 + 225 = 625$$

$$\sqrt{625} = 25 \text{ ns.}$$

One major drawback to this rule is that the coaxial cables required to interconnect the test equipment are subject to "skin effect." As frequency increases, the signals tend to travel on the outside or skin of the conductor. This decreases conductor area and increases resistance. In turn, this increases cable loss. The losses of cables do not add properly when applied to the root-sum-squares method, except as an approximation.

The second rule or method states that if the equipment or pulse being measured as a rise time is 10 times lower than the test equipment, the error is 1%. This is small and can be considered as negligible. If the equipment being measured has a rise time three times slower than the test equipment, the error is slightly less than 6%.

1-6 TESTING TRANSISTORS IN CIRCUIT

The normal forward-bias characteristics of transistors can be used to test transistor circuits without removing the transistor and without using an in-circuit tester. Germanium transistors normally have a voltage differential of 0.2 to 0.4 V between emitter and base. Silicon transistors normally have a voltage differential of 0.4 to 0.8 V. The polarities of voltages at the emitter and base depend upon the type of transistor (NPN or PNP). The voltage differential between emitter and base acts as a forward bias for the transistor. That is, a sufficient differential or forward bias turns the transistor on, resulting in a corresponding amount of emitter–collector flow. Removal of the voltage differential, or an insufficient differential, produces the opposite results. That is, the transistor is cut off (no emitter–collector flow or very little flow).

The following sections describe two methods of testing transistors in-circuit. One method involves removal of the forward bias; the other method introduces an external forward bias to the circuit.

1-6.1 *In-Circuit Transistor Test by Removal of Forward Bias*

Figure 1-11 shows the test connections for an in-circuit transistor test by removal of forward bias. The procedure is simple. First, measure the emitter-collector differential voltage under normal circuit conditions. Then, short the emitter-base junction and note any change in emitter-collector differential. If the transistor is operating, the removal of forward bias causes the emitter-collector current flow to stop, and the emitter-collector voltage differential increases. Typically, the collector voltage rises to or near the power-supply value.

For example, assume that the power-supply voltage is 10 V and that the differential between the collector and emitter is 5 V when the transistor is

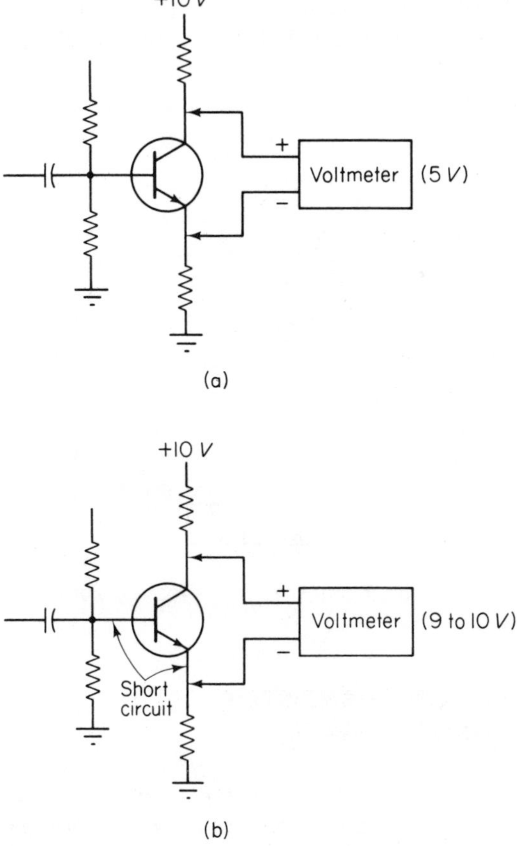

FIGURE 1-11 In-circuit transistor test by removal of forward bias.

operating normally (no short between emitter and base). When the emitter-base junction is shorted, the emitter–collector differential should rise to about 10 V (probably somewhere between 9 and 10 V).

1-6.2 In-Circuit Transistor Test by Application of Forward Bias

Figure 1–12 shows the test connections for an in-circuit transistor test by the application of forward bias. The procedure is equally simple. First, measure the emitter–collector differential under normal circuit conditions. As an alternative, measure the voltage across R_E, as shown in Fig. 1–12.

Next, connect a 10-kΩ resistor between the collector and base, as shown, and note any change in emitter–collector differential (or any change in voltage across R_E). If the transistor is operating, the application of forward bias will cause the emitter–collector current flow to start (or increase), and the emitter–collector voltage differential will decrease, or the voltage across R_E will increase.

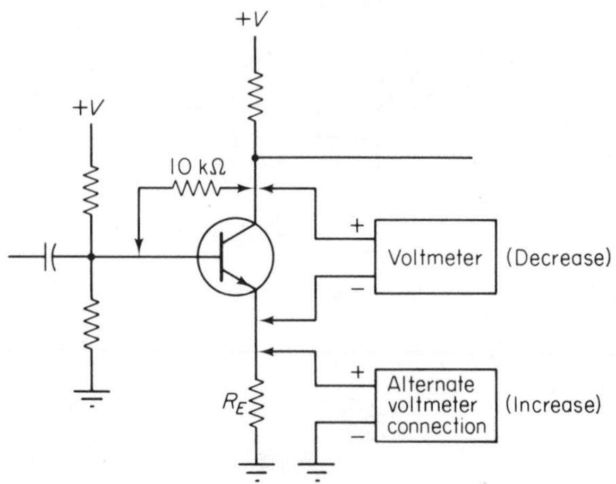

FIGURE 1–12 In-circuit transistor test by application of forward bias.

1-7 TWO-JUNCTION TRANSISTOR TESTS USING A CURVE TRACER

The most practical means of measuring transistor characteristics in the laboratory is to display the characteristics as *curve traces* on an oscilloscope. Since the oscilloscope screen can be calibrated in voltage and current, the transistor characteristics can be read from the screen directly. If a number of curves are made with an oscilloscope, they can be compared with the curves drawn on transistor datasheets. (Some datasheet curves are direct reproductions of those obtained by tracing curves on an oscilloscope).

There are a number of oscilloscopes (or oscilloscope adapters, generally known as curve tracers) manufactured specifically to display transistor characteristic curves. Some curve tracers are for one type of transistor. However, most present-day tracers will display the characteristics of many types of transistors (two-junction, FET, UJT, etc.) as well as other semiconductor devices (diodes, zeners, SCRs, triacs, diacs, etc.).

Figure 1–13 shows the functional block diagram of a typical curve

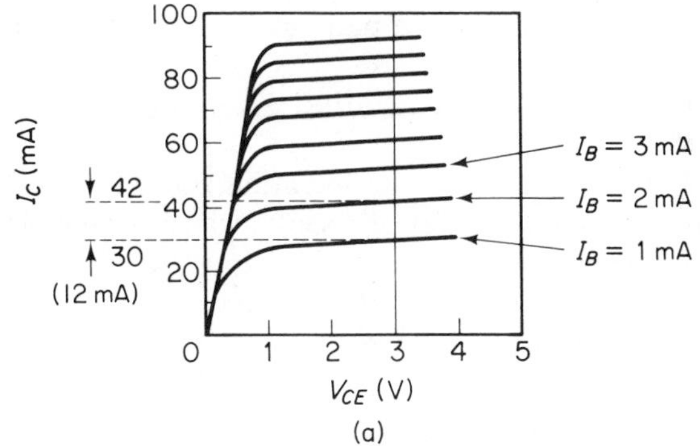

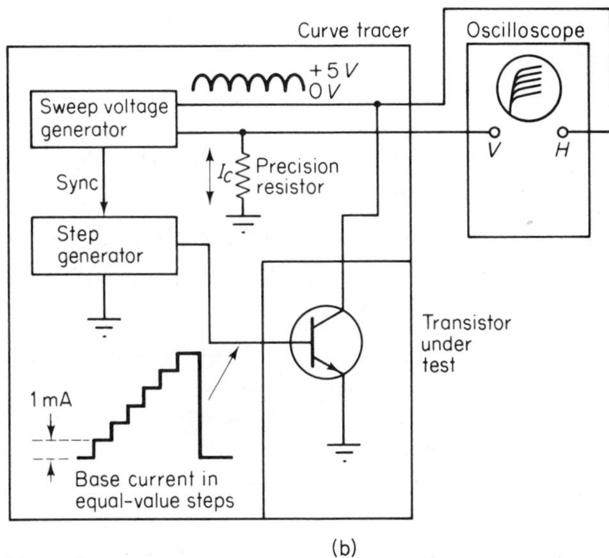

FIGURE 1–13 Basic test connections and typical oscilloscope displays for transistor curve tracer.

tracer, such as the B&K Precision 501A. When testing a two-junction transistor, the tracer introduces changes in base current in the form of equal-value steps (steps of selectable, known value). These steps occur at the same rate as the collector supply voltage is swept between 0 V and some peak value and back to zero. This produces a separate curve corresponding to each different value of base current.

When curves show collector current versus collector voltage (for different values of base current), the change in collector current induced by one step of the base current is proportional to the vertical distance between adjacent curves. This change can be read directly from the scale. Which vertical line is chosen for the scale depends on what collector voltage is specified (since each vertical line corresponds to a particular collector voltage).

1-7.1 Basic Curve-Tracer Operating Procedure

The following is a typical or basic operating procedure for a curve tracer. More detailed procedures for test of specific transistor characteristics using a curve tracer are given in the remaining sections of this chapter. The use of curve tracers to test characteristics of other devices (FETs, UJTs, SCRs, etc.) are given in the appropriate chapters.

1. As shown in Fig. 1–13, the transistor is connected into a grounded-emitter circuit. Precise 1-mA current steps are applied to the base. Voltage sweeps from 0 to approximately 5 V are applied to the collector. The oscilloscope vertical deflection is obtained from the resultant collector current. All of this is done by setting of switches on the curve tracer.
2. On some curve tracers, each sweep must be initiated individually, whereas other instruments produce a series of 10 (or more) curves in sequence automatically.
3. Once the curves are made, they are interpreted as follows:
4. Choose the vertical line corresponding to the specified collector voltage. For example, the 3-V vertical line is chosen in Fig. 1–13.
5. Note (on that line) the distance between the two curves that appears above and below the specified base current (or specified collector current). For example, assume that the transistor of Fig. 1–13 is to be operated with a collector current of 30 to 40 mA, with 3 V at the collector. The two most logical curves are then the 1- and 2-mA base current curves.
6. Note that the distance between these two curves represents a difference of 12 mA in collector current. (The 1-mA curve intersects the 3-V line at 30 mA, and the 2-mA curve intersects the 3-V line at 42 mA.)

7. Divide the collector current difference by the base current difference that caused it. Since the base current per step is 1 mA and the difference in collector current is 12 mA, the gain (beta) is 12. If the base current per step is 0.1 mA (as it is for some curve tracers), and all other conditions are the same, the beta is 120 (12/0.1 = 120).

1-7.2 NPN versus PNP Transistors

On a typical curve tracer, the curves of an NPN transistor are in a positive direction. That is, zero volts is at the left and zero current is at the bottom of the display. The curves sweep to the right and upward as collector voltage and current increases from zero. The collector sweep voltage is of positive polarity.

The family of curves of a PNP transistor is in the negative direction. That is, zero volts is at the right and zero current is at the top of the display. The curves sweep to the left and downward as collector voltage and current increase from zero. The collector sweep voltage is of negative polarity.

All transistor tests described in this chapter apply equally to NPN and PNP transistors. Any examples showing only an NPN or PNP type should be understood to apply to the counterpart as well. Basic characteristics of both types are the same, although the displays are inverted with reference to one another. Therefore, any test that can be made for NPN transistors can also be made for PNP transistors, and vice versa.

1-8 TRANSISTOR CURRENT GAIN TESTS USING A CURVE TRACER

The current gain of a transistor is the single most important characteristic and is usually measured before any other tests are performed. The general condition of a transistor can most often be determined while testing for current gain. As discussed in Section 1-4, current gain can be measured in two ways: direct current gain (or d-c beta) and alternating current gain (a-c beta). The following paragraphs describe both test procedures using a curve tracer.

1-8.1 D-C Beta Tests

D-c beta can be defined as the ratio of collector current to base current measured at *one specific point* of collector voltage and current. The validity of d-c beta depends upon the point of measurement, and this point is usually specified on manufacturers' datasheets. The measurement point usually centers about the typical operating range for which the transistor is designed. For example, a typical d-c beta is specified as h_{FE} = 40 (minimum), 125 (typical), and 400 (maximum), with an I_C of 5 mA and a V_C of 10 V.

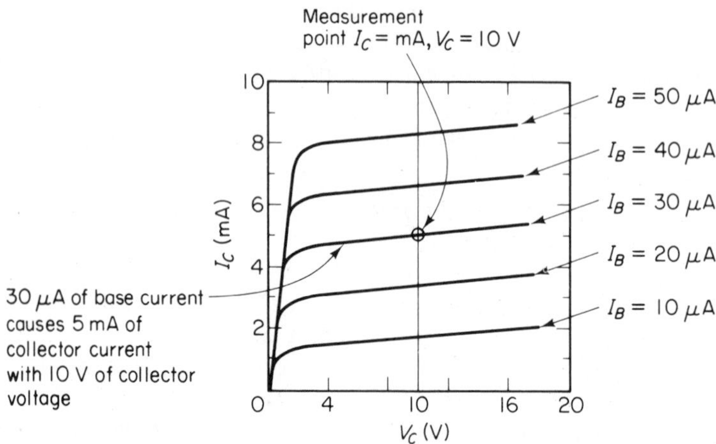

FIGURE 1-14 D-c beta measurement using a curve tracer.

Figure 1-14 shows the technique of d-c beta measurement using a curve tracer. The point of measurement is conveniently centered on the display by using 1 mA/division vertical and 2 V/division horizontal calibration. With each step at 10 μA, the third curve (30 μA) passes through the measurement point. Thus, 30 μA of base current produces 5 mA of collector current at the 10-V collector point. Using these values, d-c beta = I_C/I_B = 5 mA/ 0.03 mA = 166.

1-8.2 A-C Beta Tests

The a-c or dynamic current gain of a transistor can be defined as the *ratio of change* in collector current to the *change* in base current at a specified collector voltage. A-c beta is generally more useful than d-c beta since the transistor is tested under actual operating conditions. This provides a better basis for predicting transistor performance. Fig. 1-15 shows the technique of a-c beta measurement using a curve tracer.

1. Measure the difference in collector current (ΔI_C) between the two curves on the oscilloscope display. (The settings of the curve tracer and/or oscilloscope controls determine the amount of collector current represented by each vertical division of the display scale. In Fig. 1-15, each vertical division represents 2 mA of collector current.) Be sure that *both curve readings* are taken at the same collector voltage. (In Fig. 1-15, both readings are taken at 5 V.)
2. Note the change in base current (ΔI_B) that produces each curve. In Fig. 1-15, each curve is produced by a change of 10 μA. Thus, ΔI_B is 10 μA for any of the curves in Fig. 1-15.

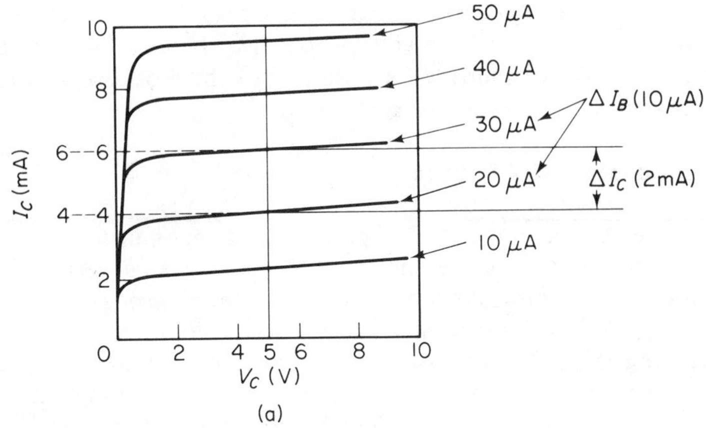

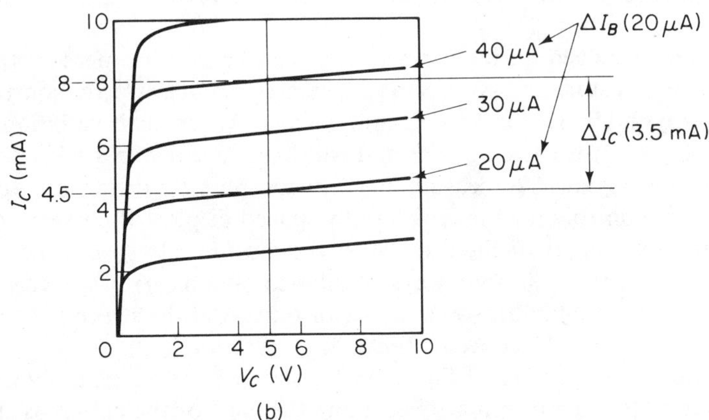

FIGURE 1-15 A-c beta measurement using a curve tracer.

3. Calculate beta by dividing ΔI_C by ΔI_B. For example, if ΔI_C is 2 mA and ΔI_B is 10 μA, as shown in Fig. 1–15(a), beta is 200 (2/0.01 = 200).

It is generally easier to use the two centermost curves of the display to measure a-c beta. However, this is not always practical. If so, the ΔI_C measurement may be made between two nonadjacent curves. For example, the difference between the collector current of the second and fourth curves may be used for measurement of ΔI_C, as shown in Fig. 1–15(b). If this method is used, make certain to use *two steps* of base current for determining ΔI_B when calculating beta. Using the values of Fig. 1–15(b), a-c beta = 3.5 mA/ 20 μA = 175 (at a V_C of 5 V).

If the transistor datasheet is available, measure the a-c beta at the approximate collector current and voltage specified. If no values are specified, adjust the curve tracer controls for a display of the most evenly and widely spaced curves.

1-8.3 Inconsistent A-C Beta

Note that the beta (a-c or d-c) of any transistor is not constant. Beta always depends upon the point of measurement. The distance between all curves is seldom equal, which means that ΔI_C is not the same at various regions of base current and, to a lesser extent, at various collector voltages. When the curves are closer together, the transistor is operating in a region of lower gain. Generally, gain is the highest at the normal operating region of a transistor and is lower at points above and below the normal operating region.

1-8.4 Testing Transistor Linearity and Distortion

The curves produced during a-c beta tests can be used to measure transistor linearity and distortion. As discussed, beta is not necessarily constant, but may vary with changes in voltage and current. When such variation of gain occurs, the transistor is nonlinear and will introduce distortion if operated in the nonlinear region. To measure linearity and possible distortion, adjust the curve tracer controls for the most evenly spaced display of curves, or so that the centermost curves of the display are the most evenly spaced, as shown in Fig. 1-16. Note that it is not always possible to obtain evenly spaced curves on any part of the display for some transistor types. Set the sweep voltage to approach, but not reach, collector breakdown (Section 1-9).

Note that the curves of Fig. 1-16 are somewhat shorter at higher collector currents. Plot an imaginary line along the ends of the curves as shown in Fig. 1-16. This line is called the *test load line*. Plot an *operating load line* in parallel with the test load line, but intersecting the zero I_C line at the rated operating voltage (at about 13 V in Fig. 1-16). Pick three successive curves which represent the normal operating base current region for the transistor (probably those curves with the most even spacing, 10, 20, and 30 μA in Fig. 1-16). Measure and compare the change in collector current (ΔI_C) between the curves along the operating load line.

As an example, first measure between the 10- and 20-μA curves, then between the 20- and 30-μA curves. If the ΔI_C values are equal, the transistor is operating in a linear region and is free of distortion. If the ΔI_C values are not equal, distortion will be introduced. The greater the difference in ΔI_C values, the greater the distortion.

Measuring Percentage of Distortion. It is possible to measure the approximate percentage of distortion using a family of curves such as shown in

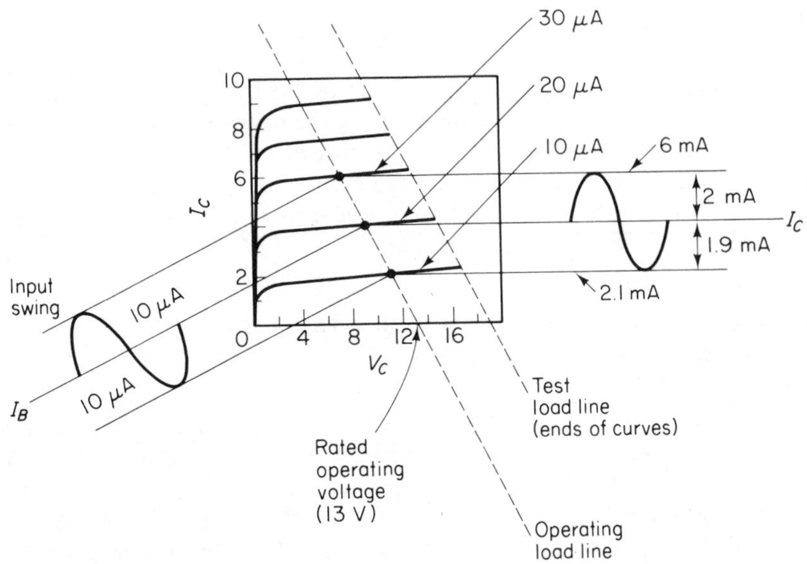

FIGURE 1-16 Testing transistor linearity and distortion using a curve tracer.

Fig. 1–16. The theoretical input and output sine waves superimposed on the curves assume that the transistor is operating in a basic common-emitter amplifier circuit, that the static or no-signal base current I_B is 20 μA, and that the no-signal collector current I_C is 4 mA. A theoretical sine wave signal of 20 μA peak to peak is applied to the base, causing the base current to swing from 10 μA to 30 μA. (In reality, the swing is caused by the 10-μA current steps applied to the base.)

Assuming a beta of 200 and no distortion, the output collector current should then swing from 2 mA to 6 mA, 2 mA above and below the no-signal point of 4 mA. However, as shown in Fig. 1–16, the output collector current swings from 2.1 mA to 6 mA, producing a total swing of 3.9 mA. Thus, there is a 0.1-mA imbalance (the output negative swing is 1.9 mA instead of 2 mA). This amounts to approximately 2.5% distortion (0.1/3.9 = 0.0256). Such distortion might be quite acceptable for some applications (class C amplifiers, frequency multipliers, switching transistors, etc.) but not for other applications (class A and B audio amplifiers).

Using the Operating Load Line. The advantage of using the operating load line rather than some specific value of collector voltage is because the operating load line nearly duplicates the dynamic conditions of operation. The transistor operates with a fixed load and with a fixed supply voltage, but with a signal present the collector voltage does not remain fixed. An increase in collector current reduces collector voltage, and vice versa. However, the load line follows the variation in collector voltage produced by the changing signal.

If a particular transistor shows nearly horizontal collector current curves, there is little difference between using the operating load line or some specific collector voltage. However, in transistors with more slope to the collector current curves (which is usually the case), the collector voltage affects the collector current, and can produce considerably different results in making linearity and distortion measurements.

1-9 TRANSISTOR BREAKDOWN VOLTAGE TESTS USING A CURVE TRACER

The variable collector sweep voltage of a curve tracer can be used to test for breakdown voltage. As sweep voltage is increased, a collector breakdown will be reached. Of course, the value at which this occurs depends upon the transistor type. A typical curve tracer, such as the B&K Precision 501A, tests breakdown up to 100 V, which is sufficient to test all but the high-voltage-rated transistors. At the collector breakdown voltage, the collector current becomes independent of base current, and rises sharply to the current-limiting protection limit of the curve tracer. (If the curve tracer is not provided with a current-limiting feature, the transistor can be destroyed if the collector voltage is maintained at or above the breakdown point for any length of time.)

Figure 1-17 shows a typical family of curves with the sweep voltage set high enough to cause collector breakdown. In the examples shown, breakdown occurs at a collector voltage of approximately 35 to 40 V for the transistor with abrupt breakdown. Note that base current has little effect upon the point at which the increase in collector current occurs, particularly with transistors having an abrupt breakdown characteristic. As shown in Fig. 1-17, it is sometimes difficult to determine the precise breakdown voltage for transistors with a gradual breakdown characteristic.

To perform the measurement, first adjust the curve tracer controls for a normal family of curves on the display. For best results, the normal display should not fill the scale horizontally. Next, increase the sweep voltage control until the upturn in collector current at the tail of the curves is observed. As shown in Fig. 1-17, this upturn is very sharp for most transistors, but can be gradual for a few transistor types.

Read the collector voltage value at which the upturn occurs, using the horizontal scale. If a breakdown voltage specification for the transistor is available, use the figure to determine if the transistor is acceptable for a particular application. If no specifications are available, a simple guideline is that the transistor should withstand approximately twice the collector supply voltage of the circuit in which the transistor is to be used.

When making any breakdown test, *keep the test as short as possible.* Even with the current limiting found in most curve tracers, the current value is

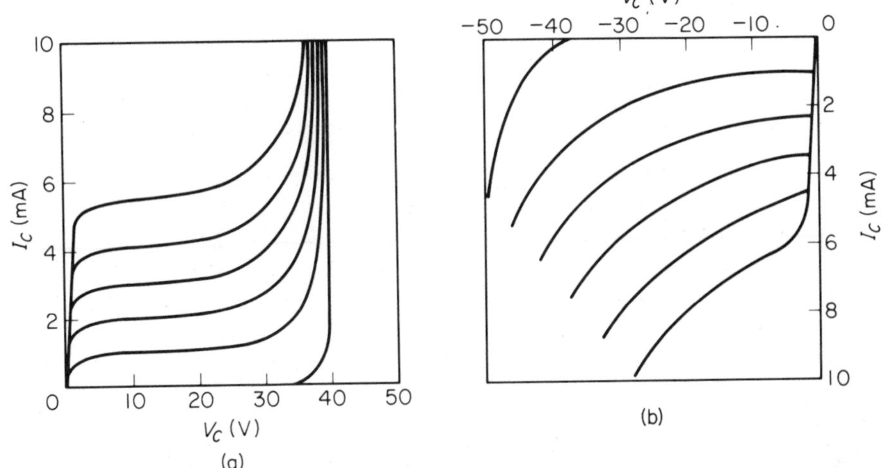

FIGURE 1-17 Transistor breakdown voltage tests using a curve tracer: (a) abrupt breakdown; (b) gradual breakdown.

much higher than normal and causes a temperature increase in the transistor. This can result in damage to the transistor.

1-10 TRANSISTOR LEAKAGE TESTS USING A CURVE TRACER

The zero-base-current curve of a curve tracer can be used to test the collector leakage of a transistor. Collector leakage current is the collector-to-emitter current that flows when the transistor is supposed to be completely off, and is listed as I_{CEO} on most datasheets. If the transistor is leaking, increasing collector voltage causes the collector current to increase independently of the base current. Some collector leakage is normal for most germanium transistors, but not for silicon transistors. Excessive leakage in silicon transistors is usually a sign of a defect. When measured in relation to datasheet specifications, all leakage tests should be made at the specified collector voltage and temperature. Leakage current generally increases with increases in temperature.

Although any sloping line or curve can be used for measurement, leakage can be measured best by observing the zero-base-current line, as shown in Fig. 1-18. This is because there should be no change in collector current (the curve should remain horizontal) when the collector voltage is increased, if there is zero base current and no leakage. Thus, any slope of the zero-base-current curve indicates some leakage. In the example of Fig. 1-18, there is approximately 0.5 mA of leakage at 10 V and about 0.25 mA of leakage at 5 V.

When making a transistor leakage test using a curve tracer, it is possible

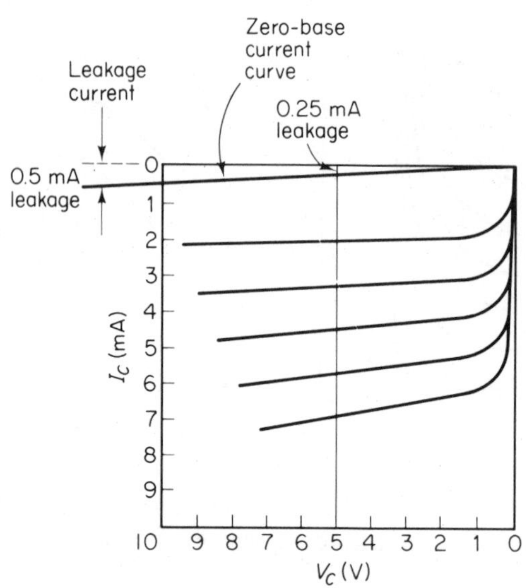

FIGURE 1-18 Transistor leakage tests using a curve tracer.

that some leakage may be indicated, even though the transistor is not leaking. This is because the horizontal input impedance or resistance of the oscilloscope is connected across the collector–emitter of the transistor, as shown in Fig. 1-13. To eliminate any doubt, switch the transistor in and out of the circuit while observing the zero-base-current curve. If the curve remains the same with the transistor switched in and out, the indicated leakage is caused by the oscilloscope impedance.

1-11 TRANSISTOR SATURATION VOLTAGE TESTS USING A CURVE TRACER

A family of curves produced on a curve tracer can be used to test saturation voltage of a transistor. The collector saturation region of a transistor is that portion of the family of curves in the area of low collector voltage and current below the knee of each curve, and is listed as $V_{CE(\text{sat})}$ on most datasheets. This is shown in Fig. 1-19. Notice that the knee of each curve occurs at approximately the same collector voltage, regardless of base current. Also, collector voltage above the knee has little effect upon collector current. Instead, base current has the predominant effect.

Saturation voltage is the collector voltage at the knee of the curve. For measurement in comparison to specifications, both base current and collector current should be stated. The specification value is the *maximum value* at which the knee should occur. Thus, if the specification value is on or above the knee, the transistor is acceptable.

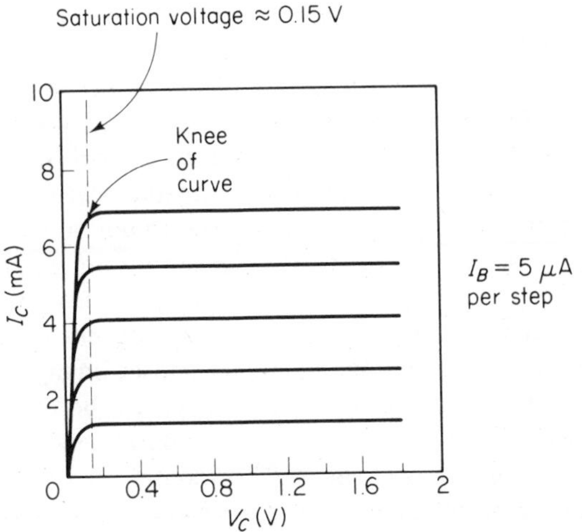

FIGURE 1-19 Transistor saturation voltage tests using a curve tracer.

To measure saturation voltage on a curve tracer, only the saturation region need be displayed. This is the low collector voltage portion up to and including the knee of each curve. If curve tracer controls permit, the display should be spread out using a low-voltage horizontal calibration (collector voltage) such as 0.2 V/division, as shown in Fig. 1-19. This provides the most accurate means of measuring the low collector voltage value that produces saturation.

If desired, you can calculate the *saturation resistance,* listed as $r_{CE(sat)}$ on some datasheets. Saturation resistance equals the collector voltage divided by the collector current for a given value of base current in the collector saturation region. Using the saturation voltage of 0.15 V shown on Fig. 1-19, and a collector current (I_C) of 4 mA, the saturation resistance is 37.5 Ω (0.15/0.004 = 37.5).

Generally, saturation resistance is stated as a maximum acceptable limit. *Dynamic resistance* is found by calculating or plotting the average saturation resistance over a range of base current.

1-12 TRANSISTOR OUTPUT ADMITTANCE AND IMPEDANCE TESTS USING A CURVE TRACER

Output admittance of a transistor can be measured from the same family of curves as displayed for gain or beta measurement. The dynamic output admittance of a transistor, listed as h_{oe} on some datasheets, is the measurement of the change in collector current (ΔI_C) resulting from a specific change in collector voltage (ΔV_C) as a constant base current, as shown in Fig. 1-20. Admittance is measured in mhos. Using the values of Fig. 1–20, the output admittance or h_{oe} is 750 μmhos when the base current is a constant 150 μA.

Output impedance of a transistor is the reciprocal of the output admittance and is measured in ohms. Output impedance can be calculated by transposing the current and voltage values used in determining output admittance. Using the values of Fig. 1–20, the output impedance is 1.333 kΩ (4/0.003 = 1333). Note that a transistor provides maximum power transformer when the load impedance equals the output impedance.

A change in collector voltage normally causes change in collector current. For some transistors, the effect is quite apparent because the curves have a noticeable slope. Such transistors have a comparatively high output admittance. Other transistors show a nearly horizontal curve with a very small change in collector current. Such transistors have a low output admittance.

Both output impedance and admittance measurements are taken at a constant base current (along one of the display curves). If datasheet information is used for reference, use the base current specified. Otherwise, select a

$$\text{Output admittance} = \frac{\Delta I_C}{\Delta V_C} = \frac{3\text{ mA}}{4\text{ V}} = \frac{0.003}{4} = 0.00075 = 750\ \mu\text{mhos}$$

$$\text{Output impedance} = \frac{\Delta V_C}{\Delta I_C} = \frac{4\text{ V}}{3\text{ mA}} = \frac{4}{0.003} = 1333\ \Omega$$

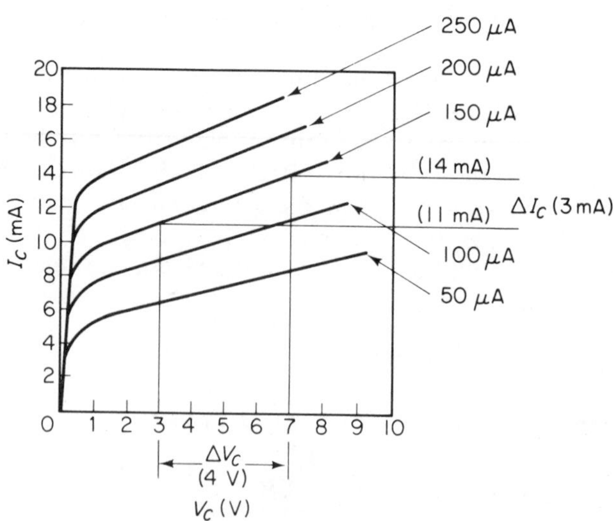

FIGURE 1-20 Transistor output admittance and impedance tests using a curve teacher.

base current curve that is typical for the normal operating range of the transistor being tested. Make the measurement between two specific collector voltages (between 3 V and 7 V on the 150-μA curve, in Fig. 1-20). When testing per datasheet specifications, use the specified voltages. Without specifications, select two voltages, one just above the saturation knee of the curve and one somewhat below collector breakdown.

If you must measure very small collector current values (ΔI_C), it may be necessary to use a higher vertical gain for the oscilloscope. However, the oscilloscope vertical scale *must* remain calibrated.

1-13 TESTING EFFECTS OF TEMPERATURE ON TRANSISTORS USING A CURVE TRACER

When current is conducted by a transistor, some heat is generated. The amount of heat increases with the value of collector voltage and current. An excessive heat buildup can result if the transistor cannot dissipate the heat generated. If excessive heat is generated while testing with a curve tracer, the results are easily detected in the display.

A high temperature can produce a noticeable loop in the curves, as shown in Fig. 1-21(a). Collector capacitance or inductance can also cause a loop in the curves but can be distinguished from a temperature loop. *For a temperature loop, the loop size decreases and disappears when the base current steps or collector voltage are reduced.* The reason for the loop in the curves is that collector current does not increase and decrease at the same rate when the sweep voltage is applied.

Under normal conditions, the transistor is cool when sweep voltage starts from zero. As the sweep voltage increases to maximum, a collector current sweep is made. Meanwhile, the temperature is increasing. The temperature increase causes an additional amount of collector current. As the sweep voltage returns to zero, the collector current decreases, but with a lag. During the return sweep, the temperature drops from maximum to normal, but a time lag is required for cooling to occur. Thus, the top portion of the loop is the increasing sweep current, and the bottom portion of the loop is the decreasing sweep current.

Another effect in some transistors is that collector current droops at the high end of the curve. In this case, the increase in temperature causes a decrease in collector current, as shown in Fig. 1-21(b).

Excessive current to the extent that *thermal runaway* begins is observed as a "vertical roll" effect, as shown in Fig. 1-21(c). Thermal runaway results when an increase in temperature, caused by increased current, results in further current increases (and a further temperature increase). If thermal runaway continues, the transistor will be burned out or permanently dam-

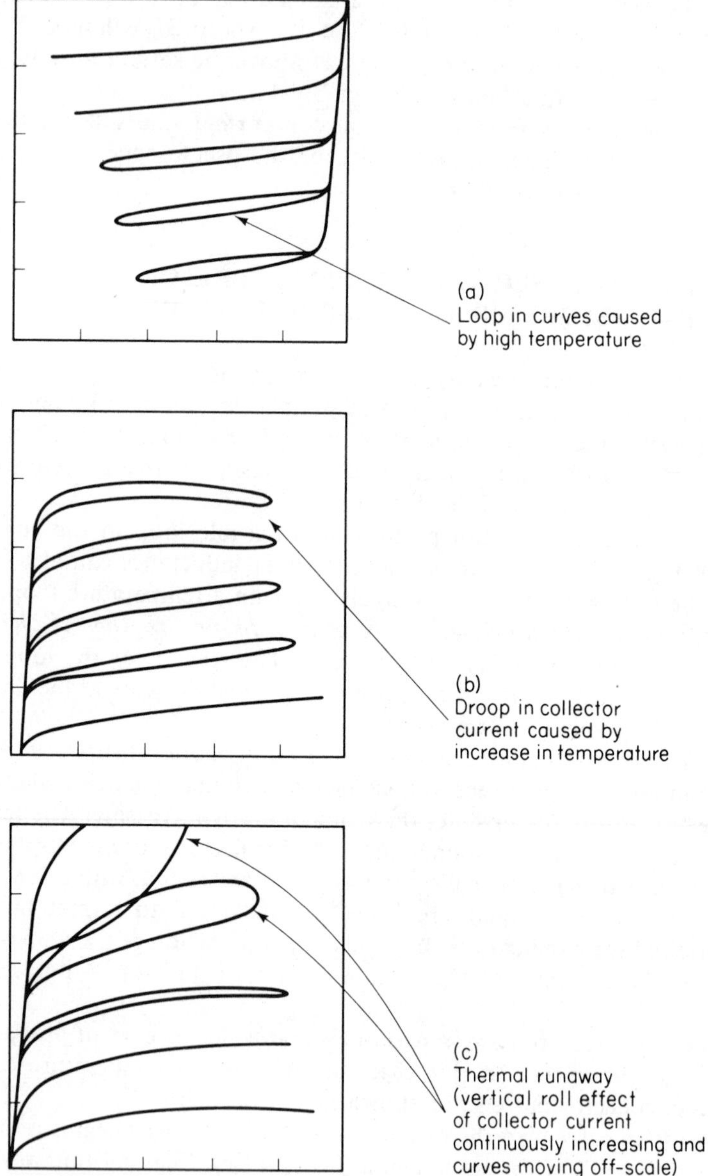
(a)
Loop in curves caused by high temperature
(b)
Droop in collector current caused by increase in temperature
(c)
Thermal runaway (vertical roll effect of collector current continuously increasing and curves moving off-scale)

FIGURE 1-21 Testing effects of temperature on transistors using a curve tracer.

aged. When thermal runaway starts as a transistor is being tested on a curve tracer, the entire family of curves moves in the direction of high collector current (as temperature buildup continues).

CAUTION: Turn off the curve tracer immediately if you observe any signs of thermal runaway (a vertical roll of all the curves, usually starting with roll of the highest base current curves).

2

Field-Effect Transistor Tests

This chapter is devoted entirely to test procedures for field-effect transistors (FETs). The first sections of the chapter describe FET characteristics and test procedures from the practical standpoint. The information in these sections permits you to test all important FET characteristics using basic shop equipment, and also help you understand the basis for such tests. The remaining sections describe how the same tests, and additional tests, are performed using more sophisticated equipment, such as curve tracers and signal generators.

2-1 FET OPERATING MODES

As in the case of two-junction transistors (Chapter 1), FETs must be tested on both a static and a dynamic basis. Static tests indicate the FET response to d-c variations. Dynamic tests show the response of a FET to alternating currents or signals. Keep in mind that FETs operate in three modes (type A, depletion only; type B, depletion/enhancement; and type C, enhancement only). Figure 2–1 shows the relationships of the three types. Also, although both JFETs and MOSFETs operate on the principle of a "channel" current controlled by an electric field, the control mechanisms for the two are different, resulting in considerably different characteristics. The main difference between the JFET (junction FET) and the MOSFET (metal-oxide-semiconductor FET) is in the

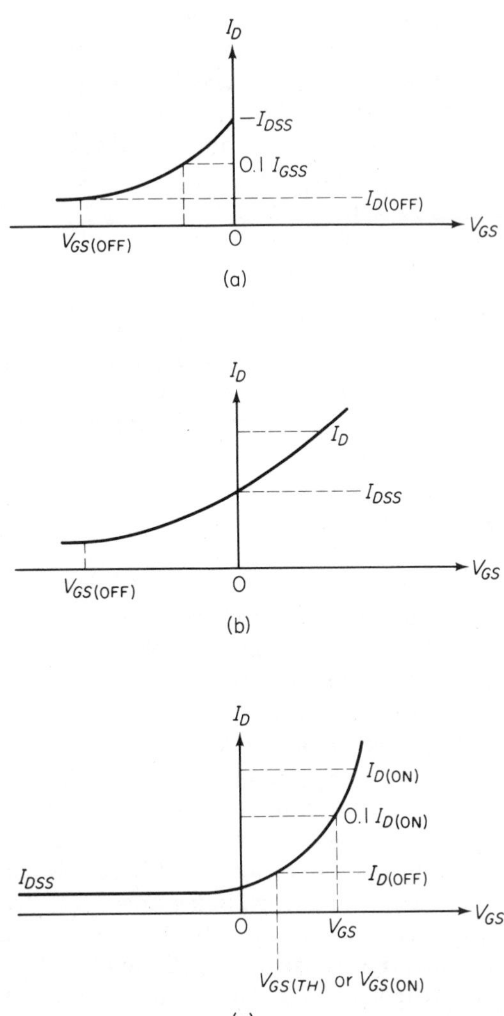

FIGURE 2-1 FET operating modes and simplified transfer characteristics: (a) type A, depletion mode; (b) type B, depletion/enhancement mode; (c) type C, enhancement mode.

gate characteristics. The input of the JFET acts like a reverse-biased diode, whereas the input of a MOSFET is similar to a capacitor.

The basic differences among the three FET operating modes can be understood most easily by examining the transfer characteristics of Fig. 2-1.

The *depletion-only FET* is classified as type A and has considerable drain-current (I_D) flow for zero gate voltage. No forward gate voltage is used. Maximum drain-current flow occurs when the gate-source voltage (V_{GS}) is zero. Drain current is reduced by applying a reverse voltage to the gate terminal. That is, I_D decreases when reverse V_{GS} is increased.

The depletion/enhancement FET is classified as type B and also has considerable drain-current flow for zero V_{GS} (but not as much as type A). Drain current is increased by application of a forward gate voltage and reduced by application of a reverse gate voltage. For a JFET used as type B, drain current can be increased by gate voltage only until the gate–source PN junction becomes forward-biased. At this point, a further increase in forward gate voltage does not produce an increase in drain current.

The *enhancement-only FET* is classified as type C and has little or no current flow for zero gate voltage. Drain current does not occur until a forward gate voltage is applied. This voltage is known as the threshold voltage and is indicated in Fig. 2–1 as $V_{GS(TH)}$. Once the threshold voltage is reached, the transfer characteristics for a type C FET are similar to those of a type B.

2-2 HANDLING MOSFETS

Electrostatic discharges can occur when a MOSFET is picked up by its case and the handler's body capacitance is discharged to ground through the series arrangement of the bulk-to-channel and channel-to-gate capacitances of the device. For this reason, MOSFETs require proper handling, particularly when the MOSFET is out of the circuit for test. In-circuit a MOSFET is just as rugged as any other solid-state component of similar size and construction.

MOSFETs are generally shipped with the leads all shorted together by means of a spring to prevent damage in shipping and handling (the short prevents static discharge between leads). The shorting spring, or similar device, should not be removed until after the MOSFET is connected into the test circuit. An alternative method for shipping or storing MOSFETs is to apply a conductive foam between the leads. Such a foam is manufactured by Emerson & Cuming, Inc., and is listed as ECCOSORB LS26.

Note that polystyrene insulating "snow" is not recommended for shipment or storage of MOSFETs. Such "snow" can acquire high static charges that can discharge through the device.

When it becomes necessary to test an installed MOSFET, consider the following points.

1. When removing or installing a MOSFET, first turn the power off. If the MOSFET is to be removed, your body should be at the same potential as the unit from which the device is removed. This can be done by placing one hand on the chassis before removing the MOSFET. If the MOSFET is to be connected to an external tester, put the hand holding the MOSFET against the tester panel or chassis

and connect the leads from the tester to the MOSFET leads (or plug in the MOSFET). This procedure prevents possible damage from static charges on the tester (or external test circuit).

CAUTION: Make certain that the chassis or external unit is at ground potential before touching it. In some obsolete or defective equipment, the chassis or panel is above ground (typically by one-half of the line voltage).

2. When handling a MOSFET, the leads should be shorted together. Generally, this is done in shipment by a shorting ring or spring as discussed. When testing MOSFETs, connect the tester lead to the MOSFET (preferably the source lead first); then remove the shorting ring.
3. When soldering or unsoldering a MOSFET, the soldering tool tip should be at ground potential (no static charge). Connect a clip lead from the barrel of the soldering tool to the chassis or tester case. The use of a soldering gun is generally not recommended for MOSFETs.
4. Remove power to the circuit before inserting or removing a MOSFET (or a plug-in module containing a MOSFET). The voltage transients developed when terminals are separated may damage the MOSFET. This same caution applies to circuits with conventional transistors. However, the chances of damage are greater with MOSFETs.

2-3 MOSFET PROTECTION CIRCUITS

Because of the static-discharge problem described in Section 2–2, manufacturers provide some form of protection for a number of their MOSFETs. Generally, this protection takes the form of a diode incorporated as part of the MOSFET substrate material. Figure 2–2 shows the relationship of the protective diode to other elements in the MOSFET. Resistors R_1 and R_2 represent the channel resistance. Capacitor C_{IN} represents the gate-to-channel capacitance.Typically, the gate-voltage-handling capability of a MOSFET (without protection) is not over 30 V, although some MOSFETs can withstand gate-to-substrate voltages of about 100 V and not result in breakdown. With any MOSFET, however, once the oxide insulation breaks down, the device is destroyed.

The ideal situation in gate protection is to provide a signal-limiting configuration that permits a typical sine wave to be handled without clipping or distortion. The signal-limiting devices must limit all transient voltages that exceed the gate breakdown voltage. For example, if the gate breakdown is 30 V, transient voltage must be limited to something less (say, 25 V). The diode con-

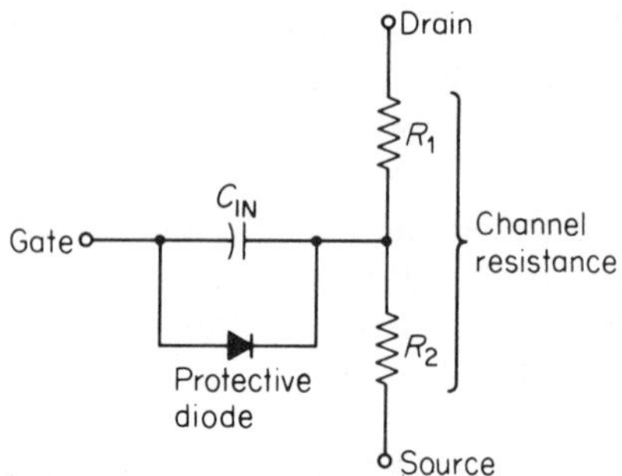

FIGURE 2-2 Static-discharge protection for MOSFETs provided by single-gate diode.

nected in parallel with C_{IN} (Fig. 2-2) is one method of limiting input voltages to a safe level and is particularly popular with single-gate MOSFETs. However, the method shown in Fig. 2-2 has certain limitations. One problem is that the single diode of Fig. 2-2 clips the positive peaks of a sine wave when the MOSFET is operated near zero bias. This problem can be overcome by using two diodes back to back as shown in Fig. 2-3. The back-to-back method is generally used with dual-gate MOSFETs, and each gate is provided with a separate pair of back-to-back diodes.

There are other methods for protecting MOSFETs, each with its own advantages and disadvantages. However, in testing a MOSFET for breakdown, the main point to remember is that on a MOSFET with gate protection, the resultant breakdown voltage is that of the protective device (diode), not the MOSFET. Also, on those FETs with protective diodes, the datasheets often list the protective diode clamp voltage. This voltage is often referred to as the *knee* voltage, or V_{KNEE}. The input signal voltages cannot exceed this value.

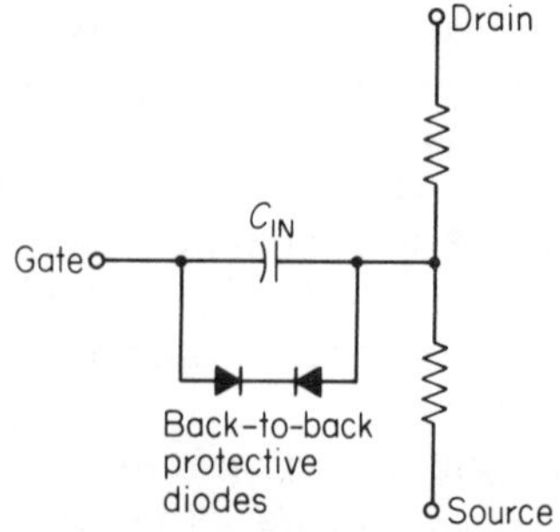

FIGURE 2-3 Static-discharge protection for MOSFETs provided by back-to-back gate diodes.

2-4 FET CONTROL VOLTAGE TESTS

The gate–source voltage or V_{GS} is considered to be the control voltage (or signal) for a FET. That is, the amount of V_{GS} controls the amount of source–drain current. $V_{GS(OFF)}$ is the gate voltage necessary to reduce the drain–source current (generally specified as I_D) to zero or to some specified value near zero. Often, $V_{GS(OFF)}$ is defined as the gate voltage required to reduce the drain current to 0.01 or preferably 0.001 of the minimum I_{DSS} value. For example, assume that 1 mA flows when V_{GS} is zero. This means that I_{DSS} is 1 mA. Under such conditions, $V_{GS(OFF)}$ is the gate voltage required to produce a drain current of 0.01 mA (or 0.001 mA).

Although the term $V_{GS(OFF)}$ can be applied to all operating modes, it is best applied to the depletion and depletion/enhancement modes (Section 2–1). In the enhancement-only mode, the term $V_{GS(TH)}$ or *threshold gate voltage* is more descriptive. $V_{GS(TH)}$ is the gate voltage of an enhancement-only-mode FET where I_D just starts to flow. Some datasheets specify $V_{GS(OFF)}$ as the gate voltage that produces some given value of I_D, such as 1 picoampere (pA) or 1 nA. The term $V_{GS(ON)}$ is sometimes used in place of $V_{GS(TH)}$ for enhancement-only FETs.

No matter what term is used, the value of $V_{GS(OFF)}$ determines the minimum operational drain–source voltage V_{DS} (Section 2–5). As a guideline, V_{DS} must be a minimum of 1.5 times $V_{GS(OFF)}$ [or $V_{GS(TH)}$ or $V_{GS(ON)}$].

2-4.1 $V_{GS(OFF)}$ Test

The basic circuit for the $V_{GS(OFF)}$ test is shown in Fig. 2–4. This test applies to depletion-only- and depletion/enhancement-mode FETs. As shown, V_{DS} is set at some fixed value and reverse-bias V_{GS} is adjusted until I_D is at some specific negligible value. This is essentially a cutoff value test. As an example, the N-channel 3N128 specifies a V_{DS} of 15 V and an I_D of 50 μA. The V_{GS} should be between -0.5 and -8 V for the 50 μA of I_D. As a guideline, V_{DS} should be a minimum of 1.5 times $V_{GS(OFF)}$ to provide proper circuit operation.

2-4.2 $V_{GS(TH)}$ or $V_{GS(ON)}$ Test

$V_{GS(TH)}$ or $V_{GS(ON)}$ are essentially the same specification as $V_{GS(OFF)}$, except that $V_{GS(ON)}$ and $V_{GS(TH)}$ are generally applied to enhancement-only-mode devices. The basic circuit of Fig. 2–4 is used, except that V_{GS} is connected for forward bias (gate positive for N channel). A forward bias is required since I_D flow is an enhancement-mode device only when the gate is forward-biased. During test, V_{GS} is increased from zero volts until the I_D is at a specific value.

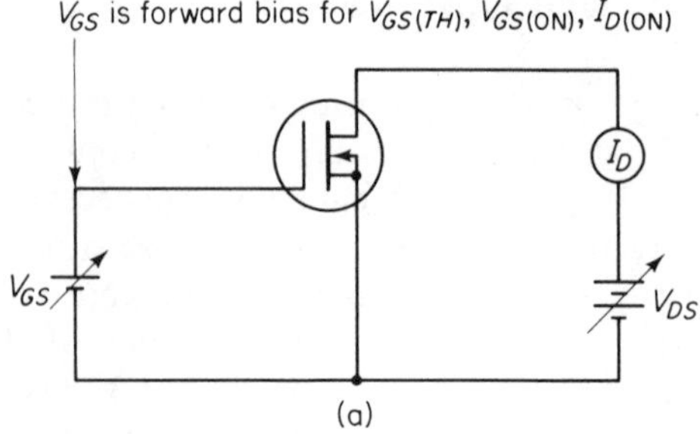

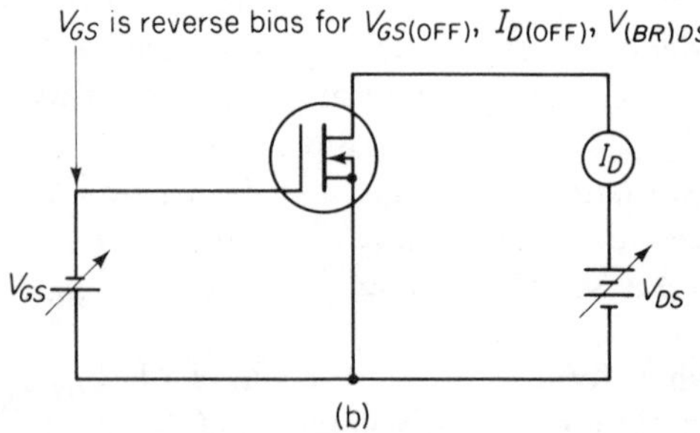

FIGURE 2-4 Test circuit for $V_{GS(OFF)}$, $V_{GS(TH)}$, $V_{GS(ON)}$, $I_{D(ON)}$, $I_{D(OFF)}$, and $V_{(BR)DSX}$. Polarities shown are for N-channel MOS devices. Reverse all polarities for P-channel.

2-5 FET OPERATING VOLTAGE TESTS

The drain–source voltage V_{DS} is considered to be the operating voltage for a FET. V_{DS} is the equivalent of collector voltage in a two-junction transistor or plate voltage in a vacuum tube. Generally, the only concern is that the maximum datasheet value for V_{DS} must not be exceeded. In most applications, however, V_{DS} must be a minimum of 1.5 times V_{GS}.

When FETs are used as switches and choppers, the terms $V_{DS(ON)}$ and $V_{DS(OFF)}$ sometimes appear on the datasheets. $V_{DS(ON)}$ is similar to saturation voltage in two-junction transistors. $V_{DS(OFF)}$ is the drain–source voltage at which the I_D increases very little for an increase in drain–source voltage, with V_{GS} held at zero. Sometimes, the term V_p, the pinch-off voltage, is used instead. V_p generally applies to JFETs, however, rather than to MOSFETs.

The various values of V_{DS} are dependent on source resistance R_S, drain resistance R_D, or drain-to-source resistance R_{DS}. All of these are static, d-c values and rarely appear on datasheets. Dynamic values are generally far more important. Although FETs do not usually have an operating voltage between drain and gate, the term V_{DG} usually appears on most FET datasheets. Generally, this is a maximum voltage and is usually the same as V_{DS} maximum.

2-5.1 $V_{DS(OFF)}$ and $V_{DS(ON)}$ Saturation Tests

The basic circuit for saturation tests is shown in Fig. 2-5. Saturation tests are generally of concern only when the FET is used for switch or chopper applications. During either test, V_{GS} is held at some fixed value, and V_{DS} is increased until I_D is at maximum, or increases very little for further increases in V_{DS}. Take care not to exceed the maximum V_{DS} when making this test. Generally, $V_{DS(ON)}$ is made with some specific value of V_{GS}, whereas $V_{DS(OFF)}$ is made with V_{GS} at zero (gate shorted to source).

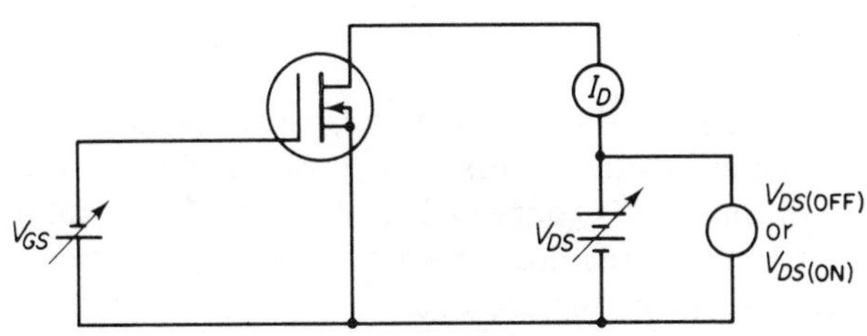

FIGURE 2-5 Test circuit for $V_{DS(OFF)}$ and $V_{DS(ON)}$.

2-6 FET OPERATING CURRENT TESTS

The drain–source current, or I_D, is considered as the operating current for a FET. I_D is the equivalent of collector current in a two-junction transistor, or plate current in a vacuum tube. Generally, the only concern is that the maximum datasheet value for I_D must not be exceeded.

In addition to I_D, the terms $I_{D(ON)}$ and $I_{D(OFF)}$ are used on some datasheets. $I_{D(ON)}$ is an arbitrary current value (usually near the maximum rated current) that locates a point in the enhancement operating mode, as shown in Fig. 2-1. $I_{D(OFF)}$ is the current value that flows when $V_{GS(OFF)}$ is applied. On some FETs operating in the enhancement-only mode, the datasheet value of $I_{D(OFF)}$ is the current when $V_{GS(TH)}$ is applied.

2-6.1 I_{DSS} Test

The basic circuit for the I_{DSS} test is shown in Fig. 2-6. This test is applied to depletion-only- and depletion/enhancement-mode devices. As shown, V_{DS} is set at some fixed value, and V_{GS} is zero (gate shorted to source). I_{DSS} is a zero-

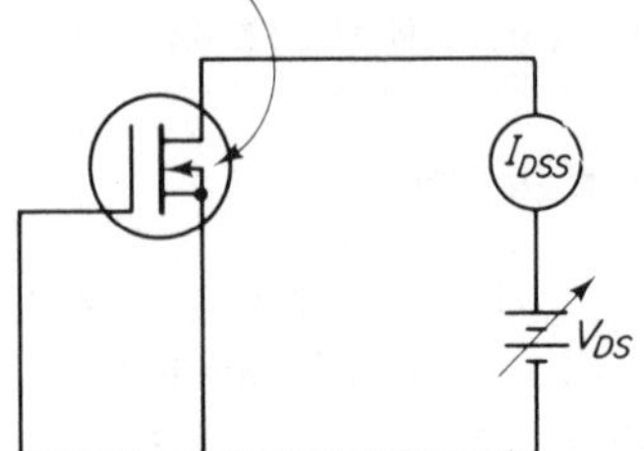

FIGURE 2-6 Test circuit for I_{DSS} and $V_{(BR)DSS}$.

bias current flow test. For example, the 3N128 specifies a V_{DS} of 15 V and V_{GS} of zero. Under these conditions, the I_D should be between 5 and 25 mA.

2-6.2 $I_{D(ON)}$ Test

$I_{D(ON)}$ is essentially the same specification as I_{DSS}, except that $I_{D(ON)}$ is generally applied to enhancement-mode devices (type B or C). The basic circuit of Fig. 2-4 is used, except that V_{GS} is connected for forward bias (gate positive for N channel). Forward bias is required since I_D flows in an enhancement-mode device only when the gate is forward-biased. Some I_D flows in a class B device without forward bias. However, $I_{D(ON)}$ is the current near the maximum (or saturation current). During test, V_{GS} and V_{DS} are set to some fixed value, and the resulting value of $I_{D(ON)}$ is measured.

2-6.3 $I_{D(OFF)}$ Test

The test circuit for $I_{D(OFF)}$ is the same as for $V_{GS(OFF)}$, as shown in Fig. 2-4. This is because $I_{D(OFF)}$ is the current that flows when $V_{GS(OFF)}$ is applied. V_{DS} and V_{GS} are set to some fixed value, and the resulting value of $I_{D(OFF)}$ is measured. For example, the 3N128 specifies a V_{DS} of 20 V and a V_{GS} of -8 V. Under these conditions, the I_D should be 50 μA maximum. On enhancement-only-mode devices (type C), the value of $I_{D(OFF)}$ is the I_D that flows when $V_{GS(TH)}$ is applied. Thus, the fixed value of V_{GS} is always some forward-bias voltage.

2-7 FET BREAKDOWN VOLTAGE TESTS

Voltage breakdown is a particularly important characteristic of FETs. There are a number of specifications to indicate the maximum voltage that can be applied to the various elements. These include:

$V_{(BR)GSS}$, *gate-to-source breakdown voltage,* which is the breakdown voltage from gate to source, with the drain and source shorted. Under these test conditions, the gate–channel junction (PN diode for a JFET; oxide layer for a MOSFET) also meets the breakdown specification since the drain and source are the connections to the channel. For the designer, this means that the drain and source may be interchanged, for symmetrical devices, without fear of the individual junction breakdown. For MOSFETs, the gate is insulated from the drain–source and channel by an oxide layer. Thus, the gate breakdown voltage is dependent on the thickness and purity of the layer. $V_{(BR)GSS}$ for a MOSFET is the voltage that will physically puncture the oxide layer.

$V_{(BR)DGO}$, *drain-to-gate breakdown voltage,* is essentially the same specification as $V_{(BR)GSS}$, except that $V_{(BR)DGO}$ represents breakdown from gate to drain. $V_{(BR)DGO}$ is more properly applied to JFETs but may appear on a MOSFET datasheet.

Drain-to-source breakdown voltage for a FET is determined by the operating mode. For enhancement-only-mode operation (type C), with the gate connected to the source (in the cutoff condition) and the substrate floating, there is no effective channel between drain and source, and the applied drain–source voltage appears (in effect) across two back-to-back-connected, reverse-biased diodes. These diodes are represented by the source-to-substrate and substrate-to-drain junctions.

During drain-to-source breakdown tests, drain current remains at a very low (typically picoampere) level as drain voltage is increased until the drain voltage reaches a value that causes reverse (*reach-through* or *punch-through*) breakdown of the diodes. This condition, generally represented on datasheets as $V_{(BR)DSS}$, is indicated by an increase in I_D.

For types A and B FETs, the $V_{(BR)DSS}$ designation is sometimes replaced by $V_{(BR)DSX}$. The main difference between the two designations is the replacement of the last subscript S with the subscript X. The S normally indicates that the gate is shorted to the source. The X indicates that the gate is biased to cutoff or beyond. To obtain cutoff in types A and B FETs, a depletion bias voltage must be applied to the gate.

In some datasheets, there may be ratings and specifications indicating a maximum voltage that may be applied between the individual gates and the drain or source, between drain and source, and so on. Not all of these specifications are found on every datasheet. Some of the specifications provide the same information in slightly different form. By understanding the various breakdown mechanisms, however, it should be possible to interpret the intent of each specification and rating.

For example, in MOSFET specifications, the breakdown voltage is generally interchangeable with the term avalanche voltage (V_A). Such avalanche voltage is the V_{DS} that causes the I_D to go "full-on," sometimes to the point of destroying the device. Note that avalanche occurs at a lower value of V_{DS} when the gate is reverse-biased than when the gate is at zero bias. This condition is caused by the fact that the reverse-bias gate voltage adds to the drain voltage, thus increasing the effective voltage across the junction. As a guideline, the maximum V_{DS} that may be applied is the drain-gate breakdown voltage, less the V_{GS}.

2-7.1 JFET Voltage Breakdown Tests

For JFETs, $V_{(BR)GSS}$ is the maximum voltage that may be applied between any two terminals and is the lowest voltage that leads to breakdown of the gate junction. As shown in Fig. 2-7, an increasingly high reverse voltage is applied between the common gates and source. Junction breakdown can be determined by the gate current (beyond normal I_{GSS}), which indicates the beginning of an avalanche (V_A) condition.

2-7.2 MOSFET Voltage Breakdown Tests

For MOSFETs, $V_{(BR)GSS}$ is the breakdown voltage from gate to source and *is a maximum voltage, not a test voltage.* For MOSFETs, $V_{(BR)GSS}$ is the voltage at which the gate oxide layer breaks down. If this full voltage is applied during a test, the oxide layer can (and probably will) be damaged. If the test is performed on a MOSFET with gate protection (Section 2-3), the resultant breakdown voltage is that of the protective diode, not the MOSFET. Thus, $V_{(BR)GSS}$ tests are usually made only during manufacture (as part of destructive testing to establish breakdown for a particular type of device). To perform a $V_{(BR)GSS}$ test, the drain and source are shorted and increasing V_{GS} is applied until gate current I_G suddenly increases, indicating puncture of the gate oxide layer. Note that $V_{(BR)DGO}$ is essentially the same specification as $V_{(BR)GSS}$, except that breakdown is from gate to drain, and the same specifications apply.

$V_{(BR)DSX}$ applies to types A and B MOSFETs. The basic circuit of Fig. 2-4 can be used. During test, reverse-biased V_{GS} is applied until the device is cut off. Usually, the $V_{GS(OFF)}$ value is used. Then V_{DS} is increased until there is some heavy flow of I_D.

$V_{(BR)DSS}$ applies to type C MOSFETs. The basic circuit of Fig. 2-6 can be used, except that the substrate must be disconnected from the source (floating substrate). Of course, the substrate cannot be disconnected on those MOSFETs with internal connections between substrate and source or other

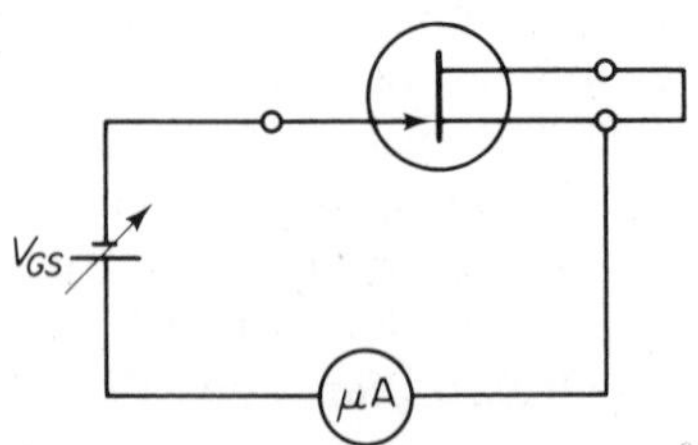

FIGURE 2-7 Test circuit for JFET breakdown voltage.

element. For type C devices, the gate and source are shorted since no reverse-biased V_{GS} is necessary for cutoff. With V_{GS} at zero, V_{DS} is increased until there is some heavy flow of I_D.

2-8 FET GATE LEAKAGE TESTS

Gate leakage is a very important characteristic of a FET since such leakage is directly related to input resistance. When gate leakage is high, input resistance is low, and vice versa. Gate leakage is usually specified as I_{GSS} (reverse-biased gate-to-source current with drain shorted to source) and is a measure of the static short-circuit input impedance.

Normally, input resistance of a FET is very high, particularly for a MOSFET, since the leakage current across a capacitor is very small. For example, at a temperature of 25°C, typical MOSFET input resistance R_{GS} is in the order of 10^{14} Ω. Of course, JFET input resistance values are lower since their input is essentially a diode junction rather than a capacitor.

Input resistance may decrease when temperature is raised to the typical maximum of +175°C. FETs are not drastically affected by temperature (particularly MOSFETs), however, and their input resistance remains high even at maximum operating temperature.

Some datasheets specify gate leakage current as I_{GDO} (leakage between gate and drain with the source open). Other datasheets use the term I_{GSO} (leakage between gate and source with the drain open). When these characteristics are used, the gate leakage current is lower than when I_{GSS} is specified. Consequently, I_{GDO} and I_{GSO} do not represent the "worst-case" condition, making I_{GSS} the preferred specification.

2-8.1 I_{GSS} Test

The basic circuit for the I_{GSS} test is shown in Fig. 2-8. This test applies to all modes of operation. As shown, V_{DS} is zero (drain shorted to source), V_{GS} is set to some given (reverse-biased) values, and the resultant value of leakage current I_G is measured. For example, the 3N128 specifies a V_{DS} of zero and a V_{GS}

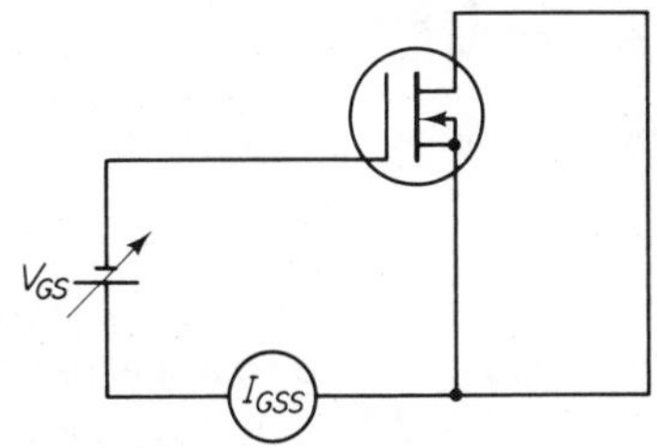

FIGURE 2-8 Test circuit for I_{GSS}.

of -8 V. Under these conditions, I_{GSS} should not exceed 50 pA (at 25°C) or 5 nA (at 125°C).

2-9 DUAL-GATE FET TESTS

In addition to the static characteristics described thus far, dual-gate FETs have certain other characteristics that may require test. For example, with dual-gate MOSFETs, both gates control I_D. As a result, there are a number of datasheet specifications that describe how the voltage on one gate affects I_D, with the other gate held at some specific voltage, or zero volts. Similarly, there are specifications that describe how gate leakage or gate current flow is affected by gate voltages. For practical situations, the most important dual-gate characteristics include *cutoff voltage, breakdown voltage,* and *gate current.*

With a dual-gate MOSFET, the gate 1-to-source cutoff voltage (with gate 2 connected to the source) is identified as $V_{G1S(OFF)}$; the gate 2-to-source cutoff voltage (with gate 1 connected to the source) is identified as $V_{G2S(OFF)}$. Note that dual-gate MOSFETs are not always symmetrical. That is, the gate voltages in different combinations produce different drain currents.

The gate 1-to-source breakdown voltage (with gate 2 connected to the source) is identified as $V_{(BR)G1SS}$; gate 2-to-source breakdown voltage is $V_{(BR)G2SS}$. The gate 1 leakage current is identified as I_{G1SS}; gate 2 leakage current is I_{G2SS}.

With a dual-gate MOSFET, gate leakage should be symmetrical, even though the gate-voltage versus drain-current characteristics may not be symmetrical. Also, gate leakage or gate current should be the same when gate voltages are reversed in polarity. Thus, dual-gate MOSFET datasheets often specify forward and reverse breakdown voltages as well as gate leakage currents. As is the case with single-gate, dual-gate FETs have a maximum recommended gate current. Gate currents for dual-gate FETs are usually identified as I_{G1SSF} and I_{G2SSF} (for forward) or I_{G1SSR} and I_{G2SSR} (for reverse).

Dual-gate FETs have a zero-bias drain-current specification I_{DS}, which corresponds to an I_{DSS} for single-gate units. The measurement for I_{DS} is usually based on 0 V at one gate, however, with a fixed voltage at the opposite gate.

2-9.1 $V_{G1S(OFF)}$ and $V_{G2S(OFF)}$ Tests

These tests apply to dual-gate devices, operating in the depletion and depletion/enhancement modes. Dual-gate devices can also be tested for $V_{GS(OFF)}$, as described in Section 2-4.1, when both gates are tied together. The basic circuit for $V_{G1S(OFF)}$ and $V_{G2S(OFF)}$ is shown in Fig. 2-9. As shown, V_{DS} is set at some specific value, one gate is forward-biased to a specific value and the other gate is reverse-biased. The reverse-biased V_{G1S} or V_{G2S} is adjusted until I_D is at some negligible value, indicating a cutoff condition.

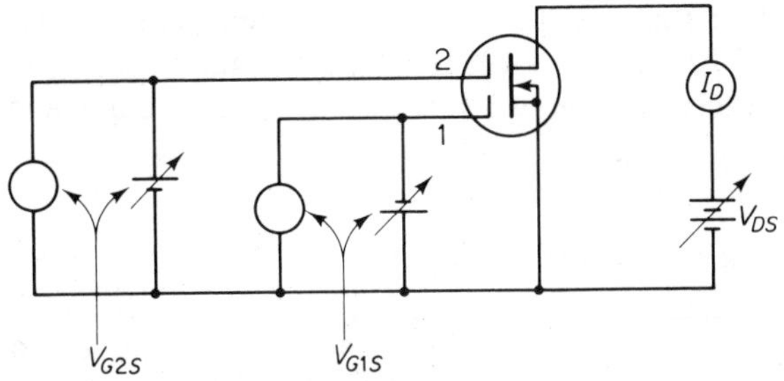

FIGURE 2-9 Test circuit for $V_{G1S(OFF)}$ and $V_{G2S(OFF)}$.

2-9.2 *Dual-Gate Voltage Breakdown Tests*

Dual-gate MOSFETs are often tested for forward and reverse gate-to-source breakdown voltage. The basic test circuit is shown in Fig. 2-10. As shown, V_{DS} and one gate are at 0 V (both shorted to source). A variable voltage is applied to the opposite gate, and the gate current is measured. For example, to measure $V_{(BR)G1SSF}$, a 40841 MOSFET specifies a V_{DS} and V_{G2S} of 0 V, and an I_{G1SSF} of 100 μA, with a typical 9 V of V_{G1} applied. $V_{(BR)G1SSR}$ is measured in the

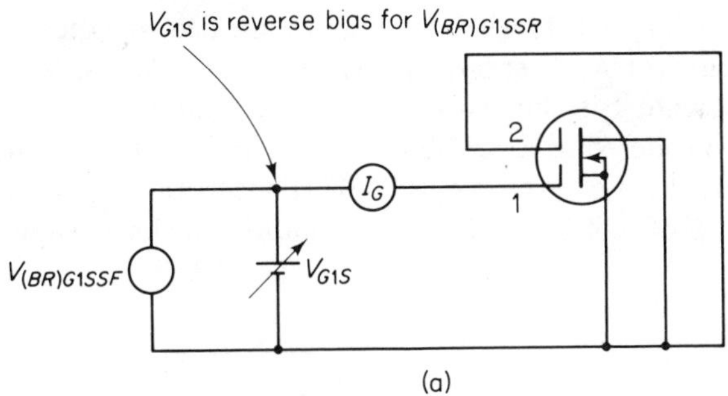

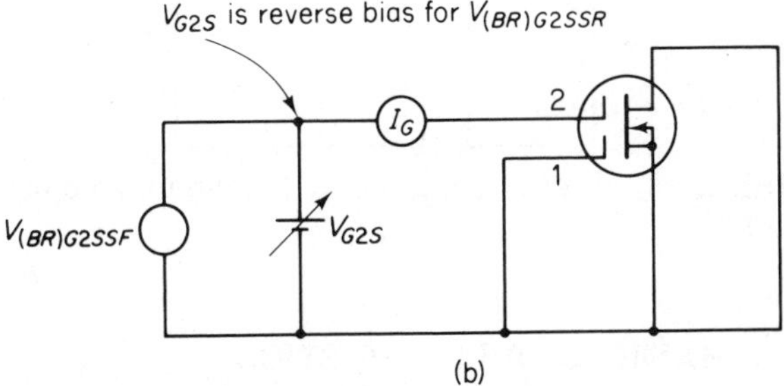

FIGURE 2-10 Test circuits for dual-gate voltage breakdown and current.

same way, except that the gate is reverse-biased. The results should be the same (100 μA of I_{G1SSR} for the 40841). $V_{(BR)G2SSF}$ and $V_{(BR)G2SSR}$ are measured in the same way, except that gate 1 is connected to the source and the voltage is applied to gate 2.

2-9.3 Dual-Gate Current Tests

Dual-gate current tests use the same basic test circuit as for gate-voltage breakdown tests, as shown in Fig. 2-10. The difference in procedure is that the gate voltage is set to a specific value and the resultant current is measured. For example, to measure I_{G1SSF}, a 40841 MOSFET specifies a V_{DS} and V_{G2S} of 0 V and a V_{G1} of 6 V. A maximum I_{G1SSF} of 60 nA should flow under these conditions. I_{G1SSR} is measured the same way, except that V_{G1S} is −6 V. I_{G2SSF} and I_{G2SSR} are measured in the same way, except that gate 1 is connected to the source and the voltage is applied to gate 2.

2-9.4 I_{DS} Test

This test applies to dual-gate devices, operating in the depletion and depletion/enhancement modes. Dual-gate devices can also be tested for I_{DSS} as described in Section 2-6.1, when both gates are tied together.

The circuit for I_{DS} is shown in Fig. 2-11. As shown, V_{DS} is set at some specific value, gate 1 is shorted to the source, and gate 2 is set to a specific value. I_{DS} is considered as a zero-bias current flow test, even though one gate has a forward bias. For example, the 40841 specifies a V_{DS} of 15 V, a V_{G1S} of 0 V, and a V_{G2S} of +4 V. Under these conditions, the I_D should be a typical 10 mA.

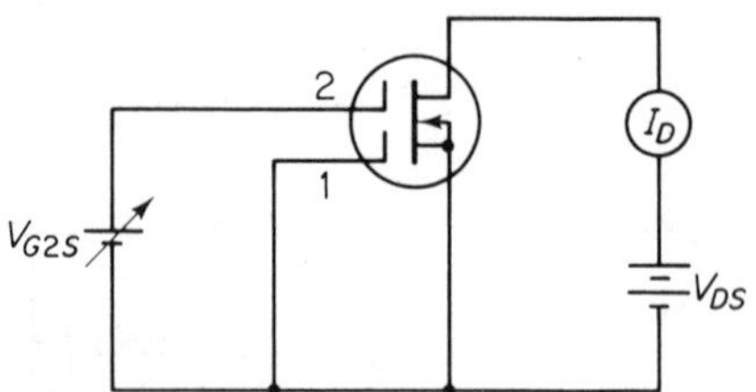

FIGURE 2-11 Test circuits for I_{DS} (dual-gate zero-bias current test).

2-10 FET DYNAMIC CHARACTERISTICS

Unlike the static characteristics described thus far, the dynamic characteristics (ac or signal) of FETs apply equally to types A, B, and C. However, conditions and presentations of the dynamic characteristics depend mostly on the intended application. We will not get into FET applications here, but will concentrate on what, why, and how of test procedures. If you want a thorough description of how FET characteristics relate to applications, your attention is invited to the author's best-selling *Handbook for Transistors* (Prentice-Hall, Inc., Englewood Cliffs, N.J., 1977).

2-10.1 Y-Parameter Tests

The *Y*-parameter tests are probably the most important dynamic tests for any FET application. *Y*-parameter tests establish the four basic admittances (forward transadmittance, reverse transadmittance, input admittance, and output admittance) required in the design of circuits using FETs.

Without going into any elaborate theory, admittance is the reciprocal of impedance and is composed of conductance (*g*, the real part of admittance) and susceptance (*b*, the imaginary part of admittance). From a test standpoint, the Y parameters can be expressed as complex numbers, where both the *g* and *b* values are given, or as a simple number in mhos. This is the same as expressing impedance (*Z*) in simple terms of ohms, or as a complex number composed of resistance (*R*) and reactance (*X*).

When it is necessary to test a FET to find a complex number (to find both the conductance and susceptance), it is necessary to use an *admittance meter* or possibly an R_X meter. An admittance meter consists essentially of a signal source and a bridge circuit. The *Y* parameter under test is connected to form one leg of the bridge. The signal source is adjusted to the frequency of interest and applied to the bridge. The bridge conductance and susceptance leg elements (usually a resistor and a capacitor) are adjusted until the bridge is balanced. The conductance and susceptance values required to produce balance are equal to the *Y*-parameter conductance and susceptance, at that frequency.

Operation of admittance meters and R_X meters is discussed in the author's best-selling *Handbook of Electronic Test Equipment* (Prentice-Hall, Inc., Englewood Cliffs, N.J., 1971), and will not be repeated here. Instead, we present FET test procedures that can be accomplished with basic test equipment to find *Y* parameters, expressed in *simple number terms.*

2-11 FET FORWARD TRANSADMITTANCE (TRANSCONDUCTANCE) TESTS

Forward transadmittance (or transconductance) Y_{fs} defines the relationship between an input signal voltage and an output signal current, with the drain-source voltage held constant, or $Y_{fs} = \triangle I_D/\triangle V_{GS}$, with V_{DS} held constant. Y_{fs} is the most important dynamic characteristic for FETs, no matter what application, and serves as a basic design parameter in audio and RF, as well as being a widely accepted figure of merit for FETs.

Because FETs have many characteristics similar to those of vacuum tubes, the symbol g_m is sometimes used instead of Y_{fs}. This is further confused since the g-notation school also uses a number of subscripts. In addition to g_m, some FET datasheets show g_{fs}, while others go even further out with g_{21}. Y_{fs} is expressed in mhos (current divided by voltage). In most datasheets, Y_{fs} is specified as 1 kHz with a V_{DS} the same as that for which $I_{D(ON)}$ or I_{DSS} is obtained. At 1 kHz, Y_{fs} is almost entirely real. Thus, Y_{fs} at 1 kHZ = Y_{fs}. At higher frequencies, Y_{fs} includes the effects of gate-to-drain capacitance and may be misleadingly high. For high-frequency operation, the real part of transadmittance $R_e(Y_{fs})$, as discussed in later sections, should be used. Since the $I_D - V_{GS}$ curves of a FET are nonlinear, Y_{fs} varies considerably with changes in I_D.

Three Y_{fs} measurements are often specified for FETs. One of these measurements, with two gates tied together, provides a Y_{fs} value for the condition where a signal is applied to both gates simultaneously. The other two measurements provide the Y_{fs} for the two gates individually. Generally, with the two gates tied together, Y_{fs} is higher and more gain may be realized in a given circuit. Because of the increased capacitance, however, the gain–bandwidth product is much lower.

For FETs used at radio frequencies, an additional value of Y_{fs} is often specified at or near the highest frequency of operation. This value is measured at the same voltage conditions as those used for $I_{D(ON)}$ or I_{DSS}. Because of the importance of the imaginary component at radio frequencies, the high-frequency Y_{fs} specification is generally a complex representation. That is, both forward transconductance (g_{fs}, g_{21}, etc.) and forward susceptance (b_{fs}, b_{21}, etc.) are given in the specification.

Some datasheets list the real part of Y_{fs} (or forward transconductance) as Re(Y_{fs}), or Re(Y_{fs})(*HF*). Such Re(Y_{fs}) is defined as the common-source forward transfer conductance (drain current versus gate voltage). For high-frequency applications, Re(Y_{fs}) is considered a significant figure of merit. The d-c operating conditions are the same as for Y_{fs}, but the test frequency is typically 100 to 200 MHz.

Rather than listing a value for the real part of Y_{fs} at a specific frequency, some datasheets show both the forward transconductance (g_{fs}) and forward susceptance (b_{fs}) over the useful frequency range of the device. *In testing FETs, the important point to remember is:* No matter what datasheet system is used, an increase in the real part of Y_{fs} produces an increase in the voltage gain.

In comparing Re(Y_{fs}) with Y_{fs} (on those datasheets that list both values), the minimum values of the two are quite close, considering the difference in frequency at which the measurements are made. At high frequencies, about 30

MHz and above, Y_{fs} increases due to the effect of gate–drain capacitance C_{gd} so that Y_{fs} is misleadingly high.

2-11.1 Y_{fs} Test

The basic circuit for the Y_{fs} test is shown in Fig. 2–12. (The same circuit applies where the value is listed as y_{21}, g_m, or even g_{fs}.) Similarly, the same circuit can apply to any single-input device, such as a two-junction transistor. The circuit for a dual-gate FET is the same, except that a forward bias is generally applied to gate 2, with the signal applied to gate 1. For example, the 40841 FET specifies a V_{G2S} of +4 V for test of forward transconductance (which the datasheet lists as g_{fs}). This forward bias produces an I_D of 10 mA when the V_{DS} is 15 V.

The value of R_L in Fig. 2–12 must be such that the drop is negligible for I_{DSS}. Just as important, the value of R_L must be such that the operating voltage point (V_{DS} in the case of a FET) is correct for a given supply voltage (V_{DD}) and operating current (I_D). For example, if I_D is 10 mA, V_{DD} is 20 V, and V_{DS} is 15 V, R_L must drop 5 V at 10 mA. Thus, the value for R_L is 5 V/0.01 A = 500 Ω. For single-gate devices, the value of I_{DSS} should be used to set the value of R_L. The value of R_G is not critical and is typically 1 MΩ.

During the test, the signal source is adjusted to the frequency of interest. The amplitude of the signal source V_{IN} is set to some convenient number such as 1 V or 100 mV. The value of Y_{fs} is calculated from the equation of Fig. 2–12 and is expressed in mhos (or millimhos and micromhos, in practical terms). As an example, assume that the value of R_L is 1000 Ω, V_{IN} in 1 V and V_{OUT} is 8 V. The value of Y_{fs} is 8/(1 × 1000) = 0.008 mho = 8 mmhos = 8000 μmhos.

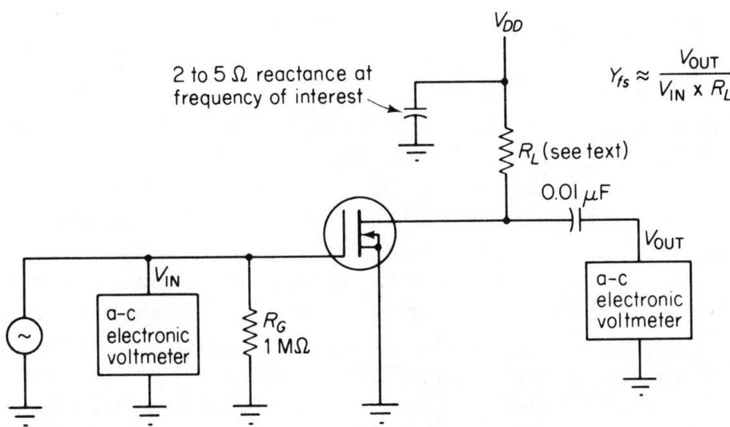

FIGURE 2-12 Test circuit for Y_{fs} (as a simple number).

2-11.2 Re (Y_{fs}) Test

For certain high-frequency applications, it is necessary to know the real part of Y_{fs}. At low frequencies, the Y_{fs} value found as described in Section 2-11.1 is sufficiently accurate. As frequency increases, Y_{fs} appears to be exceptionally high if tested by the circuit of Fig. 2-12. To obtain a true picture of the forward transconductance characteristics, it may be necessary to measure Y_{fs} as a complex number, using an admittance meter or an R_X meter. This is almost always true at 30 MHz or higher.

2-12 FET REVERSE TRANSADMITTANCE TESTS

Reverse transadmittance Y_{rs} (or Y_{12}) is not generally a critical factor in FETs. Typically, the real part of Y_{rs} (or g_{rs}) is zero over the useful frequency range. It is necessary, however, to know the values of Y_{rs} to calculate impedance-matching networks for FET RF amplifiers. For that reason, most FET datasheets list some values for Y_{rs}.

Although the real part of g_{rs} remains at zero for all conditions and at all frequencies, the imaginary part b_{rs} does vary with voltage, current, and frequency. That is, the reverse susceptance does vary and, under the right conditions, there can be undesired feedback from output to input (reverse transadmittance). This condition must be accounted for in the design of FET RF amplifier stages to prevent the feedback from causing oscillation. If it is necessary to establish the imaginary part, b_{rs}, use an admittance meter or R_X meter.

2-13 FET OUTPUT ADMITTANCE TESTS

Output admittance Y_{os} is another important dynamic characteristic for FETs. Y_{os} is also represented by various Y and g parameter, such as Y_{22}, g_{os}, and g_{22}. Y_{os} is even specified in terms of drain resistance, or r_d, where $r_d = 1/Y_{os}$.

This is similar to the vacuum-tube characteristic of output admittance, where Y_{output} is $1/r_p$ (or output admittance is equal to 1 divided by plate resistance).

No matter what symbol is used, Y_{os} defines the relationship between output signal current and output voltage, with the input voltage held constant, or $Y_{os} = \triangle I_D/\triangle V_{DS}$, with V_{GS} held constant. Y_{os} is expressed in mhos (current divided by voltage). Types A and B FETs are measured with gates and source grounded. For type C FETS, Y_{os} is measured at some specified value of V_{GS} that permits a substantial drain current to flow. Some datasheets give Y_{os} as a

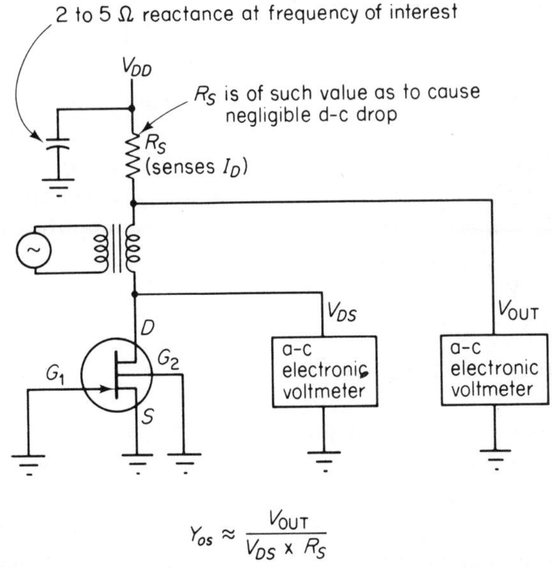

FIGURE 2-13 Test circuit for Y_{os} (as a simple number).

complex number with both the real (g_{os}) and imaginary (b_{os}) values shown by means of curves.

2-13.1 Y_{os} Test

The basic circuit for the Y_{os} test is shown in Fig. 2-13. The circuit for dual gate is the same, except that a forward bias is generally applied to gate 2. As indicated, the value of R_S must be such as to cause a negligible drop (so that V_{DS} can be maintained at a desired level, with a given V_{DD} and I_D). During the test, the signal source is adjusted to the frequency of interest. Both V_{out} and V_{DS} are measured, and the value of Y_{os} is calculated from the equation of Fig. 2-13.

2-14 FET INPUT ADMITTANCE TESTS

Input admittance Y_{is} (or Y_{11}) is not generally a critical factor in FETs. The real part of Y_{is} (or g_{is}) is nearly zero at low frequencies. It is necessary, however, to know the values of Y_{is} to calculate impedance-matching networks for FET RF amplifiers. If it is necessary to establish the imaginary part, b_{is}, use an admittance meter or R_X meter.

2-15 FET AMPLIFICATION FACTOR

The amplification factor (μ) does not usually appear on most FET datasheets. This is because amplification does not usually have a great significance in most small-signal applications for FETs. However, amplification factor is sometimes used as a figure of merit in isolated cases. Amplification factor defines the relationship between output signal voltage and input signal voltage, with the output current held constant, or amplification factor = $\triangle V_{DS}/\triangle V_{GS}$, with I_D held constant. Amplification factor can also be calculated by Y_{fs}/Y_{os}.

2-16 FET INPUT CAPACITANCE TESTS

Input capacitance C_{iss} is the common-source input capacitance with the output shorted and is used as a low-frequency substitute for Y_{is}. This is because Y_{is} is entirely capacitive at low frequencies. To find an approximate value for Y_{is} at low frequencies (below about 1 MHz), multiply C_{iss} by 6.28 farads (F). The result is b_{is}, or the imaginary part of Y_{is}. At these low frequencies, g_{is} can be considered as zero.

Note that C_{iss} is an important characteristic for FETs used in switching or chopper applications. This is because a large voltage swing at the gate must appear across the input capacitance C_{iss}.

2-16.1 C_{iss} Tests

The basic circuits for C_{iss} tests are shown in Figs. 2-14 and 2-15. Figure 2-14 is for JFETs and Fig. 2-15 is for MOSFETs. For dual-gate MOSFETs, the capacitance measurement is between gate 1 and all other terminals. For single-gate MOSFETs, the capacitance measurement is between the gate and all other terminals.

The circuit of Fig. 2-15(a) is for MOSFETs where a specific V_{GS} must be applied but V_{DS} is of no concern. The circuit of Fig. 2-15(b) is used where the capacitance is measured by means of a bridge with three terminals (high, low, and guard or ground). The reason for the two circuits is that some MOSFET datasheets specify a given V_{DS}, I_D, and V_{GS} when C_{iss} is measured. Other datasheets specify zero V_{DS} and sometimes zero V_{GS}.

2-17 FET OUTPUT CAPACITANCE TESTS

Output capacitance C_{oss} is the common-source output capacitance with the input shorted. C_{oss} does not appear on all FET datasheets. Generally, C_{oss} is found on dual-gate datasheets. The output capacitance of a typical FET is usually quite low (in the order of 1 or 2 pF).

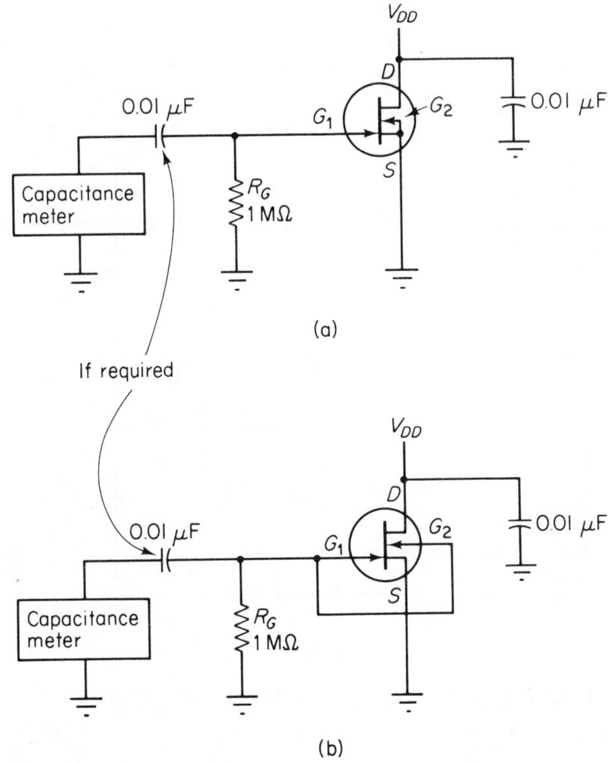

FIGURE 2-14 Test circuit for C_{iss} in tetrode JFETs: (a) with gate 2 tied to source; (b) with common gates.

2-17.1 C_{oss} Test

The basic circuit for the C_{oss} test is shown in Fig. 2-16. Generally, C_{oss} is used only in dual-gate MOSFET specifications. In a typical specification, gate 1 is shorted to the source and gate 2 has a specific voltage value applied to produce a given I_D.

2-18 FET REVERSE TRANSFER CAPACITANCE TESTS

Reverse transfer capacitance C_{rss} is defined as the common-source reverse transfer capacitance with the input shorted. C_{rss} is often used in place of Y_{rs}, the short-circuit reverse transfer admittance, since Y_{rs} is almost entirely capacitive over the useful frequency range of most FETs and is of relatively

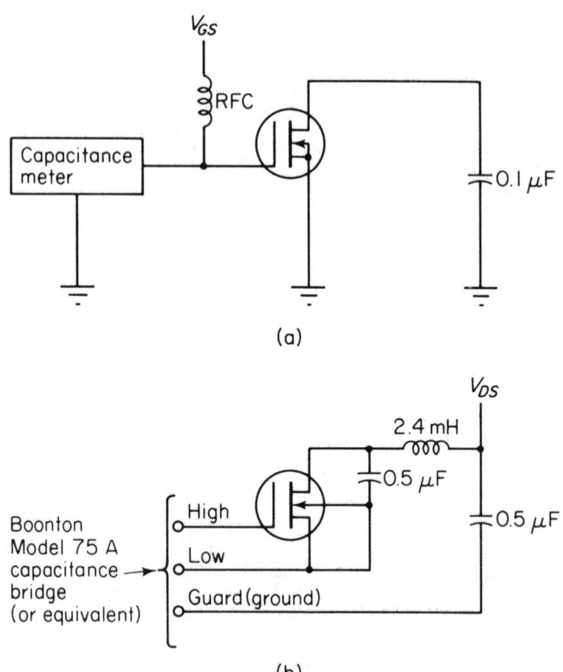

FIGURE 2-15 Test circuit for C_{iss} in MOSFETs.

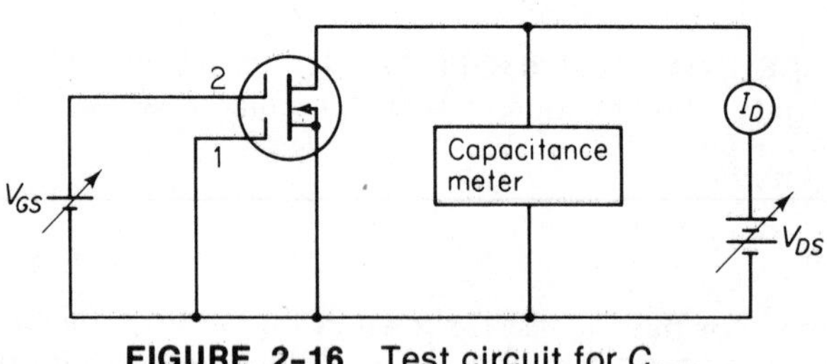

FIGURE 2-16 Test circuit for C_{oss}.

constant capacity. Consequently, the low-frequency C_{rss} is an adequate specification.

C_{rss} is also of major importance in FETs used as switches. Similar to the C_{os} of two-junction transistors, C_{rss} must be charged and discharged during the switching interval. For chopper applications, C_{rss} is the feed-through capacitance for the chopper drive. C_{rss} is also known as the Miller-effect capacitance since the reverse capacitance can produce a condition similar to the Miller effect in vacuum tubes (such as constantly changing frequency-response curves).

2-18.1 C_{rss} Tests

The basic circuits for C_{rss} tests are shown in Figs. 2–17 and 2–18. Figure 2–17 is for JFETs and Fig. 2–18 is for MOSFETs. Three-terminal measurements are used in all cases. For dual-gate MOSFETs, gate 2 and the source are

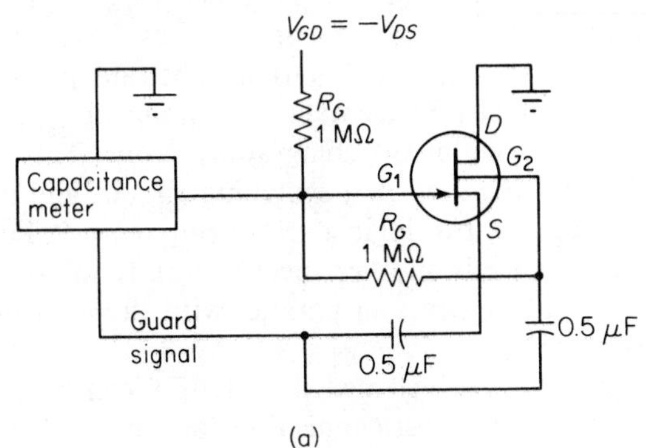

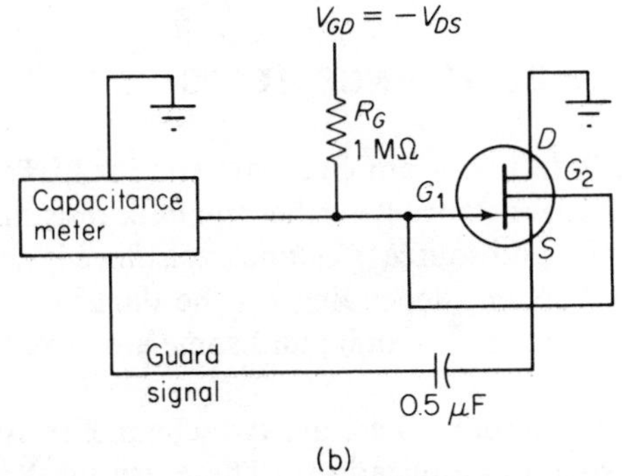

FIGURE 2–17 Test circuits for C_{rss} in JFETs.

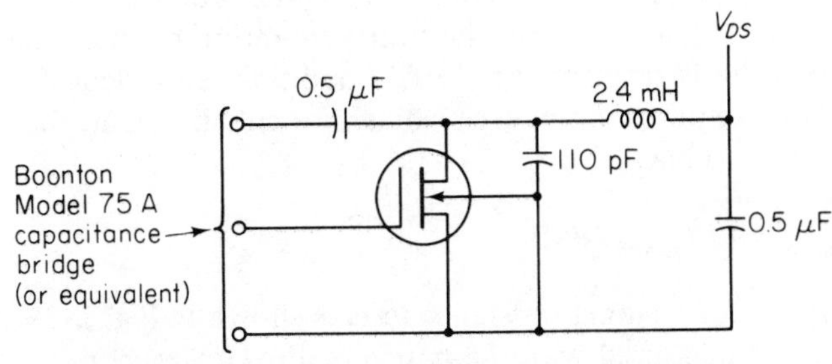

FIGURE 2–18 Test circuit for C_{rss} in MOSFETs.

returned to guard terminal and the capacitance measurement is made between gate 1 and the drain.

2-19 FET ELEMENT CAPACITANCE TESTS

The elements of a FET have some capacitance between them, as do two-junction transistors. Some of these capacitances have an effect on the dynamic characteristics of the FET. Drain-substrate junction capacitance $C_{d(sub)}$ is the most important FET element capacitance. $C_{d(sub)}$ is usually found on datasheets for FETs used in switching applications. This is because $C_{d(sub)}$ appears in parallel with the load in a switching circuit, and must be charged and discharged between the two logic levels during the switching intervals. C_{ds} drain-to-source capacitance is another specification found on some switching FET datasheets. C_{ds} also appears in parallel with the load in switching and logic applications.

The capacitance between elements of a FET can be measured with a capacitance meter. No special test connections are required. Some datasheets specify certain test connections or conditions, such as all remaining elements connected to source, or gate connected to source.

2-20 FET CHANNEL RESISTANCE TESTS

The channel resistance is an important characteristic for FETs used in switching applications. Channel resistance describes the bulk resistance of the channel in series with the drain and source. Channel resistance is described as $r_{d(on)}$, r_{DS}, R_{DS}, r_{ds}, $r_{d(OFF)}$, and so on, depending on the datasheet. Some of these descriptions are static, some are dynamic, and some are mixed (again depending on the datasheet).

From a practical standpoint, there are two channel resistance specifications of concern in switching applications. These are the "on" and "off" specifications, such as $r_{ds(on)}$ and $r_{ds(off)}$, which can be either static or dynamic. The "on" specification is the channel resistance with the FET biased "on." In a depletion-mode FET, the "on" condition can be produced by zero bias ($V_{GS} = 0$). In the enhancement mode, the "on" condition requires some forward bias. The opposite is true for the "off" condition. In a depletion FET, the "off" condition requires some reverse bias. For enhancement, the "off" condition requires zero bias.

2-20.1 $r_{ds(on)}$ and $r_{ds(off)}$ Tests

The basic circuit for channel resistance tests is shown in Fig. 2-19. Both $r_{ds(on)}$ and $r_{ds(off)}$ can be measured with the basic circuit. However, $r_{ds(on)}$ is generally the test of interest. If the FET operates in the depletion or depletion/enhance-

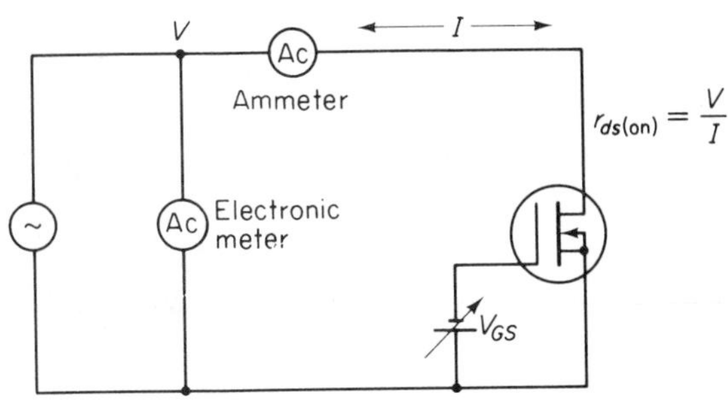

FIGURE 2-19 Test circuit for $r_{ds(on)}$ and $r_{ds(off)}$.

ment modes, $r_{ds(on)}$ is measured by adjusting V_{GS} to 0 V (or simply connecting the gate to the source). For example, the 3N128 FET specifies a V_{DS} and V_{GS} of 0 V for an $r_{ds(on)}$ of 200 Ω (at a test frequency of 1 kHz).

During test, the a-c voltage source V is adjusted to some convenient value (1 V, 10 V, etc.), and the channel current I is measured. The value of r_{ds} is found when V is divided by I. For example, if V is adjusted to 10 V and 50-mA channel current is measured, the "on" channel resistance, or $r_{ds(on)}$, is 200 Ω (10/0.05 = 200). If the FET operates in the enhancement-only mode (type C), it is necessary to forward bias the gate by adjusting V_{GS} to some specific value.

Figure 2–19 can also be used to measure the "off" channel resistance $r_{ds(off)}$. However, the bias conditions are opposite those for $r_{ds(on)}$. In the depletion and depletion/enhancement modes, the gate must be reverse-biased by adjusting V_{GS} to some specific value. In the enhancement-only mode, the gate can be connected directly to the source. Figure 2–19 can also be used to measure channel resistance of dual-gate FETs. Generally, the simplest way is to connect both gates together. However, some datasheets specify a fixed bias on one or both gates.

As a point of reference, a typical MOSFET "on" resistance is in the order of 220 Ω, whereas the "off" resistance is greater than 10^{10} Ω.

2-21 FET SWITCHING-TIME TESTS

Switching-time characteristics are of particular importance for FETs used in digital electronics. The datasheets that describe these FETs specifically designed for switching include *timing characteristics*. Typically, these include t_d (delay time), t_s (storage time), t_r (rise time), and t_f (fall time). These characteristics are measured with a pulse source and a multiple-trace oscilloscope, as are similar characteristics for two-junction transistors (Section 1-5). A typical test circuit for measurement of FET switching characteristics is shown in Fig. 2–20.

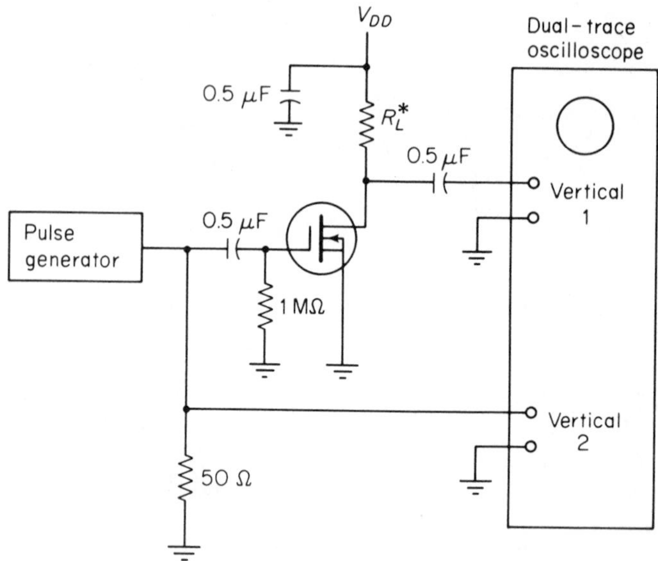

*R_L is a value that causes negligible d-c drop at I_{DSS}.

FIGURE 2-20 Test circuit for measurement of FET switching characteristics.

2-22 FET GAIN TESTS

The datasheets for FETs used as amplifier usually list some gain characteristics. Typically, these are power-gain figures, at some specified frequency and under specified test conditions (such as V_{DS} and I_D). The gain figures are expressed in decibels, sometimes as maximum or minimum values. Such terms as MAG (maximum available gain), MUG (maximum usable gain), and G_{PS} (power gain with the FET connected as common source) are used.

The datasheet notes concerning gain should always be consulted. For example, the term "MAG" usually means the maximum available gain under "ideal" or theoretical conditions. The term "MUG" usually means the maximum usable gain when the amplifier is neutralized or mismatched to produce the stability effects of neutralization.

On some datasheets, subscripts are added to identify the gain figures. For example, the subscript (*c*) can be added to MUG to indicate maximum usable gain when the FET is used as a converter.

The Y_{fs} test described in Section 2-11 can be used to establish the gain of a FET. However, the test does not establish the gain in a working circuit. The only true test of circuit gain is to operate the FET in a working circuit and to measure actual gain. The most practical method is to operate the FET in the circuit with which the FET is to be used. However, it may be convenient to have a standard or universal circuit for test of FET gain.

Generally, the main concern is with power gain in RF circuits. Most FET datasheets include a power-gain test circuit diagram. Figures 2-21 and 2-22

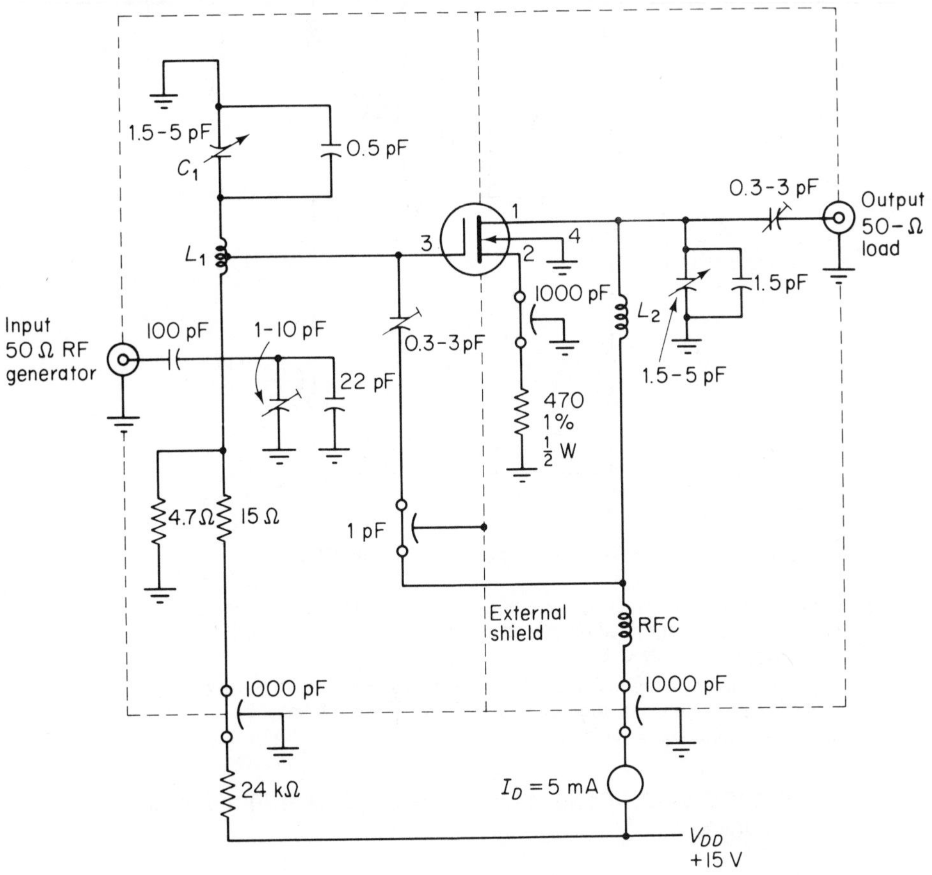

All resistors $\frac{1}{4}$-W except as noted.

L_1 = 5 turns silver-plated, 0.02 in.-thick, 0.07- to 0.08 in.-wide copper ribbon. Internal diameter of winding 0.25 in.; winding length approximately 0.65 in. tapped at $1\frac{1}{2}$ turns from C_1 end of winding.

L_2 = Same as L_1 except winding length approximately 0.7 in.; no tap.

FIGURE 2-21 Power-gain and noise-figure (NF) test circuit for 3N128.

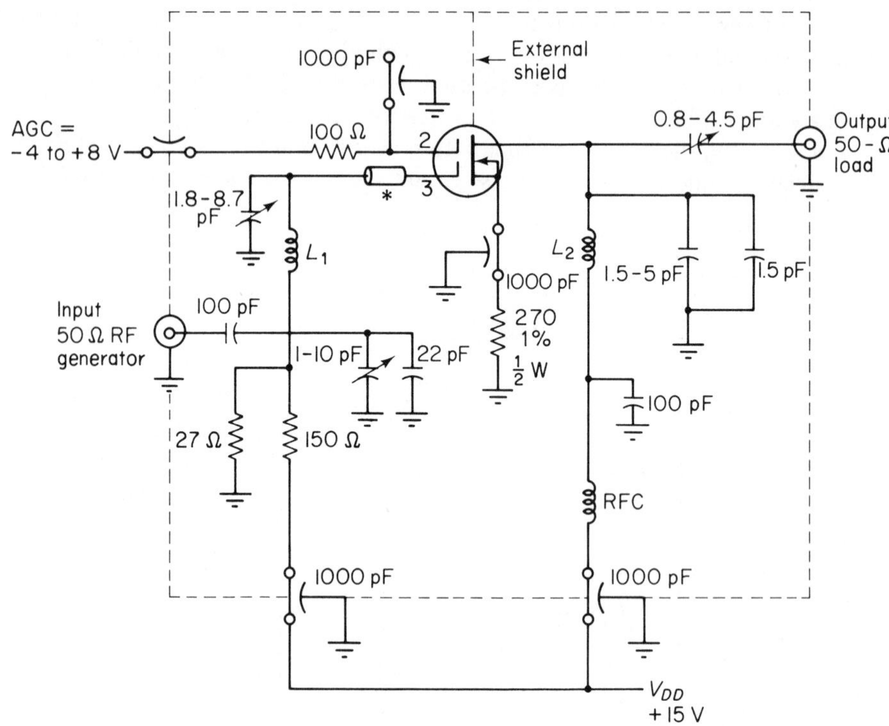

* Ferrite beads (4); Pyroferric Co. Carbonyl J, 0.09 in. OD; 0.03 in. ID; 0.063 in. thick.

L_1 = 4 turns silver-plated, 0.02-in.-thick, 0.075- to 0.085 in.-wide copper ribbon. Internal diameter of winding 0.25 in.; winding length approximately 0.80 in.

L_2 = $4\frac{1}{4}$ turns silver-plated, 0.02-in.-thick, 0.085- to 0.095-in.-wide copper ribbon. Internal diameter of winding $\frac{5}{16}$ in., winding length approximately 0.90 in.

FIGURE 2-22 Power-gain and noise-figure (NF) test circuit for 40820.

are typical examples of such power-gain test circuits. The circuit of Fig. 2-22 is for dual gate, where gate 2 is connected to a variable AGC voltage. This permits the power gain to be measured under various AGC conditions. For example, the 40821 specifies power gain between 14 and 17 decibels (dB), with a V_{DS} of 15 V, I_D of 10 mA, and V_{G2S} of +4 V (applied to the AGC line).

Keep in mind that these test circuits are for specific FETs, operating with specific loads (50 Ω) and at specific frequencies (200 MHz). However, similar test circuits can be fabricated for other load and frequencies. Once the test circuit has been fabricated, the procedures for measuring power gain are

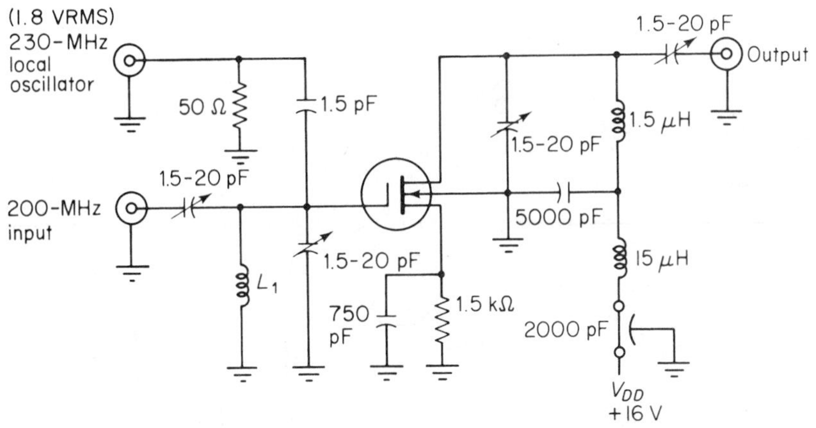

FIGURE 2-23 Conversion power-gain test circuit 3N143.

relatively simple. In brief, an RF signal voltage is applied across the input load, and the input power is calculated from voltage and load resistance (E^2/R). The amplified output voltage is measured across a similar load and the power calculated. The ratio of output power to input power is the power gain, and is usually expressed in decibels. In some cases, simple voltage gain (output voltage divided by input voltage, with identical loads at input and output) is specified instead of power gain.

For those FETs used as converters, conversion gain is of particular importance. Figures 2–23 and 2–24 are typical examples of conversion gain test circuits (supplied on the datasheet). The circuit of Fig. 2–23 is for a single-gate FET where both the RF input and local oscillator input are applied to the gate. Figure 2–24 is for dual-gate FETs where RF input is applied to gate 1 and the local oscillator input is applied to gate 2. Again, these circuits are for specific FETs operating at specific frequencies. However, the circuits serve as a design "starting point" for standard test circuits.

2-23 FET NOISE-FIGURE TESTS

The noise figure for FETs is usually listed as NF on the datasheets and represents a common-source noise figure. NF represents a ratio between input signal-to-noise ratio and output signal-to-noise ratio, and is measured in decibels. In most FET datasheets, NF includes the effects of e_n (equivalent short-circuit input noise expressed in volts-per-root cycle) and i_n (the equivalent open-circuit noise current).

The noise figure attains its highest value for a small generator resistance, and decreases for increasing generator resistance, indicating a large noise con-

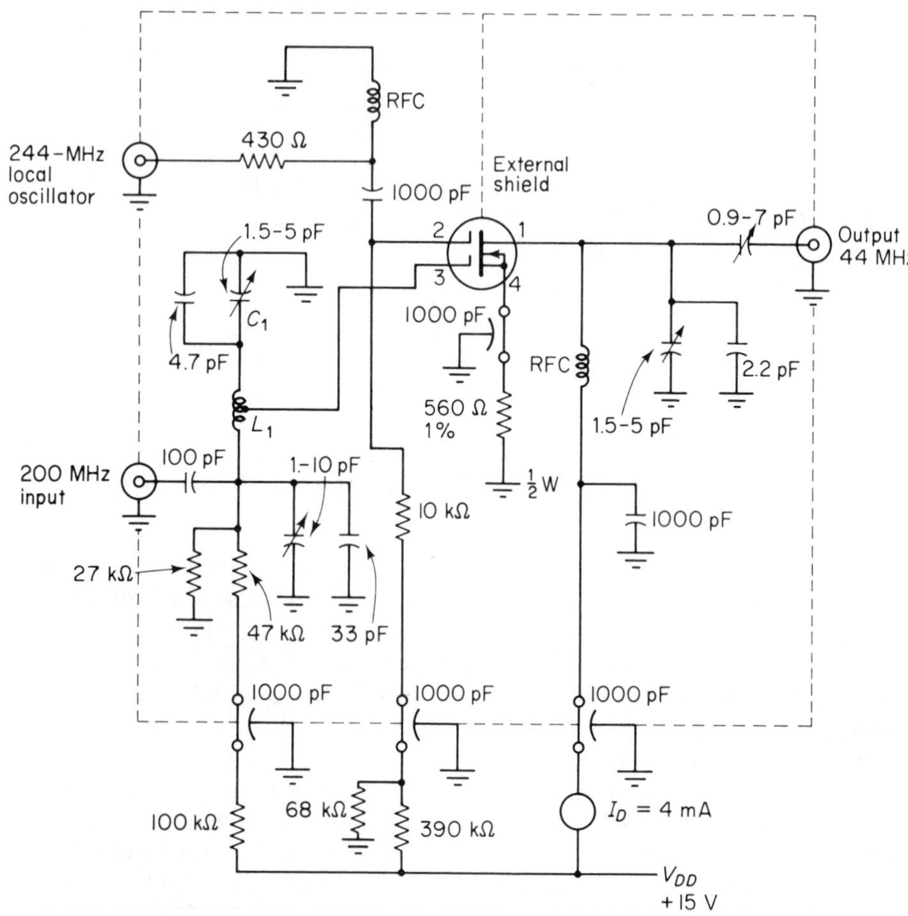

L_1 = 5 turns silver-plated, 0.02-in.-thick, 0.07-to 0.08-in.-wide copper ribbon. Internal diameter of winding 0.025 in.; winding length approximately 0.65 in. Tapped at 1½ turns from C_1 end of winding.

FIGURE 2-24 Conversion power-gain test circuit for 40821.

tribution from the noise-voltage generator. For this reason, some datasheets specify both NF and e_n but neglect i_n. In effect, NF is independent of operating current and proportional to voltage. However, the voltage effects are slight over the normal operating range of a typical FET.

Figure 2-25 is a nomograph for converting the noise figure to an equivalent input noise voltage for different generator source impedances R_S. This nomograph can be used with any FET. Since NF and e_n are frequency-dependent, Fig. 2-25 must be used with a datasheet graph or table to deter-

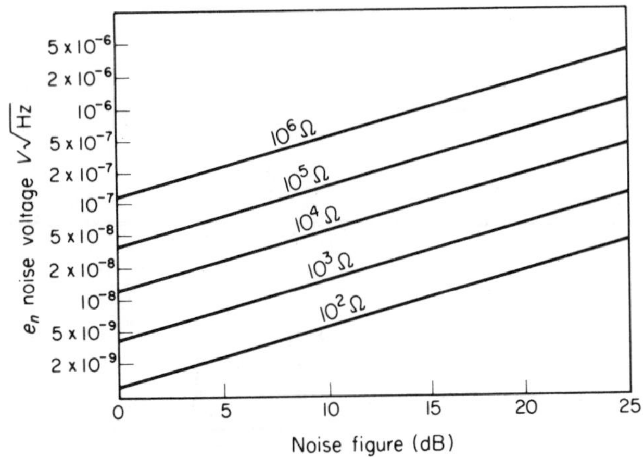

FIGURE 2-25 Noise-figure conversion chart.

mine e_n. For example, assume that the datasheet lists NF as 5 dB at 200 MHz. Also assume that the circuit with which the FET is to be used has a 1-MΩ input. Using these values, Fig. 2-25 shows e_n to be about 2×10^{-7}, or about 200 nV.

2-23.1 NF Tests

The circuits of Figs. 2-21 and 2-22 can also be used to measure the noise figure. Such noise-figure tests are best accomplished using specialized test equipment, such as a VHF noise source and VHF noise meter. The instructions supplied with the noise measurement test equipment are generally complete and provide full descriptions of noise tests. In any event, the test procedures will not be duplicated here. However, the test equipment instructions generally omit the amplifier test circuit, thus creating a need for circuits, as shown in Figs. 2-21 and 2-22.

2-24 FET CROSS-MODULATION TESTS

Cross-modulation values may or may not be listed on a FET datasheet. Figure 2-26 shows the basic block diagram, and the detailed amplifier circuits, for cross-modulation tests. These circuits and values are for a 3N128 FET being tested at frequencies of 200 MHz (desired operating frequency) and 150 MHz (interfering frequency). The circuits provide for unneutralized, neutralized, and cascade amplifier operation.

During test, signals of both frequencies are applied to the input. The test

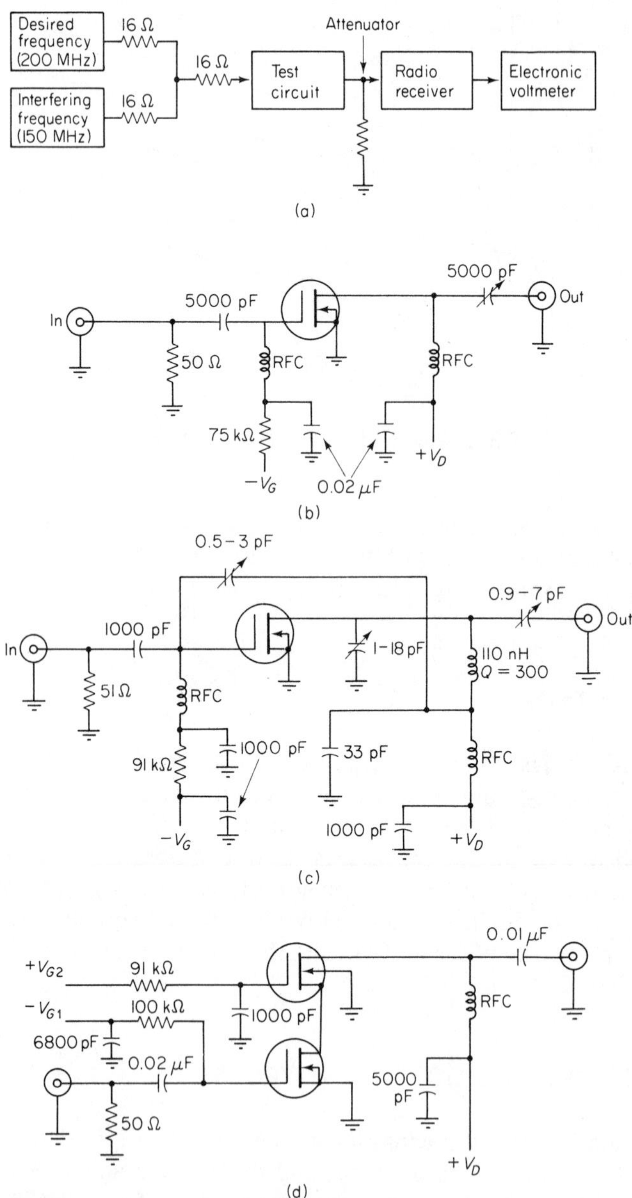

FIGURE 2-26 Test circuits used to measure cross-modulation distortion in MOSFETs: (a) block diagram; (b) unneutralized-stage test circuit; (c) neutralized-stage test circuit; (d) cascade-stage test circuit.

circuit output is monitored by the receiver. The amount of attenuation produced by the interfering signal is measured at various signal levels, and compared with the output when no interfering signal is present.

2-25 FET INTERMODULATION TESTS

Intermodulation values may or may not be listed on a FET datasheet. Figure 2-27 shows the basic circuit for intermodulation tests. This circuit and values are for a 3N128 FET being tested at frequencies of 175 MHz (F_1) and 200 MHz (F_2). If intermodulation is present, there will be signals of various frequencies at the output. For example, there will be an intermodulation signal of 150 MHz if intermodulation is present in the FET circuit.

During test, F_1 and F_2 are set to zero (amplitude) and the background noise level is measured (on the RF indicator of the receiver tuned to the intermodulation frequency, and connected at the output of the test circuit). The amplitudes of F_1 and F_2 are increased until the reading on the RF indicator is 1 mV above the noise level. The voltage levels required to produce the output indication are measured on the RF voltmeter at the test circuit input.

2-26 FET TESTS USING A CURVE TRACER

In many respects, the testing of FETs using a curve tracer is similar to testing two-junction transistors (Chapter 1). A family of curves is displayed on the oscilloscope, and the curves have a similar appearance. As shown in Fig. 2-28, N-channel FETs have a family of curves similar to NPN transistors, and P-channel FET curves are similar to PNP transistors. Two-junction transistor curves are a graph of collector current versus collector voltage, at various base currents. FET curves are a graph of drain current versus drain voltage at various gate voltages. Similarly, FET breakdown voltage may be observed and measured by the same method used for transistors.

In several other respects, testing FETs is different from testing two-junction transistors. For FETs, the *step selector* switch (or whatever the curve-tracer control may be called) is placed in a "volts per step" position. That is, the curve tracer supplies *constant-voltage* steps to the FET, rather than constant-current steps (as in the case of two-junction transistors). Also, the polarity of the step voltage is reversed in relation to the sweep voltage. While the zero-base-current step of a two-junction transistor usually provides no collector current, the zero-volt step at the gate of most FETs produces the highest drain current. Each reverse-bias voltage step results in less drain current, and when the gate voltage is sufficiently high, drain current is pinched off. The point of *pinch-off* (sometimes listed as V_P on FET datasheets) can be measured with the curve tracer.

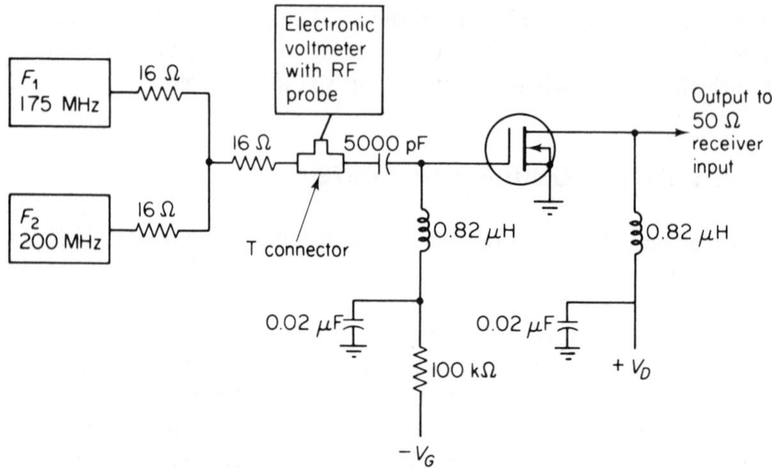

FIGURE 2-27 Test circuit used to measure intermodulation distortion in MOSFETs.

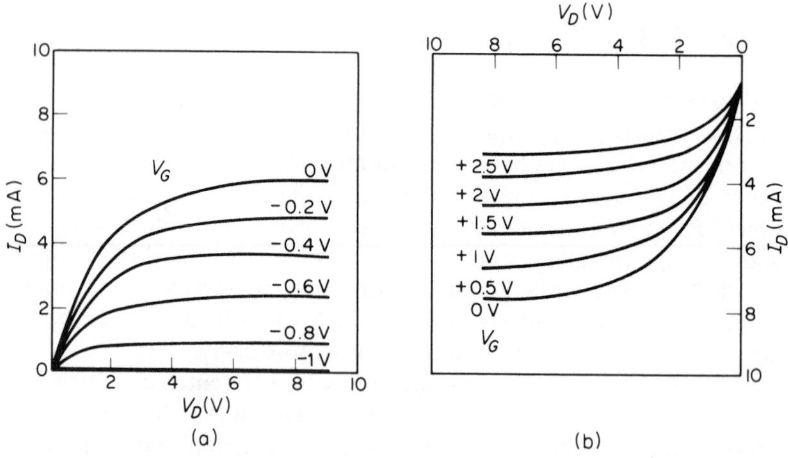

FIGURE 2-28 Typical characteristic curves: (a) N-channel; (b) P-channel.

The method of gain measurement for a FET is similar to the gain measurement of a two-junction transistor, but the forward transconductance of a FET has a voltage input characteristic which cannot be directly compared with the beta of a two-junction transistor (which has a current input characteristic).

2-26.1 *Typical FET Test Procedure*

The following steps apply to a curve tracer such as the B & K Precision 501A, but can be used as a guide for using other curve tracers to test FETs.

1. Before plugging in or connecting the FET to the curve tracer, set the controls as follows: SWEEP VOLTAGE to zero, VERTICAL SENSITIVITY to 1 mA/division, STEP SELECTOR, I_{DSS}/I_{CES}, and POLARITY to the type of FET being tested.
2. Plug the FET to be tested into the appropriate curve tracer socket. On some curve tracers, the FET is connected to *D, G,* and *S* jacks with test leads. Use the FET basing diagram, if available, to identify the gate (*G*), drain (*D*), and source (*S*) leads. On some curve tracers, the socket pins are labeled for FETs as well as two-junction transistors. If there is a separate test lead for the FET shield, clip a test lead to the shield pin and ground it to the appropriate curve tracer terminal. Do not leave the FET shield ungrounded.
3. Increase the SWEEP VOLTAGE setting a nominal amount to produce a horizontal display, but stay well below breakdown voltage (typically 5 or 10 V of sweep voltage is sufficient). A single curve should be displayed on the oscilloscope. If the FET type is unknown, try both the N CHAN and P CHAN positions of the POLARITY switch to obtain the curve.
4. Position the curve as required with the oscilloscope centering controls. If the curve extends off-scale vertically, move to the next position of the VERTICAL SENSITIVITY control.
5. Increase the SWEEP VOLTAGE as desired, but reduce the setting if breakdown is observed. (Breakdown for FETs is similar to that of two-junction transistors, as described in Section 1-9.)
6. Rotate the STEP SELECTOR control to the VOLTS PER STEP position, increasing the setting until a family of six curves is observed with the greatest attainable spacing between curves. If the setting is too low, the curves will be too closely spaced to take a reading. If set too high, some of the gate voltage steps may exceed pinch-off and result in less than six curves being displayed.

2-26.2 *Important Considerations in Testing FETs*

In general, FETs are more susceptible to damage from excess voltage or current than two-junction transistors. Starting with the recommended control setting eliminates the possibility of applying test signals that are too high. Control settings should be increased after the FET is inserted only as much as is necessary to make the tests. As discussed, some MOSFETs can be damaged by a voltage transient from a static charge carried by the person handling the FET. Safeguard against such damage and discharge any static charge by touching ground with one hand before and while handling the MOSFET with the other hand.

The I_{DSS}/I_{CES} position of the STEP SELECTOR control displays the drain current of the FET with the gate shorted to the source. (This is typical for most, but not all, FET curve tracers.) This is the zero-bias condition and produces a single curve on the display which is representative of the maximum drain current normally flowing through the FET. Most FETs normally operate in the depletion mode with a reverse bias. The constant-voltage steps of a reverse-bias polarity as generated by this curve tracer drive the FET into the depletion mode, with the curves showing lower resultant drain current with each successive step.

To test the few enhancement-mode FETs, the gate lead can be disconnected from the curve tracer and connected to a d-c bias supply which will provide the forward-bias voltage. Be sure that any such bias supply reference is common to the source of the FET by connecting a test lead between the bias supply and the *S* jack of the curve tracer.

For testing dual-gate MOSFETs, one gate should be grounded or biased while testing. One gate can be plugged into the socket of the curve tracer, and a test lead can be clipped to the other gate. To ground the gate, connect the test lead to the source (*S*) jack of the curve tracer. To bias the gate, connect the test lead to a d-c bias supply. Varying the bias supply voltage shows the effects of simultaneous inputs on the two gates of the FET. If a d-c bias supply is used, be sure to ground the d-c reference to the source (*S*) jack of the curve tracer.

2-26.3 FET Transconductance (Gain) Measurement

The most useful and common measurement to be made for a FET is the gain measurement. The dynamic gain, or gate-to-drain forward transconductance (g_m) in the common-source configuration, is the ratio of change in drain current to the change in gate voltage at a given drain voltage. Transconductance is measured in mhos. Figure 2–29 shows how typical FET curves can be used to find gain. As shown, the change in drain current ($\triangle I_D$) is 1.5 mA (from 7 mA to 5.5 mA), for a change in gate voltage ($\triangle V_G$) of 0.1 V (from 0.1 V to 0.2 V), at a V_{DS} of 6 V. This indicates a gain (transconductance, or g_m) of 15,000 μmhos (0.0015 A/0.1 V = 0.015 mho = 15,000 μmhos).

As with a two-junction transistor, the gain of a FET is not constant over the entire voltage and current range. The gain is normally calculated in the typical operating range. Distortion and linearity may be determined by the

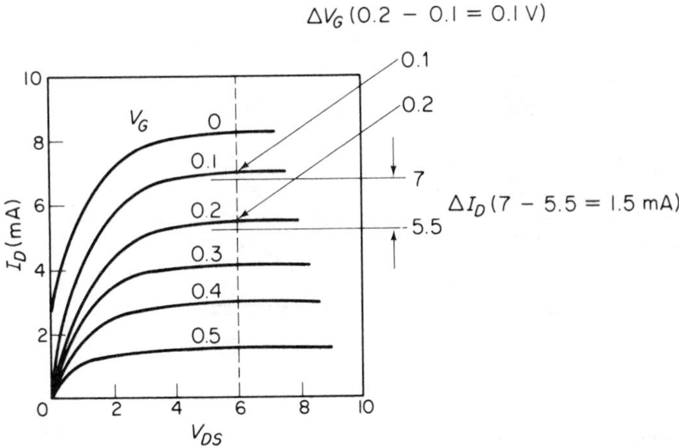

FIGURE 2-29 FET transconductance (gain) measurement.

same method as described for two-junction transistors (Chapter 1). That is, if the spacing between curves is equal, the FET is linear.

2-26.4 FET Pinch-off (V_p) Voltage Measurement

An important characteristic for depletion-mode FETs is the amount of gate voltage required to turn off drain current. This value is called the pinch-off voltage characteristic and may be measured from the family of curves as shown in Fig. 2–30(a).

Figure 2–30(a) shows the display with the STEP SELECTOR set at 0.5 V per step. Note that the entire family of curves is displayed, and that drain current continues to flow at the highest step of −2.5 V. Figure 2–30(b) shows that when the STEP SELECTOR is increased to 1 V per step, the entire family of curves is not displayed. In fact, the −3 V, −4 V, and −5 V curves are superimposed upon each other at zero drain current. From this, it can be concluded that pinch-off occurs between −2.5 and −3 V. A more precise measurement can be made, if necessary, by connecting an external d-c bias supply to the FET gate, adjusting the bias supply, and observing the exact value of pinch-off voltage on the oscilloscope.

2-26.5 FET Zero-Temperature-Coefficient Point Tests

An interesting characteristic of all FETs is their ability to operate at a zero-temperature-coefficient (OTC) point. This means that if the gate–source is biased at a specific voltage and is held constant, the drain current will not vary with changes in temperature. This effect is shown in Fig. 2–31, which illustrates the $I_D - V_{GS}$ curves of a typical FET for three different

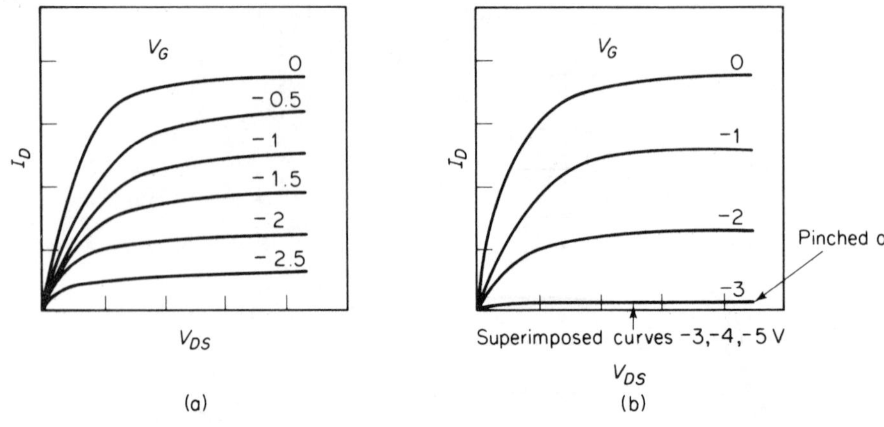

FIGURE 2-30 Determining FET pinch-off voltage using a curve tracer.

temperatures. Note that the three temperature curves intersect at a common point. If the FET is operated at this value of I_D and V_{GS} (shown as I_{DZ} and V_{GSZ}), OTC operation results. That is, drain current remains constant, even with extreme temperature changes.

Typically, JFETs show the OTC characteristic over a wide range of temperatures (approximately 150°C). MOSFETs are generally limited to a much narrower range (approximately 50°C). For that reason, OTC is not generally considered as an important MOSFET characteristic. Also, it is not always practical to operate a FET at the OTC point. For example, assume that the required V_{GS} to produce OTC is 0.3 V, and the FET is to operate as an amplifier with 0.5-V input signals. Under these conditions, a part of the input signal will be clipped. Or assume that the circuit is to be self-biased with a source resistor. An increase in bias resistance to produce OTC can reduce gain. Thus, in practical circuits, FETs are often not biased for OTC. However, the OTC characteristic should be understood, and the related test procedure is worth discussing.

The OTC point varies from one FET to another, and is dependent on I_{DSS}, the zero-gate-voltage drain current, and V_P the pinch-off voltage (Section 2-26.4). The equations shown in Fig. 2-31 provide good approximations of the OTC point. For example, if V_P is 1 V, the OTC mode is obtained if the gate-source voltage V_{GS} is 0.37 V (1 − 0.63 = 0.37).

Note that it is sometimes assumed that the forward transconductance of the FET does not vary with temperature if the FET is biased at the OTC point. However, this is not correct since the transadmittance of the FET is the slope of the $I_D - V_{GS}$ curve. Figure 2-31 shows that the slope varies with temperature at every point on the curve.

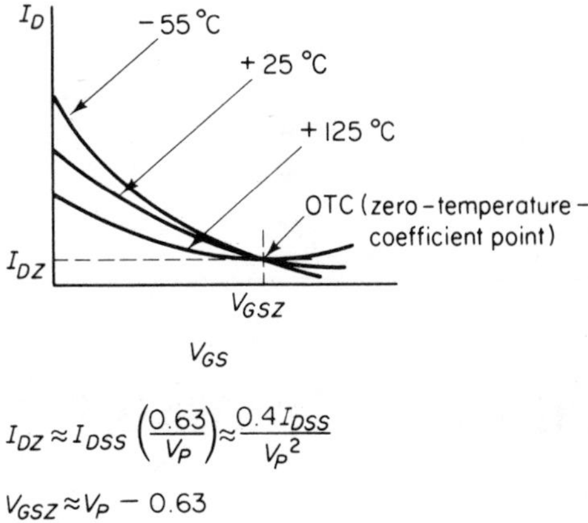

FIGURE 2-31 Zero-temperature coefficient for FETs.

As shown in Fig. 2-31, the values of I_D and V_{GS} that produce OTC can be calculated by using datasheet curves or by equations. However, such values are typical approximations. A more practical method for determining I_{DZ} (the constant drain current at the OTC point) requires a soldering tool, a coolant (a can of Freon), and a curve tracer. The curve tracer is adjusted to display a family of FET curves as described in previous sections. By alternately bringing the soldering tool near the FET, and spraying the FET with coolant, the voltage step of V_{GS} that *remains motionless* on the curve tracer can be observed. The I_D at this voltage step is the I_{DZ}. For example, if the 0.3-V step of Fig. 2-29 remains motionless when the FET is alternately cooled and heated, the I_{DZ} is 4 mA (and an 0.3-V V_{GS} produces the OTC).

Typically, FETs with an I_{DSS} of about 10 to 20 mA have an I_{DZ} of less than 1 mA. Usually, the I_{DZ} increases as I_{DSS} increases (but not always, and not in proportion). For example, the I_{DZ} of a 50-mA FET often shows an I_{DZ} below 1 mA.

3

Unijunction Transistor Tests

This chapter is devoted entirely to test procedures for unijunction transistors (UJTs) and programmable unijunction transistors (PUTs). The first sections of this chapter describe UJT and PUT characteristics and test procedures from the practical standpoint. The information in these sections permits you to test all the important UJT and PUT characteristics using basic shop equipment. The sections also help you understand the basis for such tests. The remaining sections of the chapter describe how the same tests, and additional tests, are performed using more sophisticated equipment such as the oscilloscope curve tracer.

3-1 BASIC UJT AND PUT FUNCTIONS

The UJT and PUT operates on entirely different principles from those of the two-junction transistor (Chapter 1) or the FET (Chapter 2). Therefore, the test procedures for UJTs and PUTs are quite different from other transistors. The UJT is a *negative-resistance* device where, under the proper conditions, the input voltage or signal can be decreased, yet the output or load current increases. Once the UJT is turned on, it does not turn off until the circuit is broken or the input voltage is removed. For this reason, the UJT makes an excellent trigger source. The UJT can be biased just below the "firing" point. When a small trigger voltage (either intermittent or constant) is applied, the UJT fires and produces a large output voltage pulse or signal that remains on until the circuit is broken (by switching off the base voltage).

The PUT is a four-layer device similar to an SCR (Chapter 5), except that the anode gate (rather than the cathode gate) is brought out. The PUT is normally used in conventional UJT circuits. The characteristics of both UJT and PUT devices are similar, but the triggering voltage of the PUT is programmable and can be set by an external resistive voltage divider network. The PUT is faster and more sensitive than the UJT. The PUT finds limited application as a phase-control element and is most often used in long-duration timer circuits. In general, the PUT is more versatile and is a more economical device than the UJT and can replace the UJT in many applications.

Although the basic function of the UJT is that of a switch, a relaxation oscillator is the primary building block in most UJT circuits. This basic relaxation oscillator circuit provides phase control of thyristors such as SCRs, triacs, and so on. For that reason, UJT theory and test procedures are often based on the relationship to relaxation oscillator principles.

3-1.1 UJT Symbol and Static Characteristics

Figure 3-1 shows the standard UJT symbol with appropriate terms for current and voltage. As shown, the UJT is a three-terminal device. The three terminals are emitter (E), base 1 (B_1), and base 2 (B_2). Figure 3-1 also shows the static UJT emitter characteristic curves for a single value of V_{B2B1}. Note that the emitter curve is not drawn to scale to show the different operating regions in more detail. The region to the left of the peak point is called the *cutoff region*. The emitter junction is reverse-biased in most of the cutoff region but is slightly forward-biased at the peak point. The region between the peak point and the valley point, where the emitter junction is forward-biased, is called the *negative-resistance* region. The region to the right of the valley point is called the *saturation region*. The curve for base 2 current I_{B2} equal to zero is essentially the forward characteristic of a conventional silicon diode (Chapter 4). UJT characteristics are discussed further in Section 3-2.

3-1.2 PUT Symbol and Static Characteristics

Figure 3-2 shows the symbol and a transistor equivalent circuit for a PUT. As shown, the PUT has three terminals: an anode (A), a gate (G), and a cathode (K). As seen from the equivalent circuit, the PUT is actually an anode-gated SCR. This means that if the gate is made negative with respect to the anode, the PUT switches from a blocking state to an "on" state. Since the PUT is

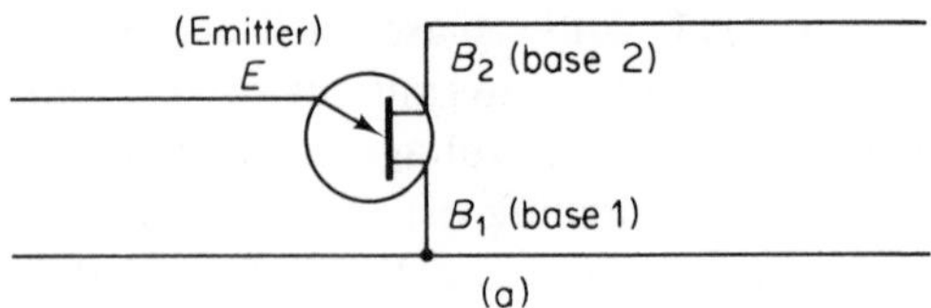

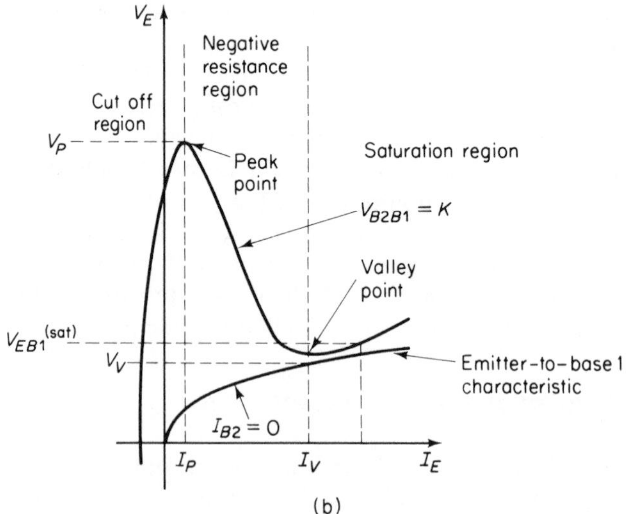

FIGURE 3-1 UJT symbol and static emitter characteristic curves.

normally used as a UJT, the UJT terminology is used to describe PUT parameters. In order to operate a PUT as a UJT, an external reference voltage must be maintained at the gate terminal.

A typical PUT relaxation oscillator circuit is also shown in Fig. 3-2, together with the characteristic curve (looking into the anode-cathode terminals). As is the case with the UJT, the peak and valley points of the PUT curve are stable operating points at either end of a negative-resistance region. The peak-point voltage (V_P) is essentially the same as the external gate reference (obtained from the voltage divider of R_1 and R_2), the only difference being the gate diode drop (V_S). Since the V_S reference is circuit-dependent (dependent upon the values of R_1, R_2, and V_1), rather than device-dependent (which is the case with a UJT), the V_S reference may be varied. Thus, V_P of a PUT is made programmable by varying the reference. This feature is the most significant difference between the UJT and PUT. PUT characteristics are discussed further in Section 3-3.

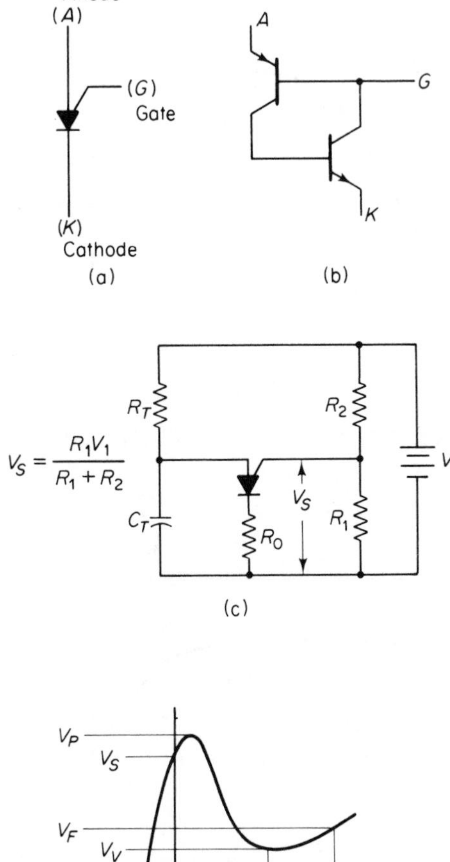

FIGURE 3-2 PUT symbol, equivalent circuit, relaxation oscillator circuit, and static characteristic curve.

3-2 UJT CHARACTERISTICS

The static emitter characteristics shown in Fig. 3-1 are not drawn to scale, in order to show all three regions (cutoff, negative resistance, saturation) of a UJT. Figure 3-3 shows a typical UJT emitter curve for a V_{B1B2} of 20 V, drawn to scale. Figure 3-3(a) shows part of the cutoff region plotted on a linear scale. When the emitter voltage is zero, the emitter current is negative. Peak-point voltage is reached at a forward-emitter current of about 10 μA. As seen

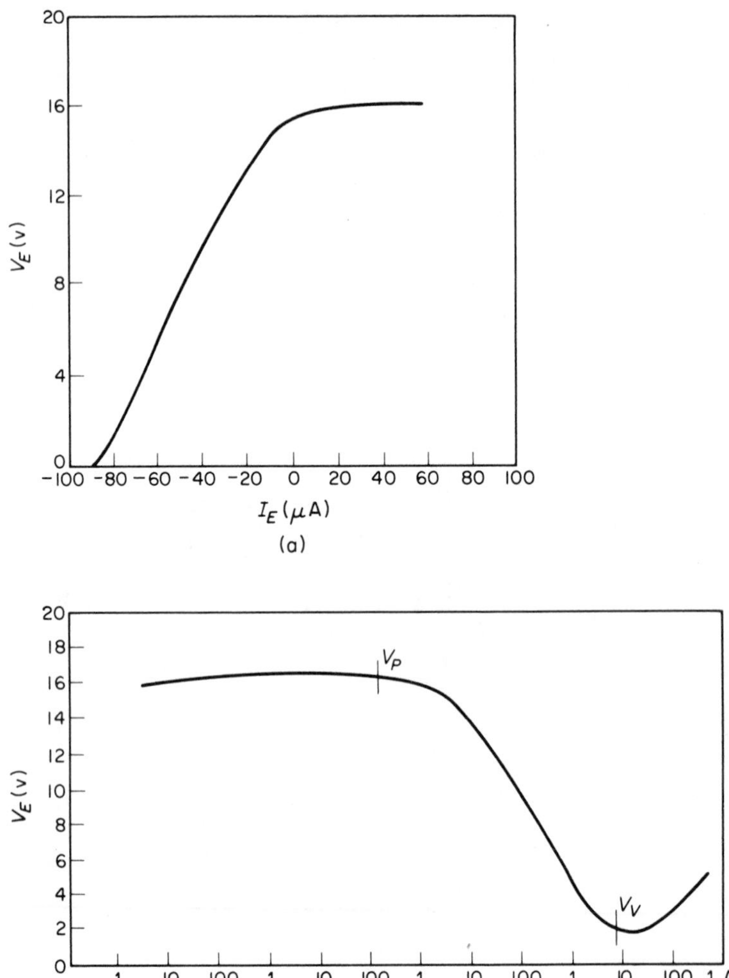

FIGURE 3-3 UJT static emitter characteristics in (a) cutoff and (b) negative-resistance regions.

in Fig. 3–3(b), the voltage remains constant until the emitter current reaches approximately 100 nA, at which point the voltage starts to decrease. Peak current I_P for the UJT graphed in Fig. 3–3 thus equals 100 nA, while peak voltage V_P equals 16 V. Referring to Fig. 3–3(b), valley voltage V_V is seen to be about 1.6 V and the valley current I_V is approximately 8 mA.

3-2.1 Saturation Resistance r_s

The saturation resistance r_s can be found from the slope of the emitter characteristic in the saturation region (above about 8 mA) and is approximately 5 Ω.

3-2.2 Diode Voltage Drop V_D

Diode voltage drop is defined as the forward voltage drop of the emitter junction. Since V_D is essentially equivalent to the forward voltage drop of a silicon diode, the value of V_D is dependent on both forward current and temperature. V_D is particularly important since it is one of the major factors in peak voltage V_P of a UJT. The equation for V_P is:

$$V_P = V_D + \eta V_{B2B1}$$

where η is the *intrinsic standoff ratio.*

In applications such as timers and oscillators, any changes in V_P result in inaccuracy since oscillator and timer circuit accuracy depend upon the repeatability of V_P. There are several ways to measure V_D. However, it is important to hold the emitter current near peak I_P when the measurement is made since it really is V_D at I_P that is required.

One simple way to measure V_D is shown in Fig. 3-4. A constant-current signal equal to I_P is applied between the emitter and base 1, and a potentiometric voltmeter is used to measure the voltage from emitter to base 2. A potentiometric voltmeter has essentially infinite input impedance when the meter is nulled, and there is no current flow in the base 2 circuit. Thus, the voltage measured is equal to V_D. Figure 3-4(b) shows V_D as a function of temperature for an emitter current of 1 μA.

3-2.3 Intrinsic Standoff Ratio

Intrinsic standoff ratio is defined by

$$\frac{V_P - V_D}{V_{B2B1}} = \frac{r_{BI}}{r_{BB}}$$

The intrinsic standoff ratio is somewhat temperature-dependent, and is also slightly dependent on V_{B2B1}. The procedure for measuring intrinsic standoff ratio is described in following sections of this chapter. It is also possible to calculate the ratio if the values of V_{B2BL}, V_P, and V_D are known. Similarly, the ratio can be measured directly if you have a very sensitive ohmmeter. First measure the base 1 to emitter resistance, then measure the base 1 to base 2 (interbase) resistance. Divide r_{B1} by r_{BB} to find the intrinsic standoff ratio.

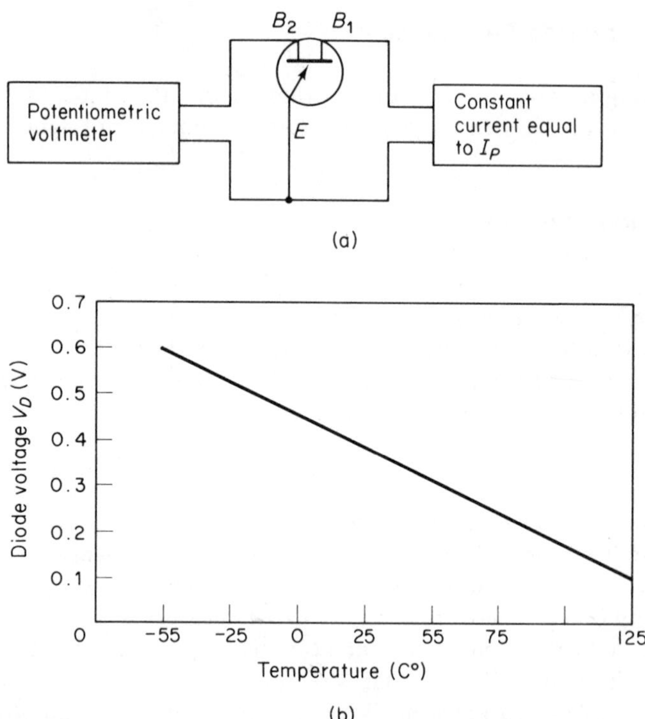

FIGURE 3-4 Basic diode voltage drop V_D test circuit and typical characteristics.

Note that r_{BB} is highly temperature-dependent and can vary with interbase voltage V_{B2B1}. Also, note that the peak-point characteristics V_P and I_P, as well as the valley-point characteristics V_V and I_V, decrease as ambient temperature is increased. When V_{B2B1} increases, both V_V and I_V increase.

3-2.4 *Interbase Characteristics*

Interbase characteristics are usually indicated by measurement of base 2 current I_{B2} as a function of interbase voltage V_{B2B1} and emitter-current I_E. Usually, interbase characteristics are measured on a sweep basis rather than with constant voltages and currents, to avoid heating effects due to power dissipation.

A circuit for sweep test of interbase characteristics is shown in Fig. 3-5(a). A constant current is applied to the emitter from time zero t_0 to time t_1. Simultaneously, a voltage ramp going from 0 to 30 V is applied to base 2.

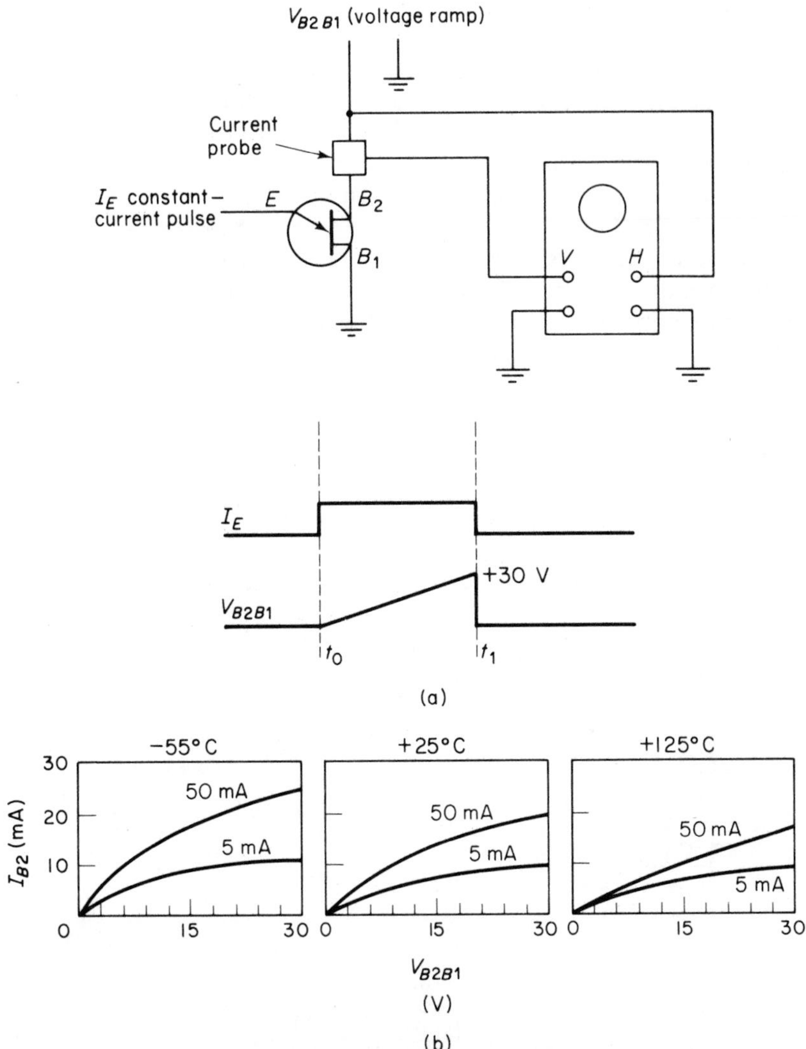

FIGURE 3-5 Interbase characteristic test circuit and typical characteristic curves.

Base 1 is grounded to complete the circuit. The current I_{B2} is measured with a current probe and applied to the vertical input of an oscilloscope. The voltage ramp, applied to base 2, is also applied to the oscilloscope horizontal input. Figure 3-5(b) shows typical interbase characteristics at ambient temperatures of −55,. +25, and +125°C. As shown, the percentage increase in I_{B2} decreases with increasing emitter current and temperature.

3-2.5 Transient Characteristics

The transient characteristics of a UJT are usually not specified in the same way as for a conventional two-junction transistor or FET. For example, switching times are usually not specified on a UJT datasheet. Instead, a parameter of f_{max} is given for most UJTs. The f_{max} indicates the maximum frequency of oscillation which can be obtained using the UJT in a specified relaxation oscillator circuit.

In some applications, such as critical timers, it may be of interest to determine turn-on and turn-off times associated with the UJT. The following paragraphs describe basic procedures for these measurements.

The circuit of Fig. 3-6(a) can be used to measure t_{on} and t_{off} for the case

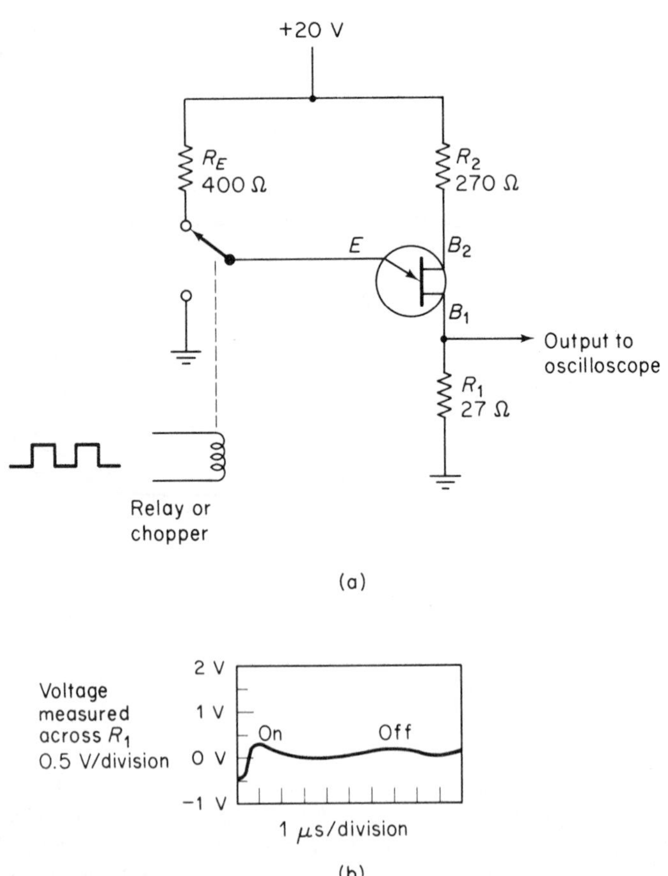

FIGURE 3-6 Test circuit for measurement of t_{on} and t_{off} where UJT emitter load is purely resistive, and typical turn-off waveform.

where the UJT emitter circuit is *purely resistive*. Typical switching-time values are t_{on} = 1 μs; t_{off} = 2.5 μs. The waveform observed at the base 1 terminal when the UJT turns off is shown in Fig. 3-6(b). Operation of the circuit in Fig. 3-6(a) is as follows. When the emitter is returned to ground, the stored charge in the junction causes a current to flow out of the emitter, and the output voltage across R_1 is smaller than the steady-state off value. Immediately following the removal of the excess charge, the voltage across R_1 goes higher than the steady-state off value because r_{b1} has still not returned to normal, following the conductivity modulation in the on-state, and I_{B2} is larger than the steady-state off value.

The circuit of Fig. 3-7 can be used to measure t_{on} and t_{off} for the case where the UJT emitter circuit has both capacitance and resistance (as usually is the case). The test circuit of Fig. 3-7(a) is a *relaxation oscillator,* with turn-on and turn-off time being measured at the base 1 terminal. The turn-on and turn-off waveforms are shown in Fig. 3-7(b).

Turn-on time is measured from the start of the turn-on to the 90%

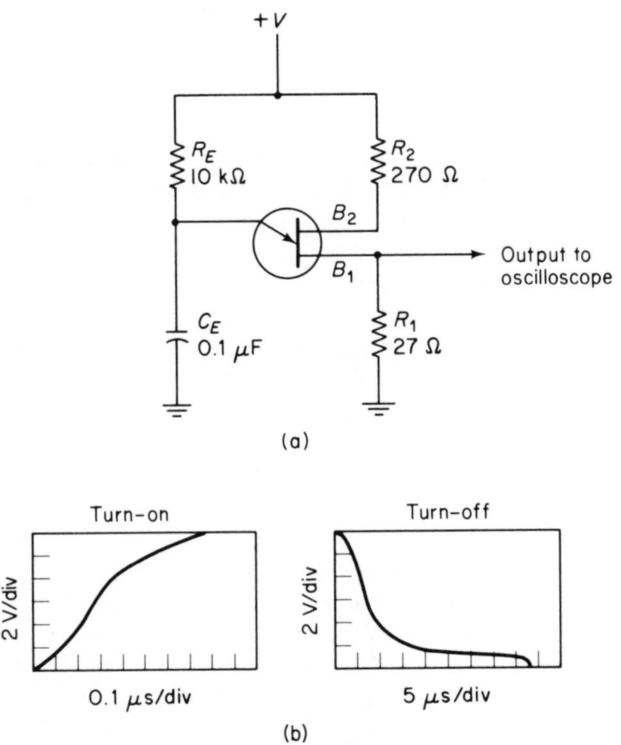

FIGURE 3-7 Test circuit for measurement of t_{on} and t_{off} where UJT emitter load is both resistive and capacitive, and typical turn-on and turn-off waveforms.

point. A typical turn-on time is 0.5 μs with the capacitance C_E shown in Fig. 3–7(a). An increase in C_E capacitance causes an increase in turn-on time. Turn-off time is measured from the start of turn-off to the 90% point, and is about 12 μs (due to the long discharge time of the capacitor). Turn-off time also increases with an increase in C_E capacitance. The effect of C_E capacitance on switching time is shown in Fig. 3–8.

3–3 PUT CHARACTERISTICS

Figure 3–9 shows some typical characteristics of selected PUTs. These characteristics are usually given on most PUT datasheets, together with the necessary test circuits and conditions. The information in Fig. 3–9 defines the static PUT curve shown in Fig. 3–2, for a 10-V gate reference V_S with various gate resistances R_G. The information also indicates currents of the PUTs and describes the output pulse. Values given in Fig. 3–9 are for 25°C unless otherwise noted. The following paragraphs define the parameters and indicate how they are measured using the various test circuits referenced in Fig. 3–9.

3–3.1 Peak-Point Current I_P

The peak point is indicated graphically by the static curve in Fig. 3–2. Reverse anode current flows with anode voltage less than the gate voltage V_S because of leakage from the bias network to the charging network. With currents less than I_P, the PUT is in a blocking state. With currents above I_P, the PUT goes through the negative resistance region to the on-state.

The charging current, or the current through a timing resistor, must be greater than I_P at V_P to ensure that the PUT switches from a blocking to an on-state in an oscillator circuit. For this reason, maximum values of I_P are given on most PUT datasheets. These values are dependent on V_S, temperature, and R_G. Typical curves on the datasheet indicate this dependence and must be consulted for most applications.

Peak-point current I_P is tested using the circuit of Fig. 3–10. This test circuit is a sawtooth oscillator that uses a 0.01-μF timing capacitor, a 20-V supply (to provide V_S), an adjustable charging current, and equal biasing resistors R. The FET circuit is used as an adjustable current source. A variable gate-voltage supply V_G is used to control the FET current source, and thus control operation of the PUT. The PUT can be adjusted to operate throughout the entire curve of Fig. 3–9 by adjustment of V_G.

The output pulse (sawtooth sweep) from the circuit is displayed on an oscilloscope as shown. The peak-point current I_P is measured across sense resistor R_S by means of a voltmeter ($I = E/R$). The value of R_S is not critical,

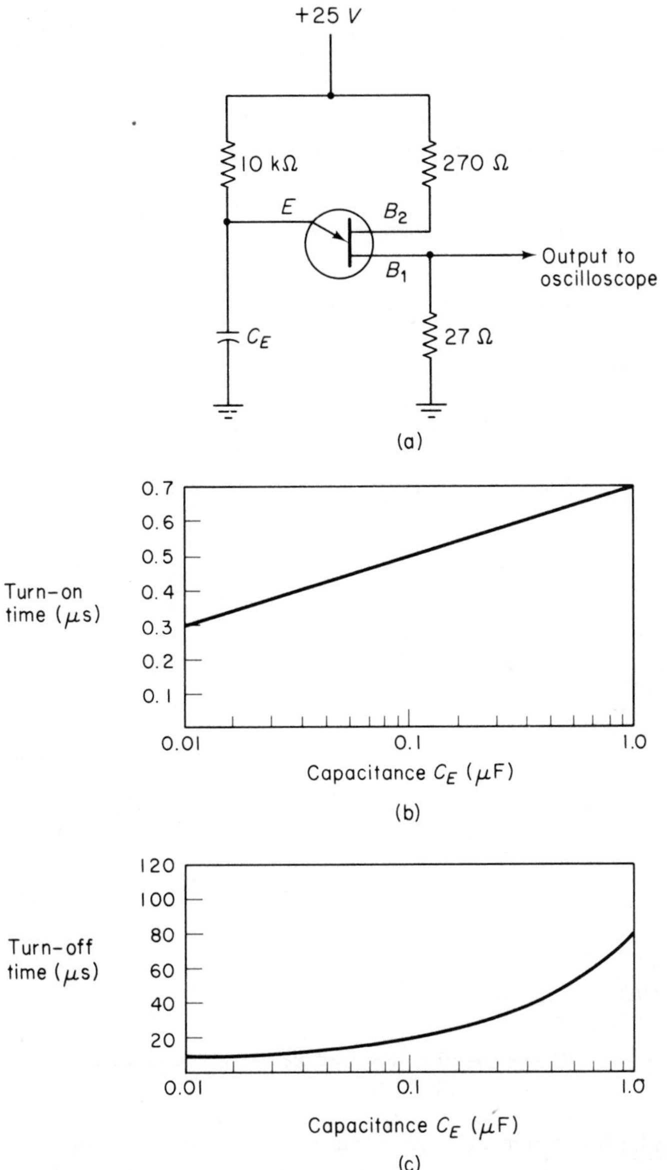

FIGURE 3-8 Turn-on and turn-off times versus emitter capacitance in relaxation oscillator test circuit.

Characterist	Test Circuit	Test Conditions	PUT under test		
			MPU131	MPU132	MPU133
I_P	3-10	R_G = 1 MΩ	1.25 μA	0.19 μA	0.08 μA
		R_G = 10 KΩ	4.0 μA	1.20 μA	0.70 μA
I_V	3-10	R_G = 1 MΩ	18.0 μA	18.0 μA	18.0 μA
		R_G = 10 KΩ	270 μA	270 μA	270 μA
V_O	3-11		16 V	16 V	16 V
t_r	3-12		40 ns	40 ns	40 ns

FIGURE 3-9 Typical PUT characteristics.

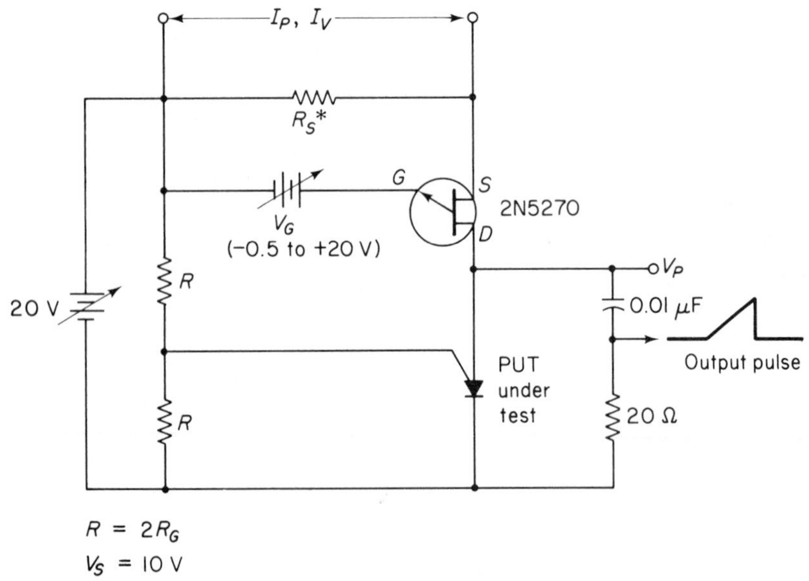

FIGURE 3-10 Test circuit for I_P, V_P, and I_V.

but should be some precision value that will keep the sense voltage in the 1-V range. To measure I_P, the PUT is set to off (just prior to oscillation) by adjustment of V_G. This condition is indicated by the absence of an output pulse to the oscilloscope. Then the voltage across R_S is measured, and the value of I_P is calculated. For example, if the indicated voltage is 1 V at I_P and R_S is 1 MΩ, I_P is 1 μA.

3–3.2 Valley-Point Current I_V

The valley point is indicated graphically in Fig. 3–2. With currents slightly less than I_V, the PUT is in an unstable negative-resistance state. A voltage minimum occurs at I_V, and with higher currents the PUT is in a stable on-state. When the PUT is used as an oscillator, the charging current or the current through a timing resistor must be less than I_V at the valley-point voltage V_V. For this reason, minimum values for I_V are generally given on the datasheet. When the PUT is used as an SCR in the latching mode, the anode current must be greater than I_V.

Valley-point current I_V is tested using the circuit of Fig. 3–10. Again, the circuit operates as a sawtooth oscillator, and operation of the circuit is controlled by V_G. Circuit output is monitored on the oscilloscope, and valley-point current I_V is measured across sensing resistor R_S in the form of a voltage (as is the case for I_P).

It is sometimes difficult to differentiate among valley-point current I_V, latching current I_L, and holding current I_H. With the PUT latched on, reducing the current (by adjustment of V_G) produces a minimum indication at the valley point (the point at which I_V is measured). However, the PUT will remain on (producing a sawtooth sweep output to the oscilloscope) at lower currents until the holding current I_H point is reached. The holding current point is detected by the absence of an output pulse (just before oscillation occurs). Latching current I_L is generally higher than I_H and is measured by increasing the current while the PUT is oscillating, then noting the value at which oscillation stops.

3–3.3 Peak-Point Voltage V_P

The unique feature of a PUT is that the peak-point voltage can be determined externally. This programmable feature gives the PUT the ability to function in voltage-controlled oscillators or similar applications. The triggering or peak-point voltage is approximated by: $V_P \approx V_T + V_S$, where V_S is the unloaded divider voltage and V_T is the offset voltage. The actual offset voltage is always greater than the anode-gate voltage V_{AG} because I_P flows out of the gate just prior to triggering. This makes $V_T = V_{AG} + I_P R_G$.

A change in R_G affects both V_{AG} and $I_P R_G$, but in opposite ways. First, as R_G increases, I_P decreases and causes V_{AG} to decrease. Second, since I_P does not decrease as fast as R_G increases, the $I_P R_G$ product increases and the actual V_T increases. These effects are difficult to predict and measure. Allowing V_T to be 0.5 V is usually sufficient for most applications.

Actual peak-point voltage V_P is tested using the circuit of Fig. 3–10. During test, V_G is adjusted until the PUT is at peak (as described in the I_P test of Section 3–3.1). Then V_P is measured across the PUT as shown. V_P can be measured with a high-impedance (10 MΩ or higher) oscilloscope.

3-3.4 *Peak Output Voltage V_0*

The peak output voltage of a PUT used in a relaxation oscillator circuit depends on many factors, including dynamic impedance, switching speed, and V_P. When the timing capacitor used in the circuit is small (less than about 0.01 μF), the effect of switching speed on V_0 is increased. This is because small-value capacitors lose part of their charge during the turn-on interval.

Peak output voltage V_0 is tested using the circuit of Fig. 3–11. This circuit is a relaxation oscillator similar to that shown in Fig. 3–2. The Fig. 3–11 circuit uses a relatively large timing capacitor (0.2 μF) to minimize the effect of losing part of the charge during turn-on. The output of the relaxation oscillator circuit is displayed on an oscilloscope across the 20-Ω resistor (in series with the cathode lead).

3-3.5 *Rise Time t_r*

Rise time is a useful parameter in pulse circuits that use capacitive coupling. Rise time can be used to predict the amount of current that flows between such circuits. As shown in Fig. 3–9, the rise time for some typical PUTs is in the order of 40 ns. This is measured using the circuit of Fig. 3–12, which is essentially a relaxation oscillator. Rise time for the PUTs shown in Fig. 3–9 is specified using a real-time oscilloscope and measuring between 0.6 and 6 V on the leading edge of the output pulse. The 0.6 and 6-V values represent the 10% and 90% points, respectively, on the leading edge with the circuit values shown in Fig. 3–12.

3-4 TESTING UJTS WITH CIRCUITS

As is the case for two-junction transistors and FETs, there are a number of commercial test sets for UJTs. Similarly, it is possible to adapt commercial

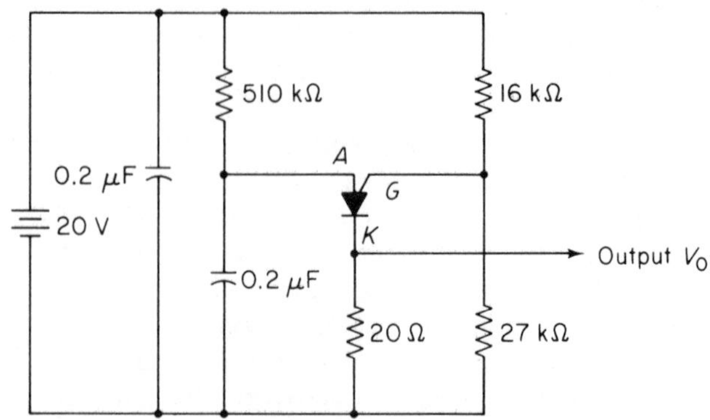

FIGURE 3-11 PUT test circuit for output voltage V_O.

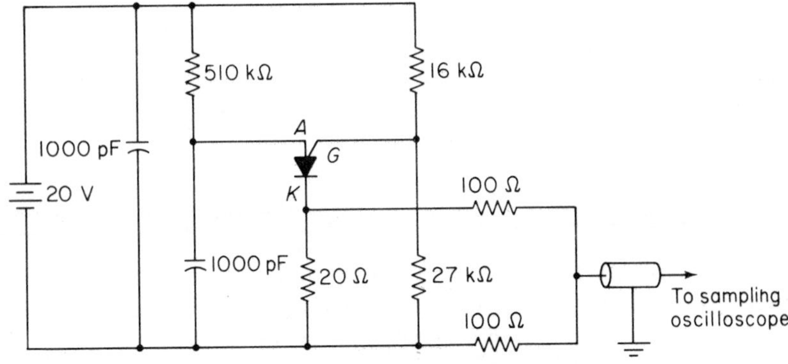

FIGURE 3-12 PUT test circuit for rise time t_r.

test sets (designed for two-junction transistors) for use with UJTs. Section 3-5 describes the procedures for testing UJTs with curve tracers. The following describes the procedures for testing all important UJT characteristics using special circuits.

3-4.1 *Simplified Test for UJTs*

Before going into a detailed test of UJT parameters, let us consider a basic test that shows operation of a UJT on a go/no go basis. The firing point of a UJT can be determined using a simple ammeter circuit shown in Fig. 3-13. The test also shows the amount of emitter-base 1 current flow after the UJT has fired. If a UJT fires with the correct voltage applied and draws the rated amount of current, the UJT can be considered satisfactory for operation in most circuits.

The base 2 voltage is shown in Fig. 3-13 as +20 V. However, any value of base 2 voltage can be used to match a particular UJT. Initially, R_l is set to 0 V (at the ground end). The setting of R_l is gradually increased until the UJT fires. The firing voltage is indicated on the voltmeter. When the UJT fires, the

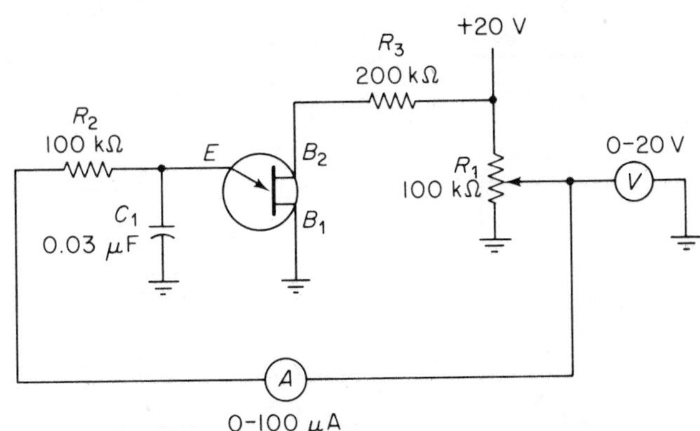

FIGURE 3-13 Simplified UJT firing test circuit.

ammeter indication suddenly increases. The amount of emitter–base 1 current is read on the ammeter. Usually, the firing point is in the range 0 to 20 V, whereas emitter–base 1 current is less than 50 μA.

3-4.2 Emitter and Interbase Curve Tracer

The circuit of Fig. 3–14 can be used to display either the emitter characteristic curves or the interbase characteristic curves on a conventional (non-curve-tracer) oscilloscope. When the switches are set to display the interbase curves, the meter indicates the emitter current. When the switches are set to display the emitter curves, the meter indicates the interbase voltage.

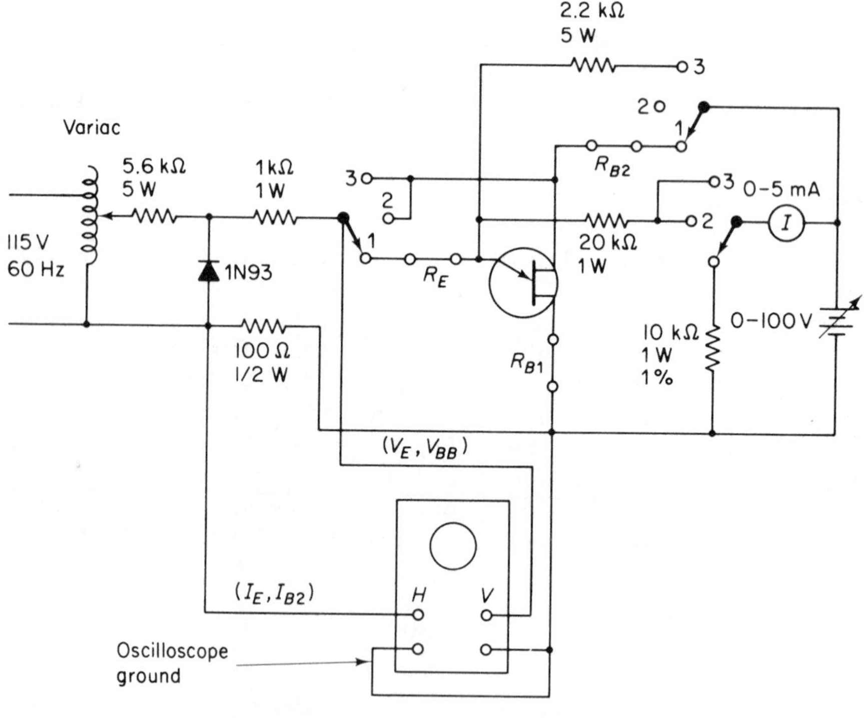

Switch position	Display	Meter
1	$V_E - I_E$	V_{BB} (0 – 50 V)
2	$V_{BB} - I_{B2}$	I_E (0 – 5 mA)
3	$V_{BB} - I_B$	I_E (0 – 50 mA)

FIGURE 3-14 Emitter and interbase curve-tracer test circuit using a conventional oscilloscope.

It is important to set the variac and the d-c supply to zero before changing the switches or inserting the UJT in the circuit so as to avoid accidental burnout. If desired, external resistors may be inserted in series with the emitter, base 1, or base 2 to determine the characteristic of the UJT in a particular circuit.

3-4.3 General-Purpose UJT Test Set

A general-purpose test set circuit for UJTs is shown in Fig. 3–15. This circuit can be used for testing six of the major UJT characteristics, as shown in the table of Fig. 3–15. Operation of the circuit is controlled by the switches. These switches can be either pushbutton or toggle. However, pushbutton switches are generally preferred. The switches are shown in the inactive position. To perform a particular test, operate all corresponding switches to the active position (the opposite position from that shown in Fig. 3–15). For example, to perform test 1 (peak-to-peak emitter voltage), operate all three switches marked 1 to the active position. This connects the meter to R_6 and the emitter to the junction of R_1-$D2$-C_3, and applies 22.5 V of power to the circuit.

The oscillator test portion of the circuit indicates if a UJT can oscillate in a relaxation oscillator circuit. Failure to pass this test usually means total failure of a UJT. The UJT is connected as an oscillator in the test 1 position, and the peak-to-peak emitter voltage is indicated on the meter. In the test 1 position, power is applied across base 1 and base 2 (through R_4), and the emitter is connected to the junction of R_1 and C_3. The meter is connected to the junction of C_1 and D_1, through R_6, and indicates the charge across C_1. If the UJT fails to oscillate, C_1 does not charge and there is no meter indication.

The intrinsic standoff ratio test (test 2) also makes use of a relaxation oscillator and a peak voltage detector. To calibrate the circuit for this test, make a temporary connection from the common point of R_4-D_3 to the common point of D_2-C_2, and adjust R_5 to give a full-scale deflection on the meter with switches for test 2 set to the active position. Remove the temporary connection before making the test. To make the test, set all switches for test 2 to the active position and check the meter to find standoff ratio. For example, if the full-scale deflection is 100 μA, and the meter indicates 80 μA during test, the standoff ratio is 0.8.

For measurement of interbase resistance (test 3), resistors R_8 and R_9 must be calibrated to give the correct meter deflection. The R_8 and R_9 values shown in Fig. 3–15 are for a typical UJT. The UJT can be disconnected from the circuit, and resistances of known value inserted between the B_1 and B_2 terminals of the circuit. Then the meter can be calibrated (by selection of R_8 and R_9 values) to produce the desired reading.

In addition to the six tests shown in Fig. 3–15, the interbase modulated current or $I_{B2(\text{mod})}$ shown on some UJT datasheets can be made by connecting an external ammeter in the B_2 shunt circuit and setting the switches for test 4 to the active position.

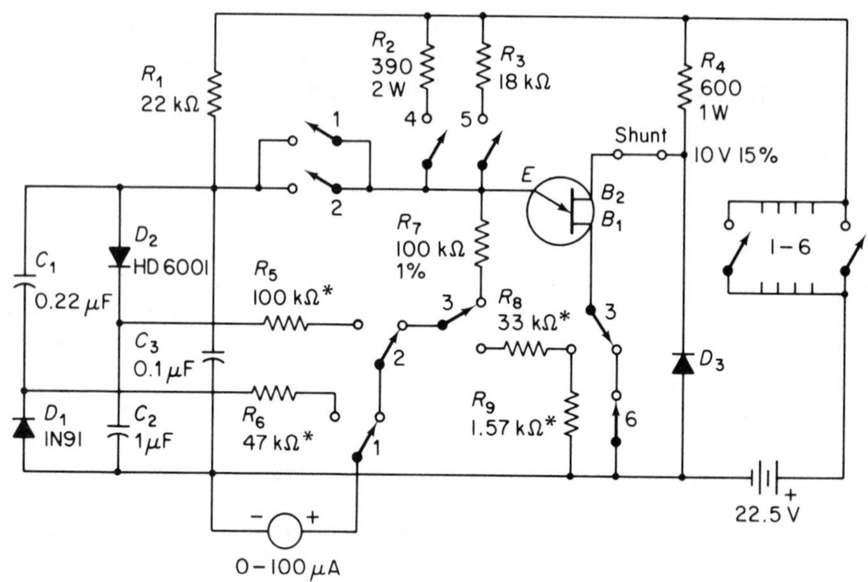

Test parameter	*Range*	*conditions*
1 Peak-to-peak emitter voltage	0-10 V	V_{B2B1} = 10 V; C = 0.1 μF
2 Intrinsic standoff ratio	0-1.0	V_{B2B1} = 10 V
3 Interbase resistance	3 kΩ to infinity	Power less than 15 mW
4 Emitter saturation voltage	0-10 V	V_{B2B1} = 10 V I_E = 50 mA
5 Emitter voltage at 1 mA	0-10 V	V_{B2B1} = 10 V I_E = 1 mA
6 Emitter leakage current	0-100 V uA	V_{EB2} = 10 V or less

All switches shown in inactive position.
All components ±5% except as noted.

*Approximate values, must be calibrated.

FIGURE 3-15 General-purpose UJT test set.

3-4.4 Intrinsic Standoff Ratio Test

The general-purpose test circuit of Fig. 3-15 is usually satisfactory for measurement of the intrinsic standoff ratio. If greater accuracy is required, or if the equivalent emitter-diode voltage V_D is to be measured, the circuit of Fig. 3-16 can be used.

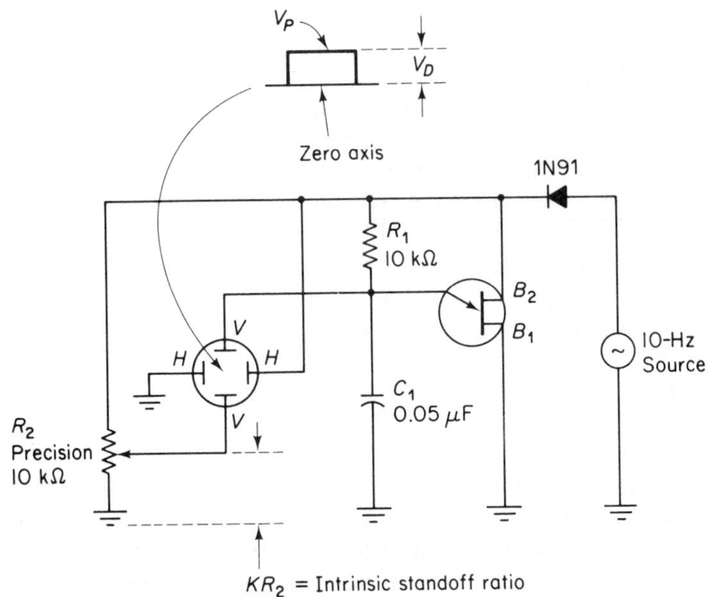

FIGURE 3-16 Test circuit for measurement of intrinsic stand-off ratio and V_D.

In the circuit of Fig. 3-16, the interbase voltage is swept by the 10-Hz oscillator, while the UJT oscillates at about 2 kHz in the basic relaxation oscillator configuration. The interbase voltage is applied to the horizontal axis of the oscilloscope. The emitter voltage is applied to one input of the oscilloscope vertical deflection amplifier. The R_2 potentiometer voltage is applied to the other input of the vertical deflection amplifier. With these connections, the oscilloscope pattern consists of a plot of V_{B2B1} on the horizontal axis against $V_P - KV_{B2B1} = V_D + (-K)V_{B2B1}$ on the vertical axis. K is equal to the fractional setting of potentiometer R_2. V_P is the upper envelope of the displayed emitter voltage.

During the test, K (potentiometer R_2) is adjusted until the upper envelope of the display is horizontal. At this point, K or the fractional setting of R_2 is equal to the intrinsic standoff ratio. For example, if R_2 is 10 kΩ, and the resistance from the contact arm of R_2 to ground is 7 kΩ, the intrinsic stand-off ratio is 0.7. If a precision potentiometer is used for R_2, the intrinsic stand-off ratio can be measured with an accuracy of better than 0.05%.

With R_2 adjusted to produce a horizontal envelope on the display, note the displacement of the upper envelope from the zero axis of the oscilloscope (in volts). This displacement is equal to V_D. If the oscilloscope vertical screen can be read accurately, V_D can be measured within about 20 mV.

3-4.5 Peak-Point Current Test

The peak-point current I_p can be measured with the circuit of Fig. 3-17. This circuit actually measures the *minimum emitter current* required for oscillation in a relaxation oscillator circuit. This minimum emitter current is a good approximation of the peak-point emitter current.

To measure I_p, set in the desired value of V_{B2B1}. Then adjust voltage V_1 until the UJT just fires, as indicated by a tone on the loudspeaker. Read the peak current on the microammeter. Take care to avoid any ripple on the V_{B2B1} supply (if a battery is not used) since ripple can reduce the apparent peak-point current considerably.

3-4.6 Valley-Point Current and Voltage Test

The valley point corresponds to the point on a particular UJT emitter-characteristic curve where the emitter voltage is at a minimum or where the dynamic resistance is zero. Because of the slight change of emitter voltage at or near the valley point, it is difficult to locate the exact position of the valley point by inspection of the emitter-characteristic curve. To overcome this problem, the circuit of Fig. 3-18 can be used to measure V_V and I_V by the *null method.*

With the circuit of Fig. 3-18, the supply voltage V_2 is adjusted to obtain a null on the electronic voltmeter. I_V and V_V can then be measured directly on the corresponding meters. The values given in Fig. 3-18 are typical. Other values of interbase voltage and B_2 series resistance can be used as desired for special tests.

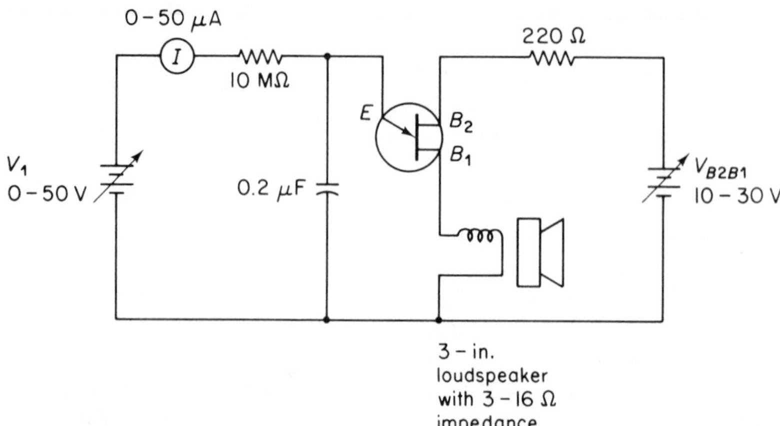

FIGURE 3-17 Test circuit for measurement of peak-point emitter current I_P.

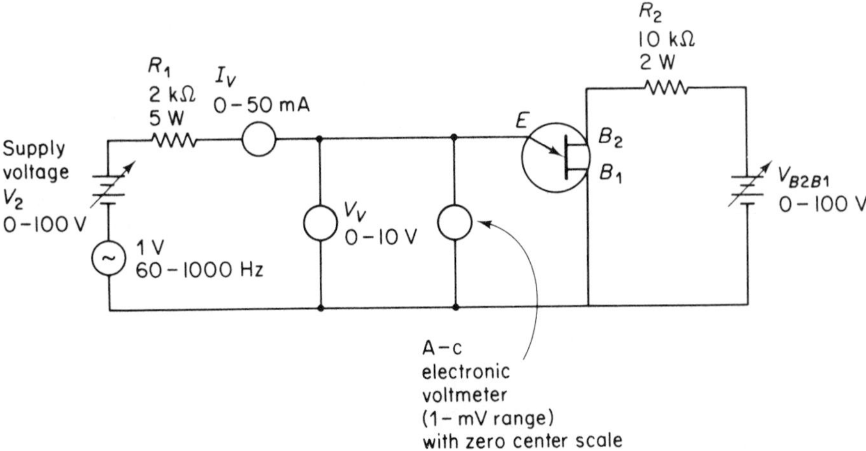

FIGURE 3-18 Test circuit for measurement of valley voltage V_V and valley current I_V.

3-4.7 *Frequency Response*

The UJT small-signal frequency-response characteristic of most value is the *emitter input impedance,* which can be measured with the circuit shown in Fig. 3-19. In this circuit, the UJT is biased in the negative-resistance region, and a null is obtained on the electronic voltmeter by adjusting R_1 and C_1. At null, the emitter input impedance is approximately equal to $R_1 + 1/(6.28C_1)$.

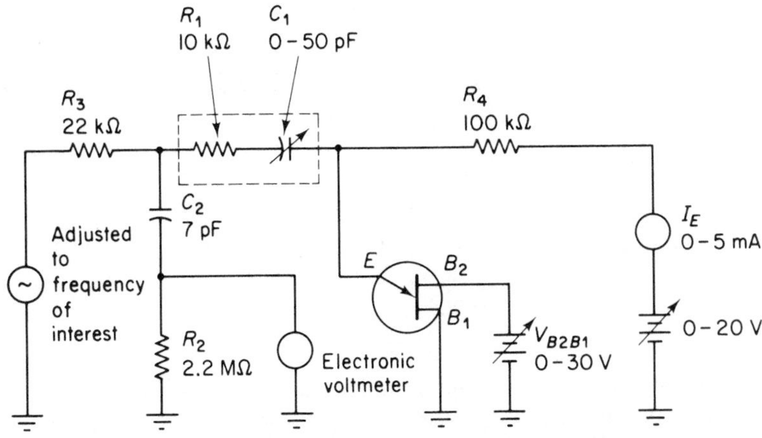

FIGURE 3-19 Test circuit for measurement of emitter input impedance of UJTs.

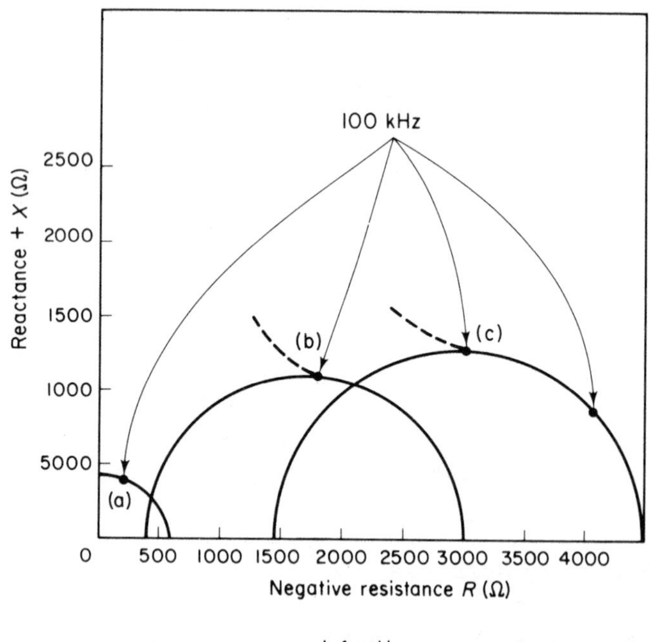

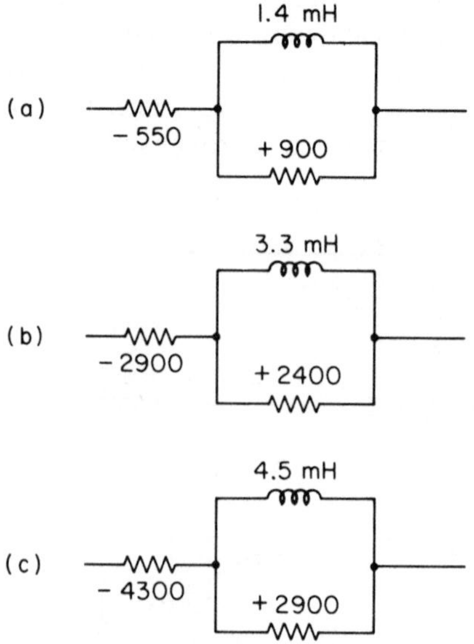

FIGURE 3-20 Emitter impedance diagram of UJT at several bias points: (a) V_{B2B1} = 25 V, I_E = 4.2 mA; (b) V_{B2B1} = 25 V, I_E = 1 mA; (c) V_{B2B1} = 25 V, I_E = 0.5 mA.

Capacitor C_2 is used to reduce the effective capacitance between the UJT emitter and ground so as to prevent oscillation in the relaxation mode.

An impedance diagram for a typical UJT, showing several bias conditions, is illustrated in Fig. 3-20. Up to 100 kHz, the simple three-element *RL* network of Fig. 3-20 provides an accurate representation of the input impedance.

The frequency at which the resistive component of the emitter input impedance is equal to zero is called the *resistance cutoff* frequency F_{RO}. This is a frequency figure of merit for a UJT since F_{RO} represents the maximum frequency at which the UJT is regenerative at a given bias point.

The circuit of Fig. 3-19 can be used to measure F_{RO} if R_1 is set equal to zero, and a null obtained on the electronic voltmeter by adjusting the frequency and C_1. Since F_{RO} varies with bias conditions, measured values are best presented by means of a plot such as that of Fig. 3-21, which gives contours of constant F_{RO} superimposed on the normal emitter-characteristic curves. F_{RO} is proportional to the magnitude of the negative resistance at a given bias point. Thus, a single measurement of F_{RO} establishes the values of F_{RO} over the entire negative-resistance portion of the emitter characteristics.

The value of F_{RO}, for a given emitter-characteristic curve, increases as

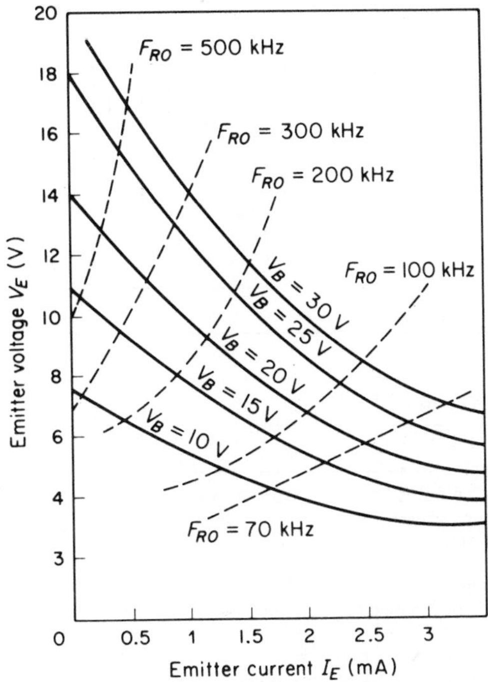

FIGURE 3-21 UJT emitter characteristic curves with contours of constant-resistance cutoff frequency.

the peak point is approached. The limiting value at the peak point corresponds to the *maximum frequency of oscillation,* f_{max}. This is the maximum frequency at which the UJT can oscillate in a relaxation oscillator circuit. For the lower range of V_{B2B1}, where the heating caused by the interbase power dissipation is not excessive, f_{max} is directly proportional to V_{B2B1}.

3-5 UJT TESTS USING A CURVE TRACER

Many curve tracers provide special circuits and connections for measurement of UJT characteristics. Most curve tracers can be converted by simple external circuits to test UJTs. For example, Fig. 3–22 shows the connections for conversion of a Tektronix 575 curve tracer to test UJTs. Note that for test of UJT interbase characteristics, the curve-tracer terminals normally used for a two-junction transistor base are connected to the UJT emitter. For the test of emitter curves, the UJT emitter is connected to the collector terminal of the test

Test	*Emitter curves*	*Interbase curves*
Circuit	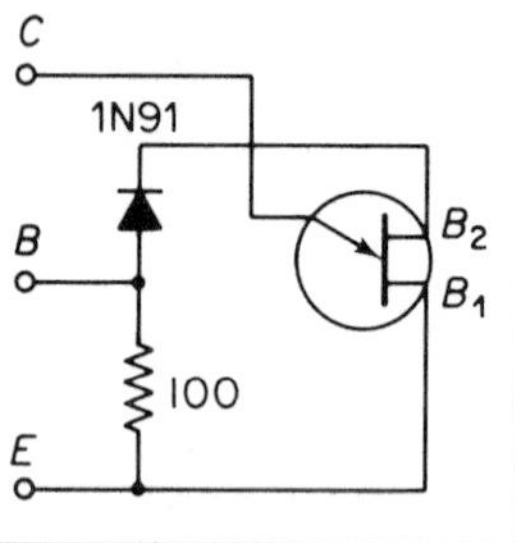 C, B, E; 1N91; 100; B_2, B_1	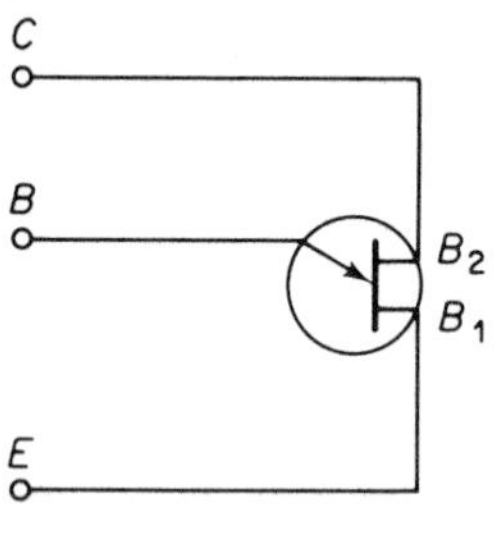 C, B, E; B_2, B_1
Collector sweep polarity	+	+
Base step polarity	+	+
Collector peak volts range	200 V	20 V
Collector limiting resistor	5 kΩ	500 Ω
Base current steps selector	20 mA/step	10 mA/step
Number of current steps	5	5
Vertical current range	2 mA/div. I_E	2 mA/div. I_{B2}
Horizontal voltage range	1 V/div. V_E	2 V/div. V_{BB}

FIGURE 3–22 Use of Tektronix 575 transistor curve tracer for display of UJT characteristic curves.

set. In a display of the emitter-characteristic curves, the interbase voltage should not exceed 12 V because of the voltage limitation of the base-current step generator.

The B&K Precision 501 curve tracer can be used without modification to display a set of UJT curves and to measure interbase resistance. The following paragraphs describe the basic steps.

3-5.1 Displaying UJT Curves on a Curve Tracer

To display UJT curves on the curve tracer, connect the UJT to the tracer as shown in Fig. 3–23. As shown, base 1 is connected to the emitter jack, base 2 is connected to the base jack, and the (UJT) emitter is connected to the collector jack. (Note that base 1 and base 2 are interchangeable.)

Set the curve-tracer POLARITY switch to NPN. Increase the SWEEP VOLTAGE from zero until the trigger voltage is exceeded. This should produce the high emitter current spike on the oscilloscope. Set the STEP SELECTOR to the "current per step" position that produces the *most curves* on the display.

The curves may appear quite close together (compared to the display of a two-junction transistor or FET). This means that you must observe the display carefully to distinguish the curves. It may be helpful to spread out the display by increasing the horizontal sensitivity of the oscilloscope, or to use expanded sweep magnification (if available) at the area of interest.

With the test configuration as shown in Fig. 3–23, the step current of the curve tracer is applied from base 1 to base 2, and the sweep voltage is applied to the emitter. Thus, the sweep voltage is the UJT trigger voltage. Note that as the sweep voltage is slowly increased from the trigger threshold producing the first current spike, the other curves are added one by one. Thus, for each base current step, the emitter trigger voltage can be measured.

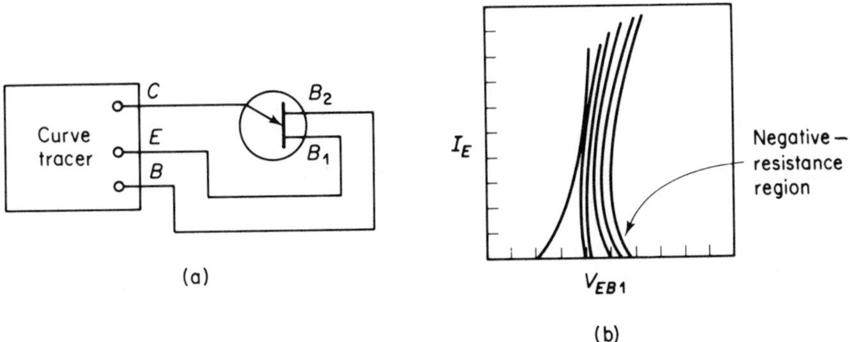

FIGURE 3-23 Test connections and typical display of UJT curves on a curve tracer.

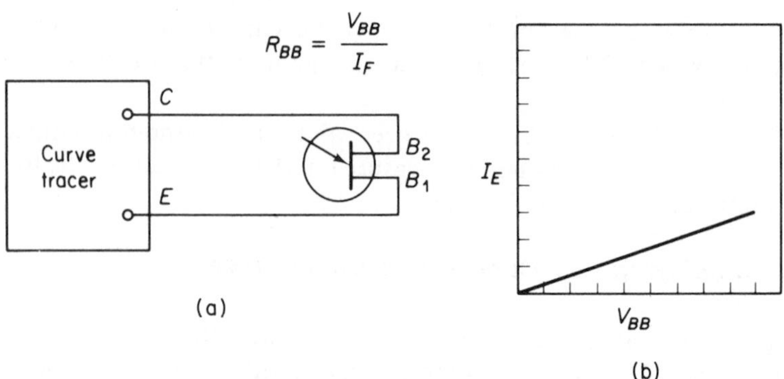

FIGURE 3-24 Test connections and typical display of UJT R_{BB} characteristics.

3-5.2 Measuring UJT Interbase Resistance with a Curve Tracer

Interbase resistance (shown as r_{BB} or R_{BB} on UJT datasheets) can be displayed using the connections shown in Fig. 3-24. As shown, base 1 and base 2 are connected to the curve-tracer collector and emitter jacks, respectively. The UJT emitter is left open-circuited (no connection). The display should be a linear trace as shown. The vertical scale of the display represents forward current (I_F), and the horizontal scale represents interbase voltage (V_{BB}). Interbase resistance equals interbase voltage divided by forward current, as shown in Fig. 3-24.

4

Solid-State Diode Tests

This chapter is devoted entirely to test procedures for solid-state diodes. These diodes include signal and rectifying diodes, zener diodes, and tunnel diodes. The first sections of this chapter describe diode characteristics and test procedures from the practical standpoint. The information in these sections permits you to test all the important diode characteristics using basic shop equipment. The sections also help you understand the basis for such tests. The remaining sections of the chapter describe how the same tests, and additional tests, are performed using more sophisticated equipment such as the oscilloscope curve tracer.

4-1 BASIC DIODE TESTS

Three basic tests are required for power rectifier diodes and small-signal diodes. First, any diode must have the ability to pass current in one direction (forward current) and prevent or limit current flow (reverse current) in the opposite direction. Second, for a given reverse voltage, the reverse current should not exceed a given value. Third, for a given forward current, the voltage drop across the diode should not exceed a given value.

All of these tests can be made with a multimeter. If the diode is to be used in pulse or digital circuits, the *switching time* must also be tested. This requires an oscilloscope and pulse generator. In addition to the basic tests, a zener diode must also be tested for the correct zener voltage point. Similarly, a tunnel diode must be tested for negative-resistance characteristics.

4-2 DIODE CONTINUITY TESTS

The elementary purpose of a diode (both power rectifier and small signal) is to prevent current flow in one direction while passing current in the opposite direction. The simplest test of a diode is to measure current flow in the forward direction with a given voltage, then reverse the voltage and measure current flow, if any. If the diode prevents current flow in the reverse direction but passes current in the forward direction, the diode meets most *basic* circuit requirements. Similarly, if there is no excessive leakage current flow in the reverse direction, it is quite possible that the diode will operate properly in all but the most critical circuits.

A simple resistance measurement or continuity check can often be used to test a diode's ability to pass current in one direction only. A basic ohmmeter can be used to measure the forward and reverse resistance of the diode. Figure 4–1 shows the basic circuit. A good diode will show high resistance in the reverse direction and low resistance in the forward direction.

If the resistance is high in both directions, with the ohmmeter connected as shown in Fig. 4–1, the diode is probably open. A low resistance in both directions usually indicates a shorted diode. If the resistance is close to the same in both directions (high or low), the diode is probably leaking.

It is possible for a defective diode to show a difference in forward and reverse resistance. The important factor in making a diode resistance test is the *ratio* of forward-to-reverse resistance (often known as front-to-back ratio). The actual ratio depends upon the type of diode. However, as a guideline, a signal diode typically has a ratio of several hundred to one,

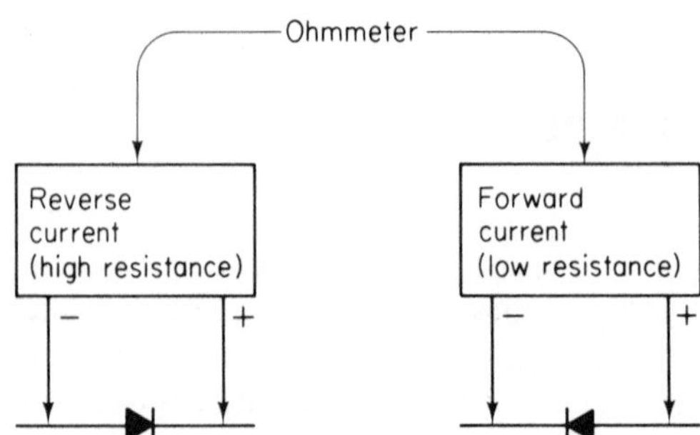

FIGURE 4–1 Basic ohmmeter test of diodes for front-to-back ratio.

whereas a power rectifier can operate satisfactorily with a ratio of about 10:1.

Diodes used in power circuits are usually not required to operate at high frequencies. Such diodes may be tested effectively with direct current or low-frequency ac. Diodes used in other circuits, even audio equipment, must be capable of operation at higher frequencies and should be so tested.

4-2.1 Diode Test Circuit

There are many commercial diode testers available for use in the shop or laboratory. Most simple diode testers operate on the continuity test principle. This is similar to testing a diode by measuring resistance, except that actual resistance value is of no concern. Instead, arbitrary circuit values are used, and the diode's condition is read out on a "good-bad" meter.

Figure 4-2 shows the basic circuit of a simple diode tester using the continuity principle. The condition of the diode under test is indicated by lamps. With an open diode, or no connection across the test terminal, lamp I_1 lights due to the current flow through CR_1, R_1, and R_2. The voltage drop across R_2 is too low to light lamp I_3. Since CR_1 is reverse-biased, lamp I_2 does not light.

If a good diode is connected across the test terminals with the polarity as shown in Fig. 4-2, lamp I_1 is shorted out and does not light. Also, lamp I_2 does not light because CR_2 is reverse-biased by the voltage developed by the diode under test. Capacitors C_1 and C_2 charge, permitting lamp I_3 to light, thus indicating that the diode is good.

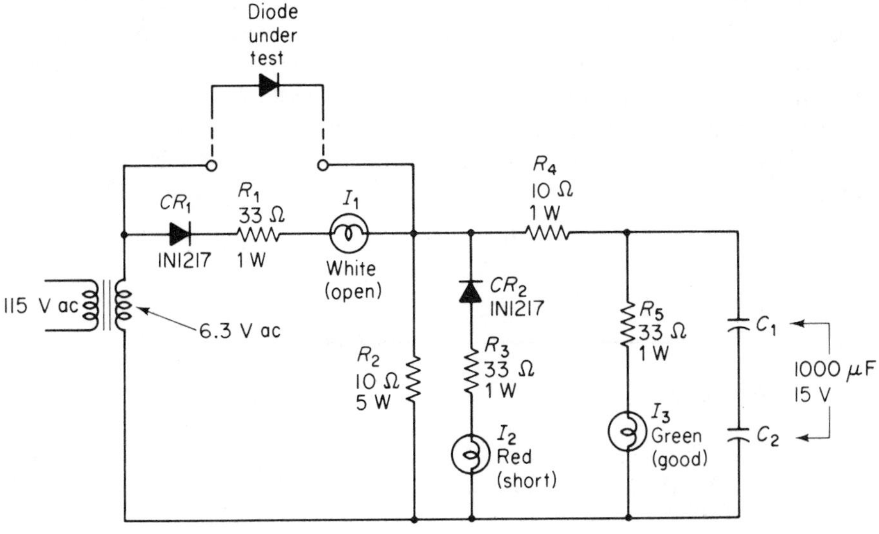

FIGURE 4-2 Continuity-type diode test circuit.

A shorted diode also shorts out lamp I_2 and allows both half-cycles of the alternating current to be applied across R_2. CR_2 conducts on the negative half-cycles (when the junction of CR_2 and R_4 is negative), causing lamp I_2 to light. Alternating current is also applied across R_4, R_5, I_3, C_1, and C_2. Capacitors C_1 and C_2 appear as a short across R_5 and I_3. Thus, lamp I_3 does not light. This leaves only lamp I_2 to light and indicate a short.

If a good diode is connected into the test terminals, but with the polarity reversed from that shown in Fig. 4–2, all lamps will light. Under these conditions, CR_1 conducts on positive half-cycles, causing lamp I_1 to light. The test diode conducts on negative half-cycles and develops a d-c voltage across R_2 (with the junction of R_2 and I_1 negative). CR_2 then conducts, and I_2 lights. Capacitors C_1 and C_2 charge, permitting I_3 to light.

4-3 DIODE REVERSE-LEAKAGE TESTS

Reverse leakage is the current flow through a diode when a reverse voltage (anode negative with respect to cathode) is applied. The basic circuit for measurement of reverse leakage is shown in Fig. 4–3. Similar circuits are incorporated in some commercial diode testers, or can be duplicated with the basic test equipment shown.

As shown in Fig. 4–3, the diode under test is connected to a variable d-c source in the reverse-bias condition (anode negative). The variable source is adjusted until the desired reverse voltage is applied to the diode as indicated by the voltmeter. Then the reverse current (if any) through the diode is measured by the ammeter. This reverse (or leakage) current is typically in the range of a few microamperes for a signal diode. Usually, excessive leakage current is undesired in any diode (signal or power), but the actual limits should be determined by reference to the appropriate datasheet. The datasheet should also specify the correct value of reverse voltage to apply.

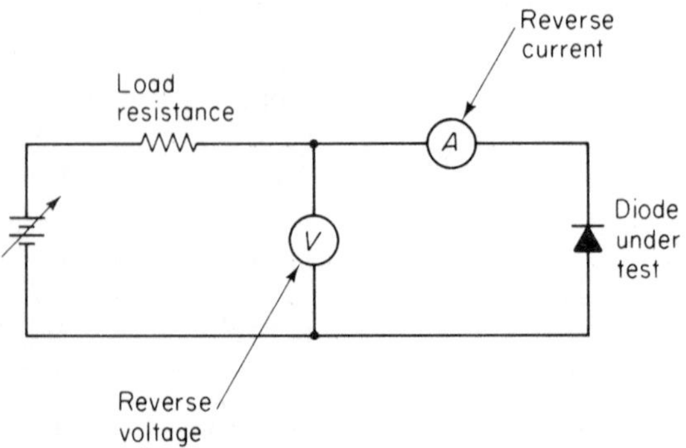

FIGURE 4-3 Diode reverse leakage test circuit.

4-4 DIODE FORWARD VOLTAGE DROP TESTS

Forward voltage drop is the voltage that appears across a diode when a given forward current is being passed. The basic circuit for measurement of forward voltage is shown in Fig. 4-4. Similar circuits are incorporated in some commercial diode testers, or can be duplicated with the basic test equipment shown.

As shown in Fig. 4-4, the diode under test is connected to a variable d-c source in the forward-bias condition (anode positive, cathode negative). The variable source is adjusted until the desired amount of forward current is passing through the diode as indicated by the current meter. Then the voltage drop across the diode is measured by the voltmeter. This is the forward voltage drop. Usually, a large forward voltage drop is not desired in any diode. The maximum limits of forward voltage drop (for a corresponding forward current) should be found in the datasheet.

Typically, the forward voltage drop for a germanium diode is approximately 0.2 V, whereas a silicon diode has a forward voltage drop of about 0.5 V, at nominal currents. Increased forward currents will produce increased voltage drops.

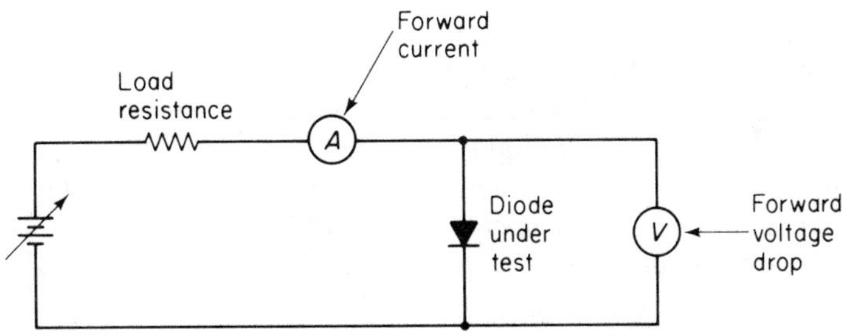

FIGURE 4-4 Diode forward voltage drop test circuit.

4-5 DIODE DYNAMIC TESTS

The circuits and methods discussed in preceding sections of this chapter provide a static test of diodes, meaning that the diode is subjected to constant direct current when the leakage and voltage drop are measured. Diodes do not usually operate this way in circuits. Instead, diodes are operated with alternating current, which tends to heat the diode junctions and change the characteristics. It is more realistic to test a diode under dynamic conditions. The following paragraphs describe two typical dynamic tests for diodes using an oscilloscope.

4-5.1 Dynamic Tests for Power Rectifier Diodes

Power rectifier diodes can be subjected to a dynamic test by using an oscilloscope to display and measure the current and voltage characteristics. To do so, a d-c oscilloscope must be used, and both the horizontal and vertical channels of the oscilloscope must be *voltage-calibrated.* Usually, the horizontal channel of most oscilloscopes is time-calibrated. However, the horizontal channel can be voltage-calibrated by using the same procedures as for voltage calibration of the vertical channel. (Such procedures are described in the oscilloscope instruction manual.) In brief, the oscilloscope sweep circuit is disconnected from the input to the horizontal amplifier (the horizontal input is set to EXTERNAL), a reference voltage is applied to the horizontal amplifier input, and the horizontal gain or width control is set for some specific deflection or width [1 centimeter (cm) of horizontal width per volt, etc.]. For best results, horizontal and vertical channels of the oscilloscope must be identical or nearly identical, to eliminate any phase difference.

As shown in Fig. 4–5, a power rectifier diode is tested by applying a controlled a-c voltage across the anode and cathode through resistor R_1. The a-c voltage (set to the maximum rated peak inverse voltage, or PIV, of the diode) alternately biases the anode positive and negative, causing both forward and reverse current to flow through R_1. The voltage drop across R_1 is applied to the vertical channel and causes the oscilloscope screen spot to move up and down. Vertical deflection is proportional to current through the diode under test. The vertical scale divisions can be converted directly to current when R_1 is made 1 Ω. For example, a 3-V vertical deflection indicates a 3-A current. If R_1 is 1000 Ω, the readout is in milliamperes.

The same voltage applied across the diode is applied to the horizontal channel (which has been voltage-calibrated), and causes the spot to move right or left. Horizontal deflection is proportional to voltage across the diode (neglecting the small voltage drop across R_1).

The combination of the horizontal (voltage) deflection and vertical (current) deflection causes the spot to trace out the complete current and voltage characteristics.

The basic test procedure is as follows:

1. Connect the equipment as shown in Fig. 4–5.
2. Place the oscilloscope in operation. Voltage-calibrate both the vertical *and* horizontal channels as necessary. The spot should be at the vertical and horizontal center with no signal applied to either channel.
3. Switch off the internal recurrent sweep of the oscilloscope. Set the sweep selector and sync selector (if any) of the oscilloscope to external. Leave the horizontal and vertical gain controls set at the (voltage) calibrate position, as established in step 2.

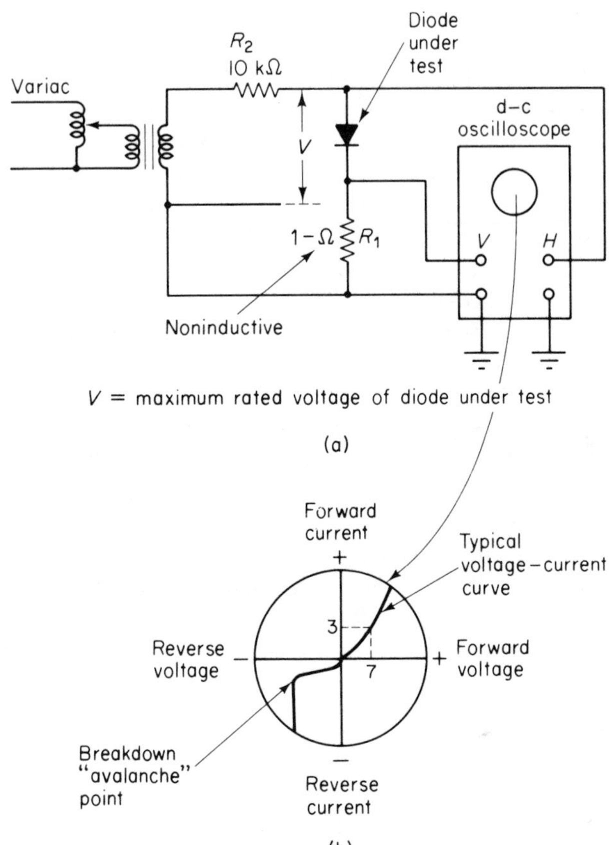

FIGURE 4-5 Test connections and typical display for power rectifier diode voltage-current characteristics.

4. Adjust the variac so that the voltage applied across the power diode under test is at (or just below) the maximum rated value (as determined from the diode datasheet).

5. Check the oscilloscope pattern against the typical curves of Fig. 4–5 and/or against the diode specifications. The curve of Fig. 4–5 is a typical response pattern. That is, the forward current (deflection above the horizontal centerline) increases as forward voltage (deflection to the right of the vertical centerline) increases. Reverse current increases only slightly as reverse voltage is applied, unless the breakdown or "avalanche" point is reached. In conventional (nonzener) diodes, it is desirable (if not mandatory) to operate considerably below the breakdown point. Some diodes will break down if operated in the reverse condition for any length of time.

6. Compare the current and voltage values against the values specified in the diode datasheet. For example, assume that a current of 3 A should flow with 7 V applied. This can be checked by measuring along the horizontal scale to the 7-V point, then measuring from that point up (or down) to the trace. The 7-V (horizontal) point should intersect the trace at the 3-A (vertical) point as shown.

4-5.2 Dynamic Tests for Small-Signal Diodes

The procedures for checking the current–voltage characteristics of a signal diode are the same as for power-rectifier diodes. However, there is one major difference. In a small-signal diode, the ratio of forward voltage to reverse voltage is usually quite large. A test of forward voltage at the same amplitude as the rated reverse voltage will probably damage the diode. On the other hand, if the test voltage is lowered for both forward and reverse directions, the voltage is not a realistic value in the reverse direction.

Under ideal conditions, a signal diode should be tested with a low-value forward voltage and a high-value reverse voltage. This can be done using a circuit shown in Fig. 4–6. The circuit of Fig. 4–6 is essentially the same as that of Fig. 4–5 (for power diodes), except that diodes CR_1 and CR_2 (Fig. 4–6) are included to conduct on alternate half-cycles of the voltage across transformer T_1. Rectifiers CR_1 and CR_2 are chosen for a linear amount of conduction near zero.

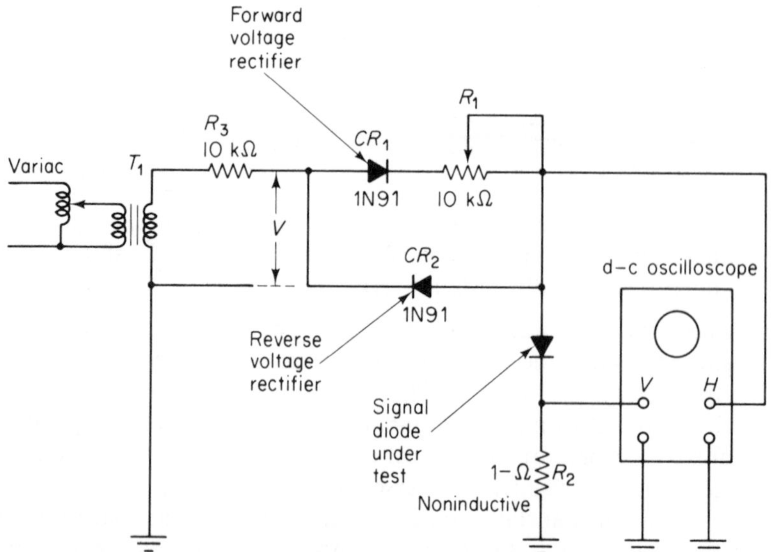

FIGURE 4-6 Test connections for measurement of signal diode voltage-current characteristics.

The variac is adjusted for maximum rated reverse voltage across the diode under test, as applied through CR_2, when the upper secondary terminal of T_1 goes negative. This applies the full reverse voltage.

Resistor R_1 is adjusted for maximum rated forward voltage across the diode, as applied through CR_1, when the upper secondary terminal of T_1 goes positive. This applies a forward voltage limited by R_1.

With resistor R_1 properly adjusted, perform the current–voltage check as described for power diodes (Section 4–5.1).

4-6 DIODE SWITCHING TESTS

Diodes to be used in pulse or digital work must be tested for switching characteristics. The single most important switching characteristic is *recovery time.* When a reverse-bias pulse is applied to a diode, there is a measurable time delay before the reverse current reaches its steady-state value. This delay period is listed as the recovery time (or some similar term) on the diode data sheet.

The duration of recovery time sets the minimum width for pulses with which the diode can be used. For example, if a 5-μs reverse voltage pulse is applied to a diode with a 10-μs recovery time, the pulse will be distorted.

An oscilloscope having a wide frequency response and good transient characteristics can be used to check the high-speed switch and recovery time of diodes. The oscilloscope vertical channel must be voltage-calibrated in the normal manner, and the horizontal channel must be time-calibrated (rather than sweep-frequency-calibrated). Most laboratory oscilloscopes are time-calibrated, and some shop oscilloscopes are frequency-calibrated.

As shown in Fig. 4–7, the diode is tested by applying a forward-biased current from a d-c supply, adjusted by R_1 and measured by M_1. The negative portion of the square-wave output from the square-wave generator is developed across R_3. The square wave switches the diode voltage rapidly to a high negative value (reverse voltage). However, the diode does not cut off immediately. Instead, a steep transient voltage is developed by the high momentary current flow. The reverse current falls to its steady-state value when the carriers are removed from the junction. This produces the approximate waveform shown in Fig. 4–7.

Both forward and reverse currents are passed through resistor R_3. The voltage drop across R_3 is applied through emitter follower Q_1 to the oscilloscope vertical channel. The coaxial cable provides some delay, so that the complete waveform is displayed. CR_1 functions as a clamping diode to keep the R_4 voltage at a level safe for the oscilloscope.

The time interval between the negative peak and the point at which the reverse current has reached the low, steady-state value is the diode recovery

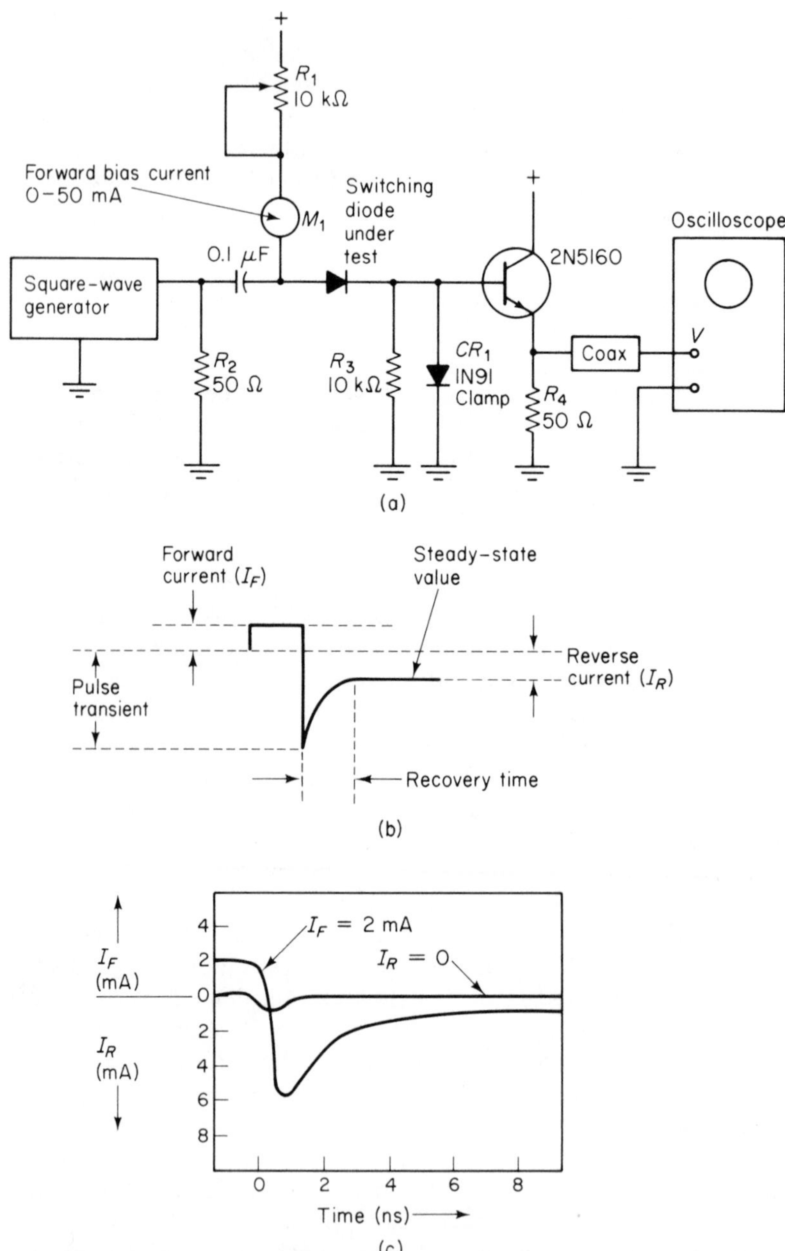

FIGURE 4-7 Test circuit and displays for switching (recovery) time of diodes: (a) theoretical display; (b) practical display; (c) approximate waveform.

time. Typically, this time is on the order of a few nanoseconds for a signal diode.

The test procedure is as follows:

1. Connect the equipment as shown in Fig. 4–7.
2. Place the oscilloscope in operation as described in the instruction manual.
3. Switch on the oscilloscope internal recurrent sweep. Set the sweep selector and sync selector to internal.
4. Set the square-wave generator to a repetition rate of 100 kHz or as specified in the diode datasheet.
5. Set R_1 for the specified forward test current as measured on M_1.
6. Increase the square-wave generator output level (amplitude) until a pattern appears.
7. If necessary, readjust the sweep and sync controls until a single sweep is shown.
8. Measure the recovery time along the horizontal (time-calibrated, probably in nanoseconds) axis.

4-7 ZENER DIODE TESTS

The test of a zener diode is similar to that of a power rectifier or signal diode. The forward voltage drop test for a zener is identical to that of a conventional diode, as described in Section 4–4. A reverse leakage test is usually not required, since a zener goes into the avalanche condition when sufficient reverse voltage is applied. In place of a reverse leakage test, a zener diode should be tested to determine the point at which avalanche occurs (establishing the zener voltage across the diode). This can be done using a static test circuit or with a dynamic (oscilloscope) test circuit.

It is also common practice to test a zener diode for impedance, since the regulating ability of a zener is related directly to the impedance. A zener diode is similar to a capacitor in this respect; as the reactance decreases, so does the change in voltage across the terminals.

4-7.1 Static Test for Zener Diodes

The basic circuit for measurement of zener voltage is shown in Fig. 4–8. As shown, the diode under test is connected to a variable d-c source in the reverse-bias condition (anode negative). (This is the way in which a zener is normally used.) The variable source is adjusted until the zener voltage is

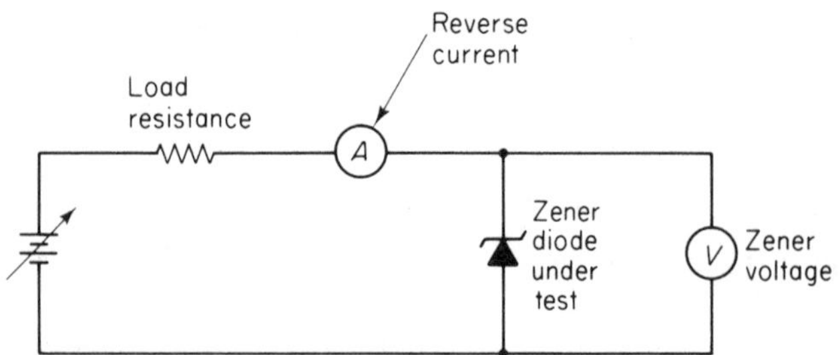

FIGURE 4-8 Static test circuit for zener diodes.

reached, and a large current is indicated through the current meter. Zener voltage can then be measured on the voltmeter. The amount of current indicated on the ammeter is usually not critical, and is partially dependent on the value of R_1. However, the voltage reading in the avalanche condition must agree with the voltage rating of the zener (within a very close tolerance).

4-7.2 Dynamic Test for Zener Diodes

The procedures and circuit for dynamic test of zener diodes are similar to those for dynamic test of conventional diodes. As shown in Fig. 4–9, the zener diode is tested by applying a controlled a-c voltage across the anode and cathode through resistors R_1 and R_2. The a-c voltage (set to some value above the zener voltage) alternately biases the anode positive and negative, causing both forward and reverse current to flow through R_1 and R_2.

The voltage drop across R_2 is applied to the vertical channel and causes the screen spot to move up and down. Vertical deflection is proportional to current. The vertical scale divisions can be converted directly to current when R_2 is made 1 Ω. For example, a 3-V vertical deflection indicates a 3-A current.

The same voltage applied across the zener (taken from the junction of R_1 and the zener under test) is applied to the horizontal channel (which has been voltage-calibrated, as described in Section 4–5.1) and causes the spot to move right or left. Horizontal deflection is proportional to voltage. The combination of the horizontal (voltage) deflection and vertical (current) deflection causes the spot to trace out the complete current and voltage characteristics.

The test procedure is as follows:

1. Connect the equipment as shown in Fig. 4–9.
2. Place the oscilloscope in operation as described in the instruction manual. Voltage-calibrate both the vertical and horizontal channels as necessary. The spot should be at the vertical and horizontal center with no signal applied to either channel.

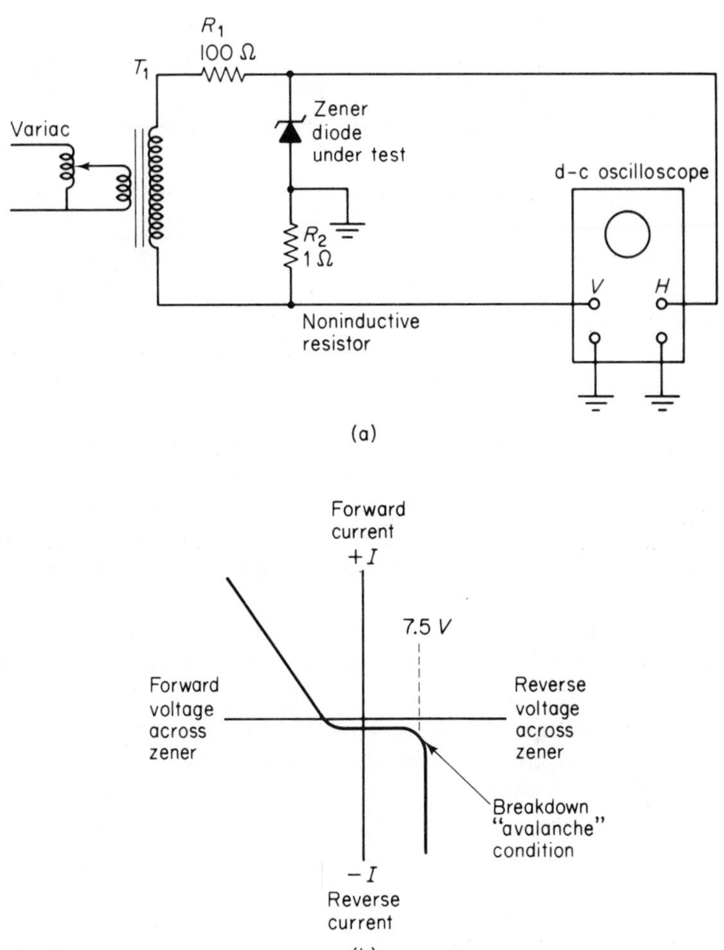

FIGURE 4-9 Test circuit and displays for dynamic test of zener diodes.

3. Switch off the internal recurrent sweep. Set sweep selector and sync selector controls to external. Leave the horizontal and vertical gain controls set at the (voltage) calibrate position established in step 2.
4. Adjust the variac so that the voltage applied across the zener diode, and resistors R_1 and R_2 in series, is greater than the rated zener voltage.
5. Check the oscilloscope pattern against the typical curves of Fig. 4-9 and/or against the diode specifications. The curve of Fig. 4-9 is a

typical response pattern. That is, the forward current increases as forward voltage increases. Reverse (or leakage) current increases only slightly as reverse voltage is applied until the avalanche or zener voltage is reached. Then the current increases rapidly (thus the term "avalanche").

6. Compare the current and voltage values against the values specified in the zener diode datasheet. For example, assume that avalanche current should occur when the reverse voltage (zener voltage) reaches 7.5 V. This can be checked by measuring to the right along the horizontal scale to the 7.5-V point.

4-7.3 Impedance Test for Zener Diodes

As discussed, the regulating ability of a zener diode is directly related to the diode impedance. Similarly, zener diode impedance varies with current and diode size. Thus, to properly test a zener diode for impedance, you must make the measurements with a specific set of conditions. This can be done using the circuit of Fig. 4-10.

With such a circuit, the diode direct current is set to approximately 20% of the zener maximum current by adjustment of R_1. The zener direct current is indicated by meter M_1. Alternating current is also applied to the zener, and is adjusted by R_2 to approximately 10% of the maximum current rating of the zener. The zener alternating current is indicated by meter M_3.

When these test conditions have been met, the a-c voltage developed across the zener can be read on meter M_2. When zener a-c voltage $V_{Z(AC)}$ and

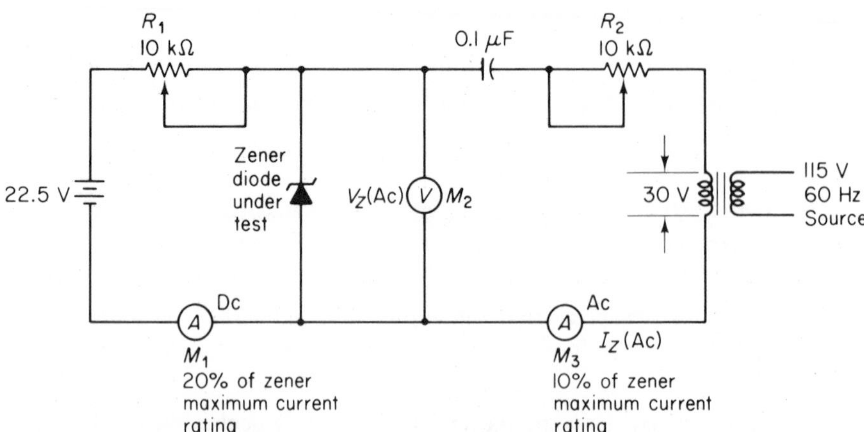

FIGURE 4-10 Test circuit for measurement of zener-diode impedance.

zener alternating current $I_{Z(AC)}$ are known, the impedance Z_Z is calculated using the equation

$$Z_Z = \frac{V_{Z(AC)}}{I_{Z(AC)}}$$

4-8 TUNNEL DIODE TESTS

The single most important test of a tunnel diode is the negative-resistance characteristic. The most effective test of a tunnel diode is to display the entire forward voltage and current characteristics on an oscilloscope, permitting the valley and peak voltages, as well as the valley and peak currents, to be measured simultaneously. It is also possible to make basic negative-resistance tests, as well as switching tests, of tunnel diodes using meters.

4-8.1 Switching Test for Tunnel Diodes (Meter Method)

The basic switching test circuit for tunnel diodes is shown in Fig. 4–11. The tunnel diode under test is connected to a variable d-c supply. Initially, the power supply is set to zero, and then is gradually increased. As the voltage is increased, there is some voltage indication across the tunnel diode. When the critical voltage is reached, the voltage indication "jumps" or suddenly increases. This indicates that the diode has switched and is operating normally. Usually, the voltage indication will be of the order 0.25 to 1 V. The power supply is then decreased gradually. With a normal tunnel diode, the voltage indication gradually decreases until a critical voltage is reached. Then the voltage indication again "jumps" and suddenly decreases.

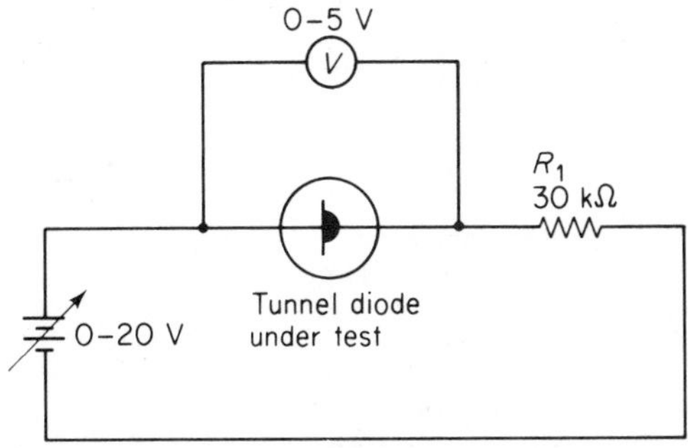

FIGURE 4–11 Switching test circuit for tunnel diodes (meter method).

4–8.2 *Negative-Resistance Test for Tunnel Diodes (Meter Method)*

Although the negative-resistance characteristics of a tunnel diode are best tested with an oscilloscope, it is possible to obtain fairly accurate results using meters connected as shown in Fig. 4–12. Note that the diode under test is connected in the reverse-bias condition (anode negative). Thus, any current indication on the ammeter is reverse current.

Initially, the power supply is set to zero and is gradually increased until the voltage reading starts to drop (indicating that reverse current is flowing and the diode is in the negative-resistance region). The negative-resistance region should not be confused with leakage. True negative resistance is indicated when further increases in supply voltage cause an increase in current reading but a decrease in voltage across the diode. The amount of negative resistance can be calculated using the equation

$$\text{negative resistance (ohms)} = \frac{\text{decrease (volts) across diode}}{\text{increase (amperes) through diode}}$$

It is not recommended that a conventional diode by subjected to a negative-resistance test unless there is a special need for the test (even though conventional diodes may show some tunnel diode characteristics). Also, do not operate a conventional diode in the negative-resistance region for any longer than is necessary. Considerable heat is generated and the diode may be damaged.

4–8.3 *Negative-Resistance Test for Tunnel Diodes (Oscilloscope Method)*

A d-c oscilloscope is required for test of a tunnel diode. Both the vertical and horizontal channels must be voltage-calibrated. Also, the horizontal and ver-

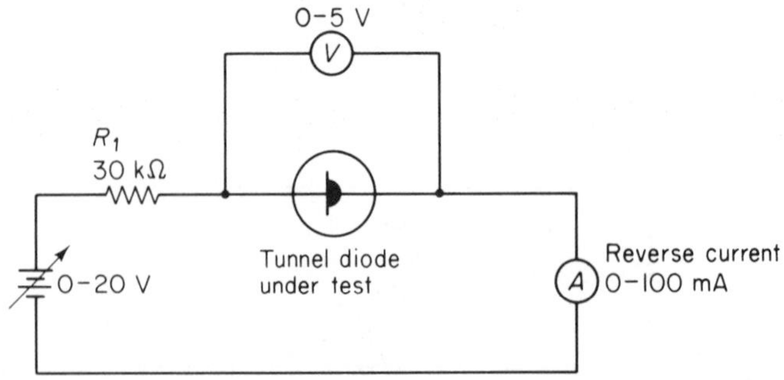

FIGURE 4–12 Negative-resistance test circuit for tunnel diodes (meter method).

tical channels must be identical, or nearly identical, to eliminate any phase difference.

As shown in Fig. 4-13, the tunnel diode is tested by applying a controlled d-c voltage across the diode through resistor R_3. This d-c voltage is developed by rectifier CR_1 and is controlled by the variac. Current through the tunnel diode also flows through R_3. The voltage drop across R_3 is applied to the ver-

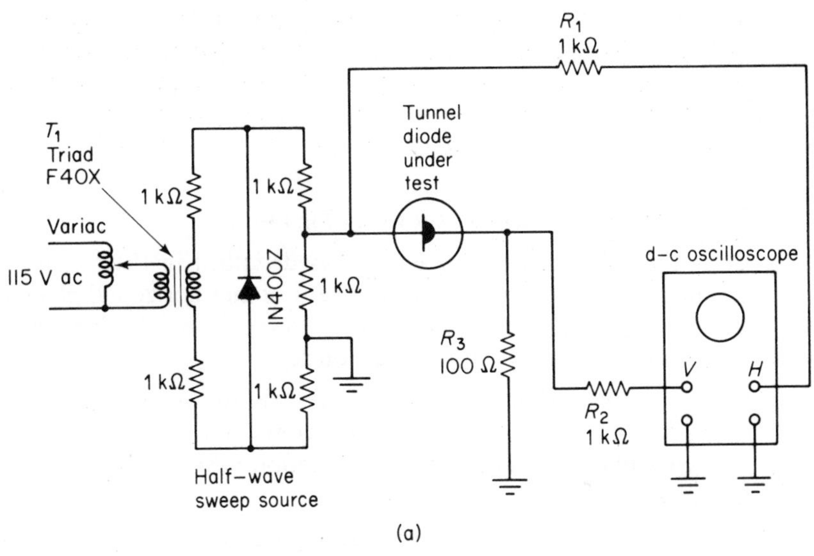

$$\frac{0.44 - 0.14}{2(0.1 - 0.01)} = \frac{0.30\ \text{V}}{0.18\ \text{A}} = 1.66\ \Omega\ \text{(approx.)}$$

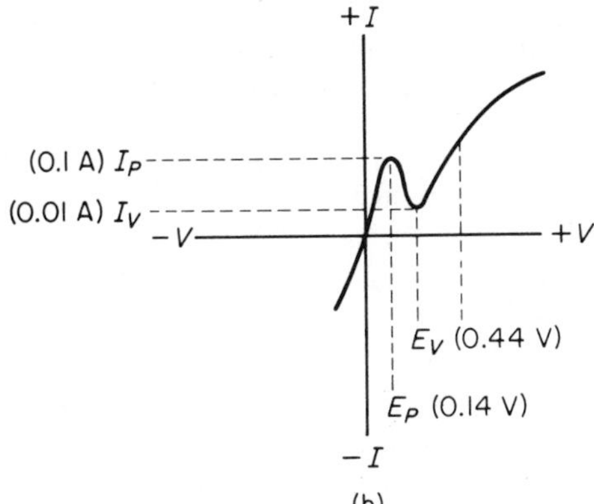

FIGURE 4-13 Negative-resistance test circuit and typical display for tunnel diodes (oscilloscope method).

tical channel and causes the spot to move up and down. Thus, vertical deflection is proportional to current. Vertical scale divisions can be converted directly to a realistic value of current when R_3 is made 100 Ω. For example, a 3-V vertical deflection indicates 30 mA.

The same voltage applied across the tunnel diode is applied to the horizontal channel (which has been voltage-calibrated) and causes the spot to move from left to right. [For a tunnel diode, the horizontal and vertical zero-reference point (no-signal spot position) should be at the *lower left* of the screen rather than in the center.] The horizontal deflection is proportional to voltage. The combination of the horizontal (voltage) deflection and vertical (current) deflection causes the spot to trace out the complete negative-resistance characteristic.

The test procedure is as follows:

1. Connect the equipment as shown in Fig. 4–13.
2. Place the oscilloscope in operation as described in the instruction manual. Voltage-calibrate both the vertical and horizontal channels as necessary. The spot should be at the lower left-hand side of center with no signal applied to either channel.
3. Switch off the internal recurrent sweep. Set the sweep selector and sync selector controls to external. Leave the horizontal and vertical gain controls set at the (voltage) calibration position, as established in step 2.
4. Adjust the variac so that the voltage applied across the tunnel diode under test is the maximum rated forward voltage (or slightly below). This can be read across the voltage-calibrated horizontal axis.
5. Check the oscilloscope pattern against the curve of Fig. 4–13 or against the tunnel diode datasheet.
6. The following equation can be used to obtain a *rough approximation* of negative resistance in tunnel diodes:

$$\text{negative resistance} = \frac{E_V - E_P}{2(I_P - I_V)}$$

where E_V = valley voltage
I_V = valley current
E_P = peak voltage
I_P = peak current

Figure 4–13 shows some typical tunnel diode voltage–current values, and the calculations for the approximate negative resistance.

4-9 SIGNAL AND POWER DIODE TEST USING A CURVE TRACER

Signal and power (rectifier) diodes conduct easily in one direction and are nonconducting in the opposite direction. These properties may be tested and observed with a curve tracer and oscilloscope. For testing diodes, the pulsating d-c sweep voltage is applied across the diode, and the diode current/voltage are plotted on the oscilloscope screen. The step current or step voltage signals used for testing transistors and FETs are not used in diode testing, and the curve tracer STEP SELECTOR has no effect on test results.

4-9.1 *Basic Diode Test Connections*

The diode under test can be plugged into the collector and emitter pins of the transistor socket, or directly into the collector and emitter jacks, depending on the type of curve tracer. Test leads can also be run from the connector and emitter jacks to the terminals of the diode under test.

Since the polarity of the sweep voltage can be reversed with the curve tracer POLARITY switch, the diode may be inserted into the socket without observing polarity. Of course, diodes connected with one polarity produce an oscilloscope display that deflects to the right and upward from the starting point, whereas the opposite polarity produces a display that deflects downward and to the left from the starting point. For a consistent display, connect the cathode of the diode to the emitter jack as shown in Fig. 4–14. With this polarity connection and the POLARITY switch set to NPN, the diode will be forward-biased, and should produce a display similar to that of Fig. 4–14. With the same polarity connection and the POLARITY switch set to PNP, the diode is reverse-biased, and should produce a display similar to that of Fig. 4–15.

4-9.2 *Diode Forward-Bias Display on a Curve Tracer*

To display forward bias, connect the diode and operate the curve tracer controls as shown in Fig. 4–14. For realistic values, the oscilloscope horizontal sensitivity should be calibrated at some low voltage. For a typical diode, a horizontal sensitivity of 0.5 V/division (or preferably less) is necessary to obtain any degree of accuracy in the voltage reading. Most oscilloscopes used with curve tracers can be calibrated for a sensitivity of 0.25 V/division in the horizontal axis.

When testing diodes as described here, only one curve is displayed, not a family of curves as displayed for transistors and FETs. Both *forward voltage drop* and *dynamic resistance* can be measured using the connections of Fig. 4–14. As shown, no current flows until the applied voltage exceeds the for-

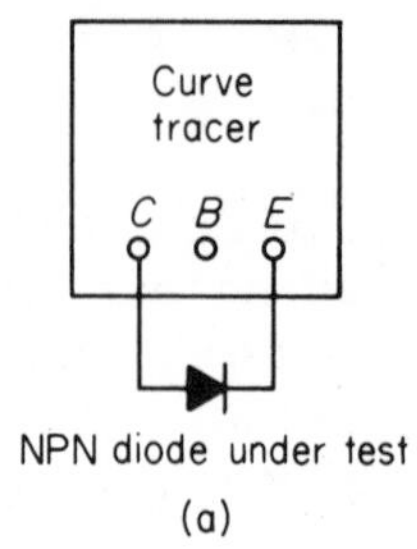

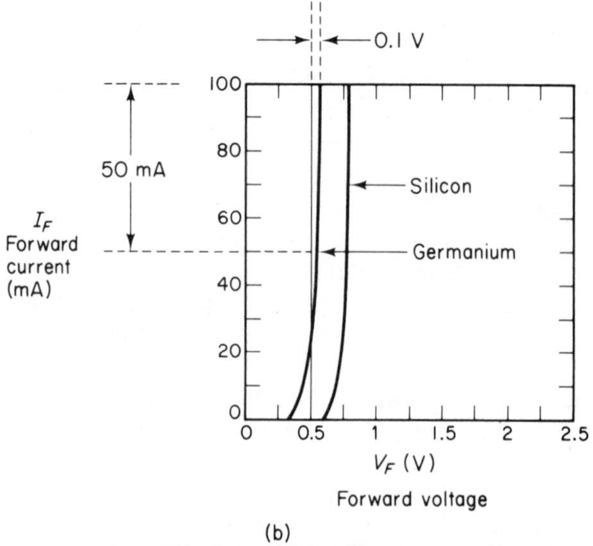

FIGURE 4-14 Diode forward bias display on a curve tracer.

ward voltage drop. As discussed in Section 4-4, the forward voltage drop of germanium diodes is in the range 0.2 to 0.4 V, whereas the forward voltage drop of silicon diodes is about 0.5 to 0.7 V. Above the forward voltage point, current increases rapidly with an increase in forward voltage. As shown, the current increases more rapidly and the "elbow" has a sharper bend for silicon diodes than for germanium diodes.

The dynamic resistance of a diode equals the change in forward voltage divided by the change in forward current. For example, the germanium diode curve of Fig. 4-14 shows an increase of about 50 mA in forward current for an increase of about 0.1 V of forward voltage. This indicates a dynamic resistance of about 2 Ω. Germanium diodes, with more slope to their curves, as shown, have higher dynamic resistance than do silicon diodes.

There is no need to increase the curve-tracer SWEEP VOLTAGE control set-

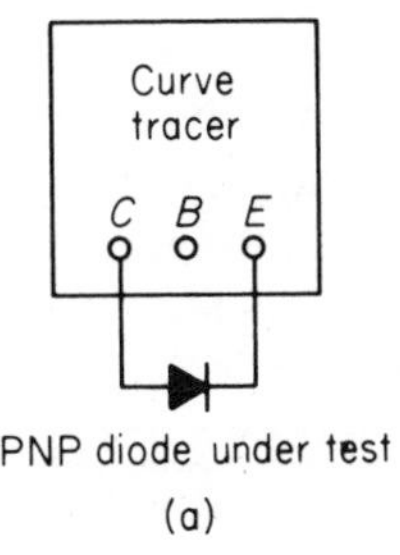

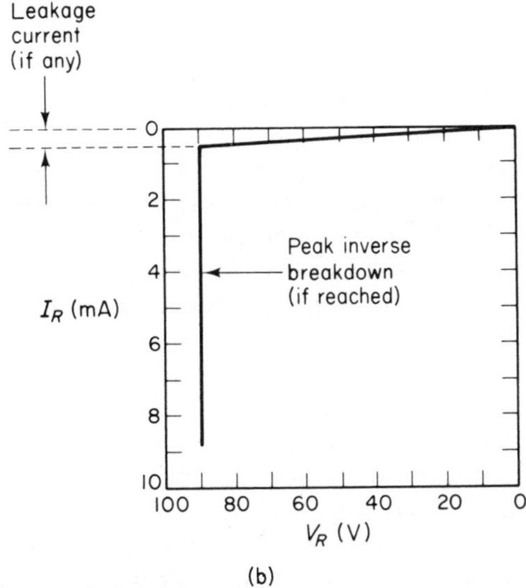

FIGURE 4–15 Diode reverse bias display on a curve tracer.

ting beyond that which gives a full-scale vertical presentation, although there is generally very little danger that a higher setting will do harm. Typically, the oscilloscope VERTICAL SENSITIVITY control may be set to the 1-mA/division position for examining the low current characteristics, or to a lower sensitivity position (such as 10 mA/division) for observing a wider range of forward current conduction.

4-9.3 Diode Reverse-Bias Display on a Curve Tracer

To display reverse bias, connect the diode and operate the curve-tracer controls as shown in Fig. 4–15. When the POLARITY switch is first set to the opposite position as used for the forward-bias tests, there should be only a horizontal line displayed. The oscilloscope centering must be readjusted because the polarity reversal causes the trace to move off-screen.

Leakage current is easier to check with a much higher voltage than is used for forward-bias testing. Typically, the oscilloscope can be calibrated for 10 V/division, which allows display of the test up to 100 V. Keep in mind that the diode must never be operated at reverse voltage greater than the peak inverse voltage (or breakdown voltage) for any prolonged period of time. To do so can damage the diode. Note that leakage current is displayed as a slope of the horizontal line, whereas breakdown voltage is displayed as a sharp vertical drop. Figure 4–15 shows a leakage current of about 0.5 mA at a breakdown voltage of 90 V, with correspondingly less leakage at lower reverse voltages.

4-10 ZENER DIODE TEST USING A CURVE TRACER

When a curve tracer is used, the procedure for testing zener diodes is almost the same as for testing signal and rectifier diodes. In fact, the forward characteristics of the diodes are essentially identical and the test procedures would be the same, except that forward voltage measurements are seldom used for zener diodes, since zeners are designed to be used in the reverse voltage breakdown mode. In this mode, a large change in reverse current occurs while the zener voltage remains nearly constant. Because of this characteristic, zener diodes are most often used as voltage regulators.

The zener voltage value (reverse voltage breakdown value) may be measured with the curve tracer and oscilloscope set up as described for reverse voltage measurement of signal and power diodes (Section 4–9). To get the most accurate voltage reading possible, calibrate the full-scale oscilloscope horizontal sensitivity to a convenient value slightly above the zener voltage. For example, for a 6-V diode, calibrate full scale at 10 V.

Make sure that the curve tracer POLARITY switch is set to display the reverse voltage condition. Increase the curve-tracer SWEEP VOLTAGE control setting to display the zener region, as shown in Fig. 4–16. No reverse current should flow until the reverse breakdown voltage value is reached. At that point, there should be a very sharp "elbow" or "knee" and a very vertical current trace. If the "knee" is rounded or the vertical current trace has a measurable voltage slope, the zener diode is probably defective. Read the zener voltage from the display.

4-11 TUNNEL DIODE TEST USING A CURVE TRACER

The characteristics of tunnel diodes may be measured with a curve tracer by connecting the diode between the collector and emitter jacks as shown in Fig. 4–17. If the cathode cannot be readily identified, try both positions of the

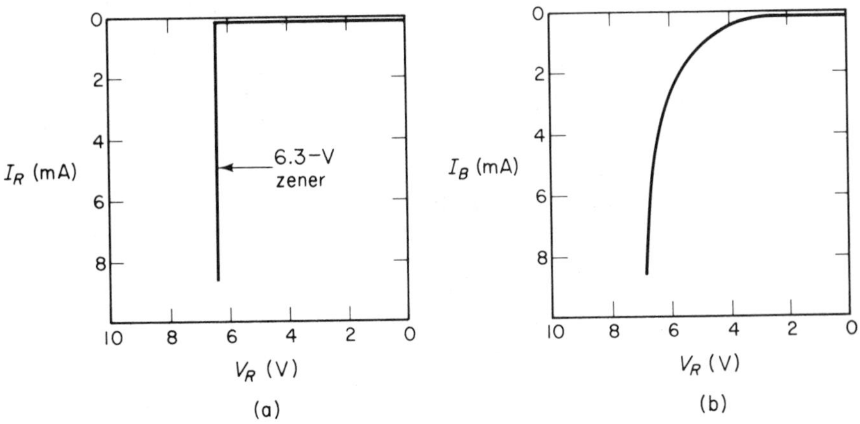

FIGURE 4-16 Typical zener diode curve tracer displays: (a) sharp zener knee; (b) sloppy zener knee.

POLARITY switch. The display shown in Fig. 4–17(a) is obtained when the tunnel diode cathode is connected to the emitter jack, and the curve tracer POLARITY switch is set to NPN.

Calibrate the oscilloscope horizontal amplifier for high sensitivity, such as 0.1 V/division. It may be necessary to use sweep magnification or horizontal scale expansion (if the oscilloscope is capable of such operation) to achieve this sensitivity. Set the curve tracer SWEEP VOLTAGE control to a low value so that the voltage just sweeps through the tunnel region (typically less than 0.5 V). Set the oscilloscope VERTICAL SENSITIVITY as required for the largest possible on-scale display.

Several characteristics of the tunnel diode can be measured directly from the display, as discussed in Section 4–8.3.

4-11.1 Tunnel Rectifier Test Using a Curve Tracer

Tunnel rectifiers are similar to tunnel diodes but do not use the negative-resistance characteristic in operation. The region that tends to tunnel is more resistive in tunnel rectifiers, and peak current is not as pronounced. Because of this characteristic, a high front-to-back ratio at low voltages allows the tunnel rectifier to be used as a very low signal voltage rectifier. The tunnel rectifier conducts very easily in one direction with very little voltage drop (actually the opposite direction from conventional diodes insofar as the N and P material is concerned; thus, tunnel rectifiers are sometimes called *back diodes*).

The reverse direction in a tunnel rectifier (direction that tends to tunnel) is resistive at low voltage values, but conducts readily at voltages that approx-

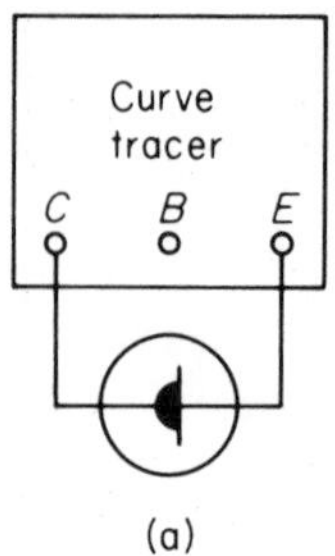

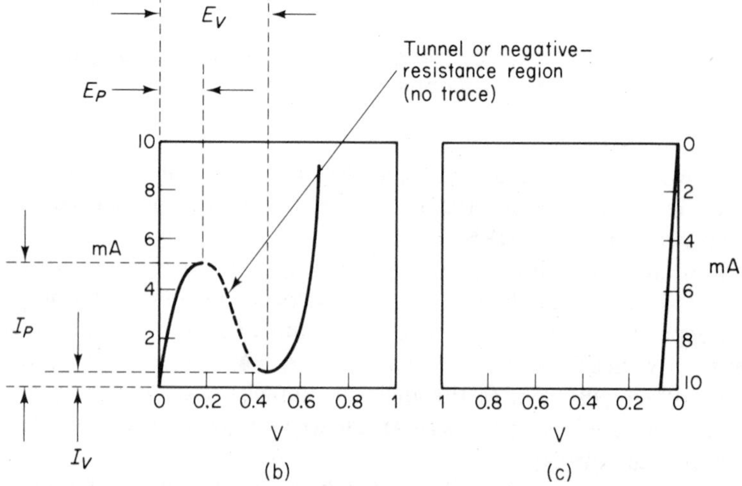

FIGURE 4-17 Typical tunnel diode and tunnel rectifier curve tracer displays: (a) tunnel diode under test; (b) tunnel diode; (c) reverse characteristics of tunnel rectifier.

imate the forward drop of a conventional diode. Therefore, the peak voltage of the signal to be rectified should not exceed the resistive region.

Tunnel rectifier characteristics should be displayed as described for tunnel rectifiers. Tests using both polarities of sweep voltage are required to examine the forward versus reverse conduction characteristics. The display of Fig. 4-17(b) shows typical reverse characteristics of a tunnel rectifier.

5

Thyristor and Control Rectifier (SCR) Tests

This chapter is devoted entirely to test procedures for control rectifiers and thyristors. The terms "control rectifier" and "thyristor" are alternately applied to many devices used in electronic control applications. The most common such devices are the SCR, SCS, PNPN switch, triac, diac, SUS, and SBS. The first sections of this chapter describe control rectifier or thyristor characteristics and test procedures from the practical standpoint. The information in these sections permits you to test all the important control rectifier and thyristor characteristics using basic shop equipment. The sections also help you understand the basis for such tests. The remaining sections of the chapter describe how the same tests, and additional tests, are performed using more sophisticated equipment such as the oscilloscope curve tracer.

5-1 THYRISTOR AND CONTROL RECTIFIER BASICS

Before going into specific test procedures, let us review some control rectifier and thyristor basics. The control rectifier (also called a controlled rectifier) is similar to the basic diode (Chapter 4), with one specific exception. The control rectifier must be "triggered" or "turned on" by an external voltage source.

The control rectifier has a high forward and reverse resistance (no current flow) without the trigger. When the trigger is applied, the forward resistance drops to zero (or very low), and a high forward current flows, as with the basic diode. The reverse current remains high, and no reverse current flows, so the control rectifier rectifies a-c power in the normal manner. As long as the forward voltage is applied, the forward current continues to flow. The forward current stops, and the control rectifier "turns off," if the forward voltage is removed.

Of the numerous control rectifiers in use, many are actually the same type (or slightly modified versions) but manufactured under different trade names or designations. The term *thyristor* is applied to many control rectifiers. Technically, a thyristor is defined as any semiconductor switch whose bistable action depends on PNPN regenerative feedback (Section 5-1.3). Thyristors can be two-, three-, or four-terminal devices, and are capable of both unidirectional and bidirectional operation. The following paragraphs describe the most common control rectifier or thyristor devices.

5-1.1 Silicon or Semiconductor Control Rectifier (SCR)

With some manufacturers, the letters SCR refer to semiconductor control rectifier and can mean any type of solid-state control rectifier. However, SCR usually refers to silicon control (or controlled) rectifiers.

If four semiconductor materials, two P-type and two N-type, are arranged as shown in Fig. 5-1(a), the device can be considered as three diodes arranged alternately in series as shown in Fig. 5-1(b). Such a device acts as a conventional diode rectifier (Chapter 4) in the reverse direction, and as a combined electronic switch and series rectifier in the forward direction. Conduc-

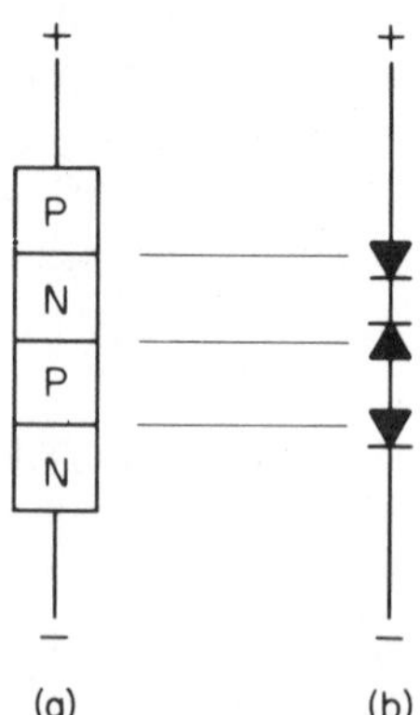

FIGURE 5-1 Semiconductor material arrangement and equivalent diode relationship of an SCR.

tion in the forward direction can be controlled, or "gated," by operation of the switch.

Figure 5-2 shows the circuit symbol and block diagram of a typical SCR. Note that there are two basic arrangements for an SCR: one with the gate terminal connected to the cathode and one with the gate terminal connected to the anode. The cathode gate is the most common arrangement.

SCRs are normally used to control alternating current but can be used to control direct current. Either a-c or d-c voltage can be used as the gate signal, provided that the gate voltage is large enough to trigger the SCR into the "on" condition.

An SCR is used to best advantage when both the load and trigger are ac. With a-c power, control of the power applied to the load is determined by the *relative phase* of the trigger signal versus the load voltage. Because the trigger control is lost once the SCR is conducting, an a-c voltage at the load permits the trigger to regain control. Each alternation of ac through the load causes conduction to be interrupted (when the a-c voltage drops to zero between cycles), regardless of the polarity of the trigger signal.

Phase Relationship between Gate and Load Voltages. Figure 5-3 shows the operation of an SCR with a-c voltages at the trigger circuit and load circuit. If the trigger voltage is in phase with the a-c power input signal as shown in Fig. 5-3(b), the SCR conducts for each successive positive alternation at the anode. When the trigger is positive-going at the same time as the load or anode voltage, load current starts to flow as soon as the load voltage reaches a value that will cause conduction. When the trigger is negative-going, the load voltage is also negative-going, and conduction stops. The SCR acts as a half-wave rectifier in the normal manner.

If there is a 90° phase difference between the trigger voltage and load voltage (say the load voltage lags the trigger voltage by 90°) as shown in Fig. 5-3(c), the SCR does not start conducting until the trigger voltage swings positive, even though the load voltage is initially positive. When the load

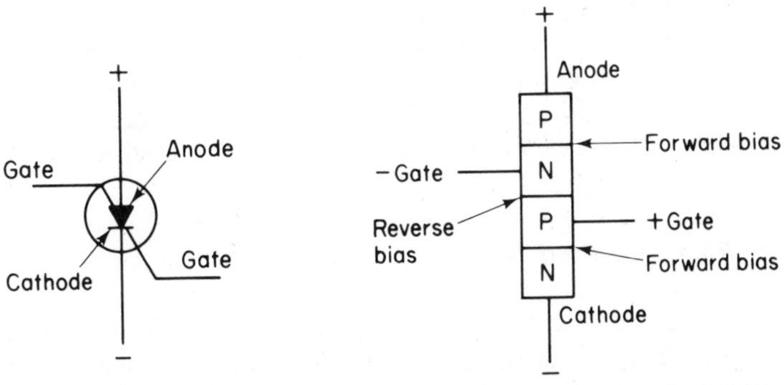

FIGURE 5-2 Symbol and block diagram of a typical SCR.

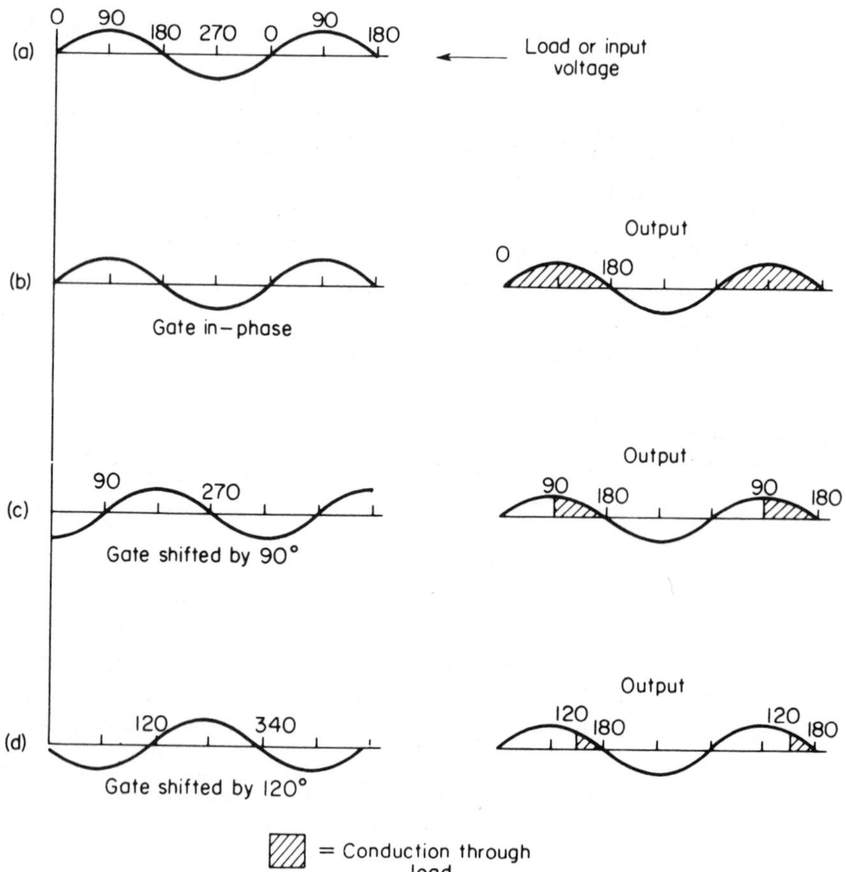

FIGURE 5-3 Operation of an SCR with ac at both load and gate or trigger.

voltage drops to zero, conduction stops, even though the trigger voltage is still positive.

If the phase shift is increased between trigger and load voltages as shown in Fig. 5-3(d), conduction time is even shorter, and less power is applied to the load circuit. By shifting the phase of the trigger voltage in relation to the load voltage, it is possible to vary the power output, even though the voltages are not changed in strength.

5-1.2 PNPN-Controlled Switch or SCS

The PNPN-controlled switch (often referred to as an SCS or silicon-controlled switch) is similar in operation to an SCR. However, the SCS is a PNPN device with all four semiconductor regions made accessible by means

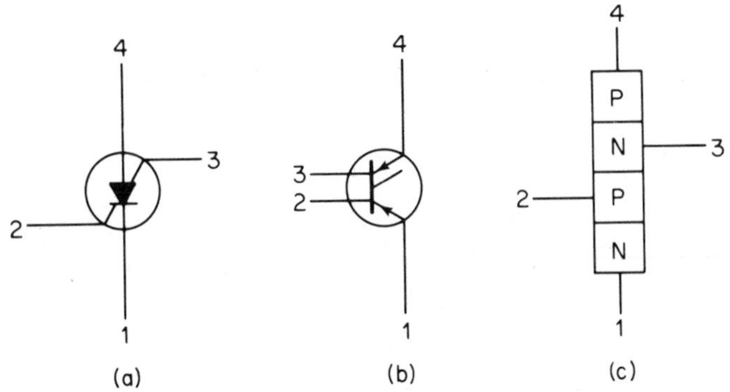

FIGURE 5-4 Symbols and block diagram of a typical SCS.

of terminals. Figure 5-4 shows the circuit symbols and block diagram of a typical SCS. Note that two circuit symbols are used; both are in common use. Often, the SCS is used as an SCR, with the extra gate terminal not connected.

For some control applications, the SCS can be considered as a transistor and diode in series. Figure 5-5 shows such an arrangement. If a negative load voltage is applied to terminal 4, with a positive voltage at terminal 1, the SCS does not turn on, no matter what trigger signals are applied. However, with a positive voltage at terminal 4 and a negative voltage at terminal 1, the SCS is turned on by either a positive voltage at terminal 2 or a negative voltage at terminal 3.

The SCS also has the ability to turn off by means of a gate signal, as shown in Fig. 5-6. Note that the gate turn-off method applies only when the conducting current is below a certain value. The SCS is often considered as

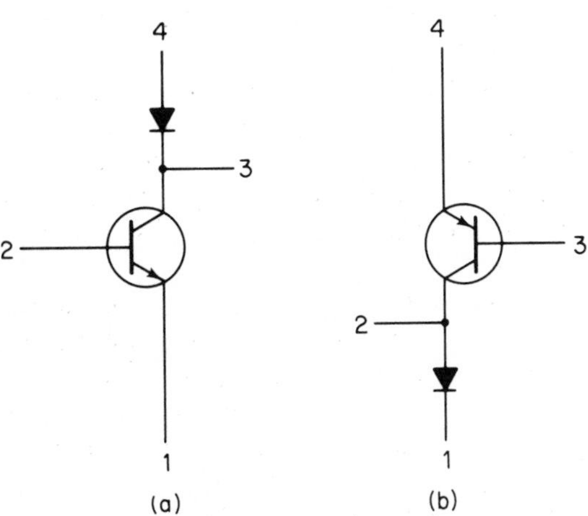

FIGURE 5-5 Equivalent diode-transistor representation of an SCS.

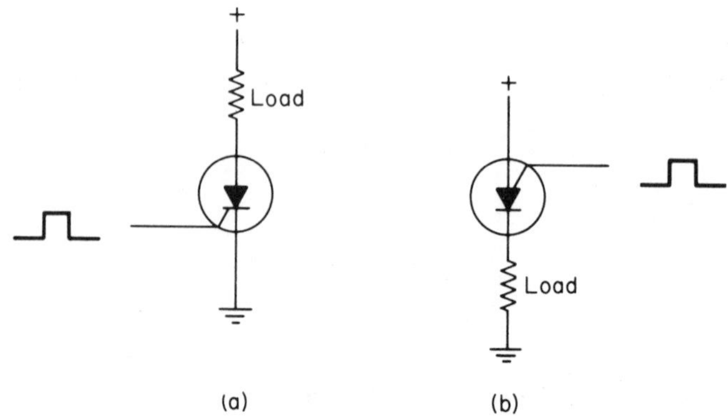

FIGURE 5-6 Gate turn-off arrangement for SCSs: (a) cathode gate pulse turn-off; (b) anode gate pulse turn-off.

two transistors (an NPN and a PNP) connected as shown in Fig. 5-7. Both transistors are connected so that the collector output of the NPN feeds into the base input of the PNP, and vice versa. If a positive trigger voltage is applied to the NPN base, the NPN is turned on and some current flows in the NPN emitter–collector circuit. Since the NPN collector feeds the PNP base, the PNP is also turned on, and the PNP collector output feeds into the NPN base, adding to the trigger voltage.

Load current then flows through the two transistors from the NPN emitter (or rectifier cathode) to the PNP emitter (or rectifier anode). Load current continues to flow even though the trigger is removed, because the current flow also keeps both transistors turned on. The same condition can be produced by

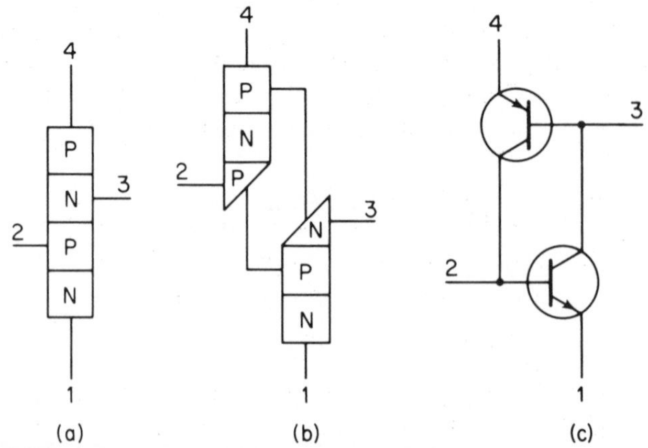

FIGURE 5-7 Equivalent two-transistor representation of an SCS.

a negative trigger applied to the PNP base. This turns on the PNP, which, in turn, turns on the NPN, until both transistors are fully conducting.

Normally, the currents do not stop until the load voltage is removed or, in the case of ac, the voltage drops to zero between cycles. However, if the load current is below a certain level (different for each type of SCS), the SCS can be turned off by a trigger voltage. For example, if a negative trigger voltage is applied to the NPN base, less current flows in the NPN emitter–collector circuit. As a result, the PNP is turned on less, and the PNP collector output drops to aid the turn-off trigger voltage at the NPN base. The feedback process continues until the load current is completely stopped.

5-1.3 *Triacs and Diacs*

Like the SCR and SCS, the triac is triggered by a gate signal. Unlike either the SCR or SCS, the triac conducts in both directions and is, therefore, most useful for controlling devices operated by a-c power (such as a-c motors). Since an SCR or SCS is essentially a rectifier, two SCRs or SCSs must be connected back to back (in parallel or bridge) to control alternating current. (Otherwise, the alternating current is rectified into direct current.) The use of two or more SCRs or SCSs requires elaborate circuits (in many cases). The elaborate control circuits can be eliminated by a triac when a-c power is to be controlled.

Figure 5-8 shows the circuit symbol of a triac. Note that the symbol is essentially one SCR symbol combined with another, complementary SCR symbol. Since the triac is not a rectifier (when turned on and conducting, current flows in both directions), the terms "anode" and "cathode" do not apply. Instead, terminals are identified by number. Terminal T_1 is the reference point for measurement of voltages and currents at the gate terminal and at terminal T_2. The area between terminals T_1 and T_2 is essentially a PNPN switch in parallel with an NPNP switch.

Like the SCS and SCR, the triac can be made to conduct when a breakdown or breakover voltage is applied across terminals T_1 and T_2, and when a trigger voltage is applied. Current continues to flow in one direction

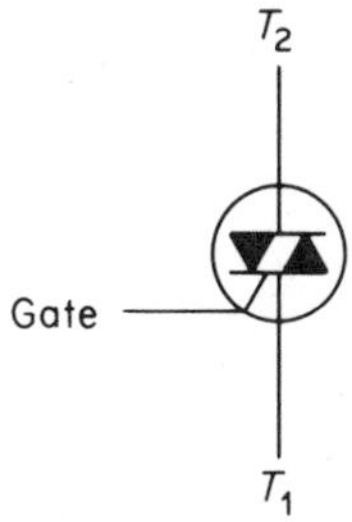

FIGURE 5-8 Typical triac symbol.

until that half-cycle of the a-c voltage (across T_1 and T_2) is complete. Current then flows in the opposite direction for the next half-cycle. The triac does not conduct on either half-cycle unless a gate-trigger voltage is present during that half-cycle (unless the breakdown voltage is exceeded).

Triac Trigger Sources. Triacs can be triggered from many sources, as can SCRs and SCSs. One of the most common trigger sources is the diac, shown in Fig. 5-9. The diac can be considered a semiconductor device resembling a pair of diodes connected in complementary (parallel) form, as shown in Fig. 5-9. The anode of one diode is connected to the cathode of the other diode, and vice versa.

Each diode passes current in one direction only, as in the case of a common diode. However, the diodes in a diac do not conduct in the forward direction until a certain breakover voltage is reached. For example, if a diac is designed for a breakover voltage of 3 V and the diac is used in a circuit with less than 3 V, the diodes appear as a high resistance (no current flow). Both diodes conduct in their respective forward bias directions when the voltage is raised to any value over 3 V. Since the diac is bidirectional, it should be tested in both polarities.

5-1.4 SUS and SBS

Figure 5-10 shows the symbols and equivalent circuits of the SUS and SBS. The SUS (silicon unilateral switch) is essentially a miniature SCR with an anode gate (instead of the usual cathode gate) and a built-in low-voltage zener diode between the gate and the cathode. The SUS operates in a manner similar to the UJT (Chapter 3). However, the SUS switches at a fixed voltage, determined by the internal zener diode, rather than by a fraction or percentage of the supply voltage. Also, the switching current of the SUS is generally higher than that of the UJT. From a test standpoint, the SUS can be tested as a UJT or as an SCR.

The SBS (silicon bilateral switch) is essentially two identical SUS structures arranged in an inverse-parallel circuit. Since the SBS operates as a switch with both polarities of applied voltage, the SBS is particularly useful for triggering triacs with alternate positive and negative gate pulses. As in the case of the diac, the SBS should be tested in both polarities.

FIGURE 5-9 Symbols for diode-type and transistor-type diacs.

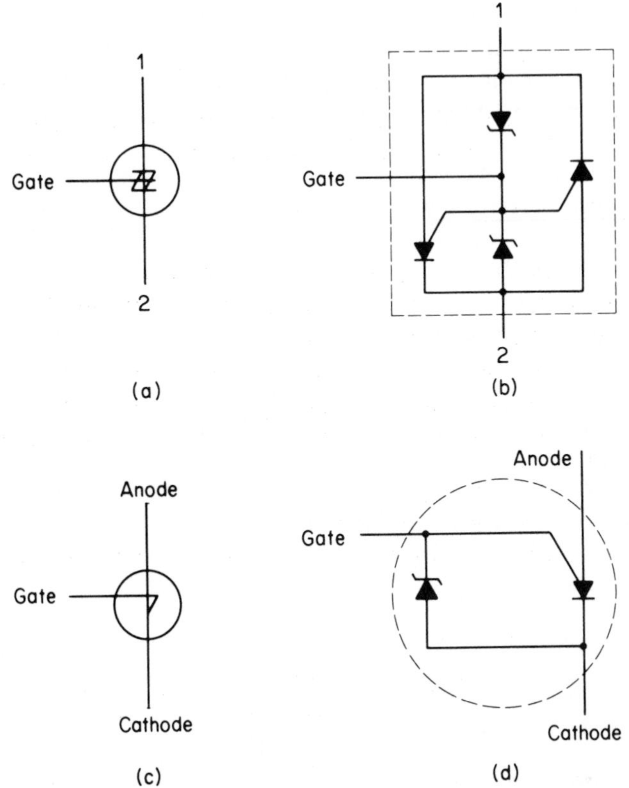

FIGURE 5-10 Symbols and equivalent circuits for SBS and SUS.

5-2 CONTROL RECTIFIER AND THYRISTOR TEST PARAMETERS

Manufacturers have their own set of symbols, letters, and terms to identify the parameters of control rectifiers and thyristors. Many of the symbols and terms are duplicates used by different manufacturers. In a few cases, special terms and letter symbols are used by a manufacturer to identify the parameters of their own particular type of control rectifier or thyristor. No attempt is made to duplicate all of this information here. However, the following paragraphs discuss the most important parameters. These parameters can then be compared with those found on the datasheets of a particular control rectifier or thyristor.

Forward voltage is the voltage drop between the anode and cathode at any specified forward anode current, when the device is in the "on" condition. *Forward anode current* is any value of positive current that flows through the device in the "on" condition. *Forward blocking voltage* is the maximum anode-cathode voltage in the forward direction that the device can withstand before conduction, at zero gate current. *Forward breakover voltage* is the value of forward anode voltage at which the device switches to the "on" state, with a shunt resistance between gate and cathode. The basic test connections for forward "on" conditions are shown in Fig. 5-11. *Forward off current* is the anode current that flows when the device is in the "off" condition (with a positive voltage applied) and is therefore sometimes listed as *forward leakage current*. Figure 5-12 shows the basic test connections for forward "off" measurements.

Reverse anode voltage is any value of negative voltage that may be applied to the anode. The rated reverse anode voltage is less than the reverse avalanche bias voltage, and is the maximum peak inverse voltage (PIV) of the device. *Reverse blocking voltage* is the maximum reverse anode-cathode voltage at zero gate current that the device can withstand before voltage breakdown, and is similar to the peak inverse voltage of a diode. The basic test connections for reverse conditions are shown in Fig. 5-13. *Reverse current*

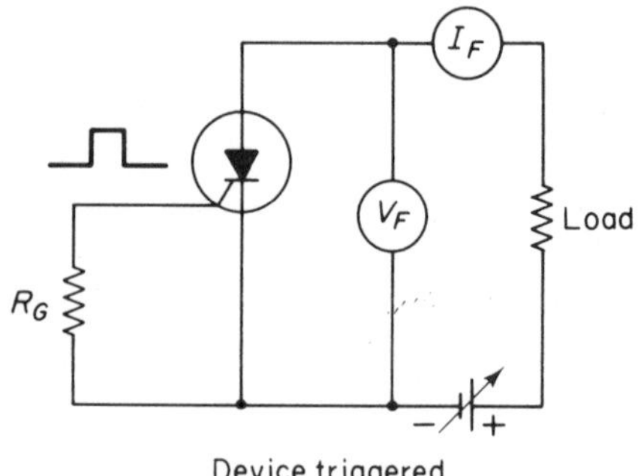

FIGURE 5-11 Basic test circuit for forward voltage and forward anode current.

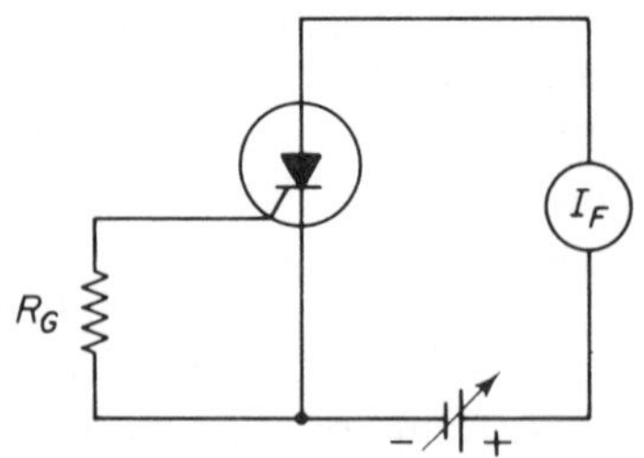

FIGURE 5-12 Basic test circuit for forward "off" current (forward leakage current).

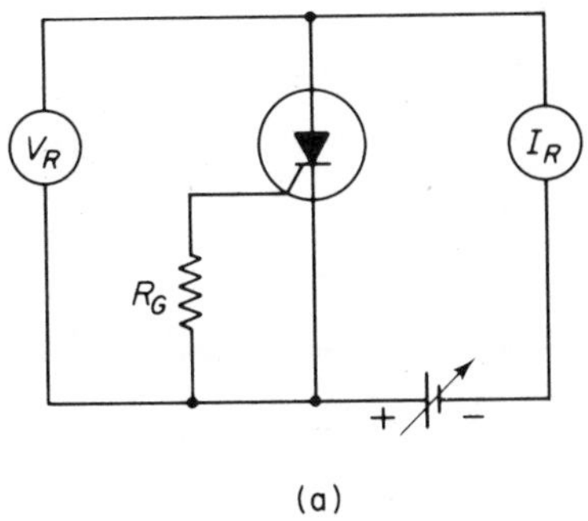

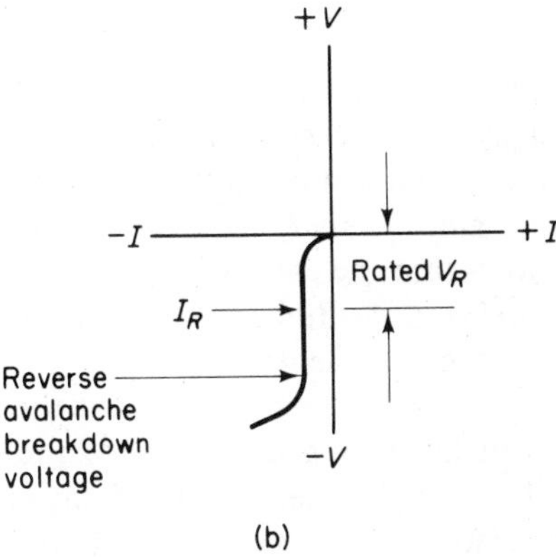

FIGURE 5-13 Basic test circuit for reverse blocking voltage and reverse current.

is the negative current that flows through the device at any specified condition of reverse anode voltage and temperature. Figure 5–13 also shows the relationship of reverse voltage and current.

Forward gate voltage is the voltage drop across the gate-cathode junction at any specified forward gate current. *Forward gate current* is any value of positive current that flows into the gate, with a shunt resistance between gate and cathode. Figure 5–14 shows the basic test connection for both forward gate voltage and current measurements.

Latching current is the minimum current that must flow through the anode terminal of thyristor for the device to switch to the "on" state, and remain in the "on" state after removal of the gate trigger pulse. *Holding current* is the minimum current that can flow through the anode terminal of a conducting thyristor without the device reverting to the off state.

Junction temperature in any PNPN device is usually considered to be a composite temperature of all three junctions. Since *all parameters* of a thyristor are temperature-dependent, operating temperature of the device

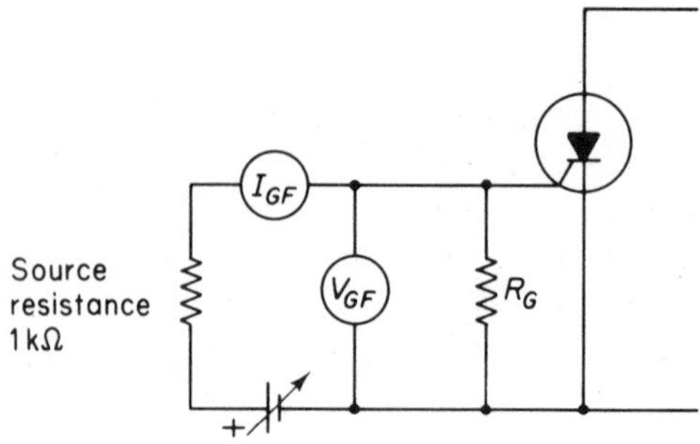

FIGURE 5-14 Basic test circuit for forward gate voltage and forward gate current.

must be considered when making any tests. This can be especially critical in holding-current and delay-time measurements.

Reverse gate voltage is the voltage drop across the gate–cathode junction at any specified *reverse gate current. Reverse gate current* is any value of negative current that flows into the gate, with a shunt resistance between gate and cathode. Figure 5–15 shows the basic test connections for both reverse gate voltage and current measurements.

Delay time is the interval between the start of the gate pulse and the instant at which the output has changed to 10% of the maximum amplitude. Figure 5–16 shows the basic test connections for delay measurements, together with a typical graph or display. As shown in Fig. 5–16, delay time is the time, following initiation of the gate pulse, required for the anode voltage to drop to 90% of the initial value. Delay time decreases with increased gate current, but at large gate currents reaches a limit of about 0.2 μs (for a typical SCR).

Rise time is the interval during which the output pulse changes from 10% to 90% of the maximum amplitude. In Fig. 5–16, rise time is the time re-

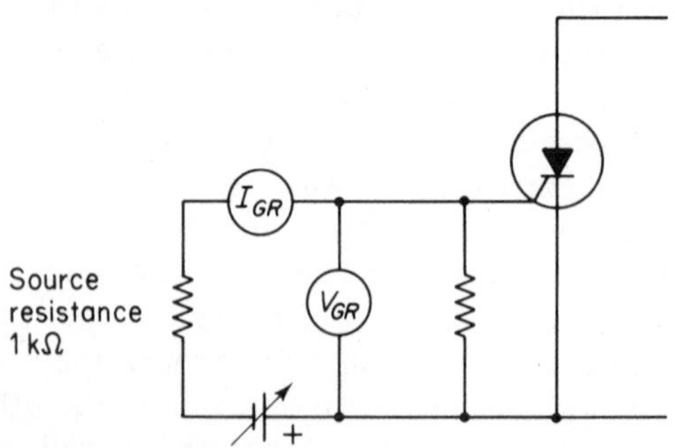

FIGURE 5-15 Basic test circuit for reverse gate voltage and reverse gate current.

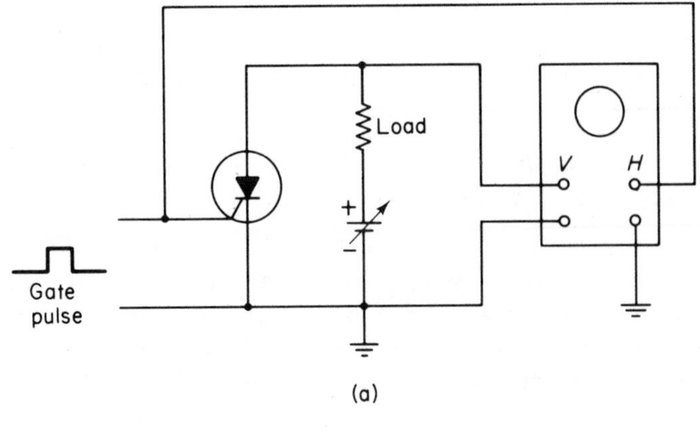

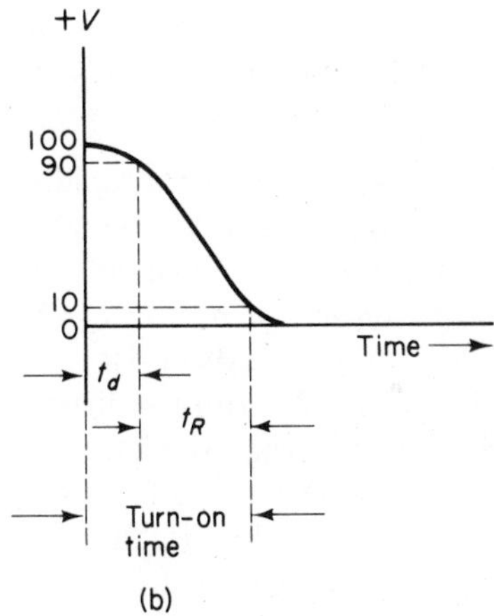

FIGURE 5-16 Test circuit and typical graph or display for delay, rise, and turn-on time.

quired for the anode–cathode voltage to drop from 90% to 10% of the initial value.

Turn-on time is usually expressed as a combination of delay time plus rise time.

Turn-off time is the interval between the start of the turn-off and the instant at which the anode voltage may be reapplied, without turning on the device. When a thyristor is triggered, and the anode voltage is positive, the device conducts. When the anode voltage swings negative, conduction stops. However, if the anode swings positive immediately, the device can conduct even though the trigger is not present. There must be some delay between the instant that conduction stops and the instant the anode can be made positive. This delay is the turn-off time.

A number of factors affect turn-off time:

Junction temperature: Turn-off time increases with increases in junction temperature.

Forward current and its rate of decay: Turn-off time increases as forward current and its rate of decay increase.

Reverse recovery current: If the device is subjected to a reverse bias (anode made negative) immediately after a condition of forward conduction (such as occurs when alternating current is used, and the anode swings from positive to negative each half-cycle), a reverse (or recovery) current flows from anode to cathode. This is essentially the same as recovery current of a conventional diode (Chapter 4). Turn-off time decreases as reverse (or recovery) current increases.

Rate of rise of reapplied forward voltage and its maximum amplitude: As the rate of rise and the amplitude of reapplied forward voltage increase, turn-off time increases.

Rate of rise (or dV/dT): When a rapidly rising voltage is applied to the anode of a thyristor, the anode may start to conduct, even though there is no trigger, and the breakdown voltage is not reached. This condition is known as *rate effect, or dV/dT effect,* or (sometimes) *dI/dT effect.* The letters *dV/dT* signify a difference in voltage for a given difference in time. The letters *dI/dT* signify a difference in current for a given difference in time. Either way, the letter combinations indicate how much the voltage (or current) changes for a specific time interval. Usually, the terms are expressed in volts of change per microsecond, or amperes of current change per microsecond.

Critical rate of rise: Every thyristor has some *critical rate of rise.* That is, if the voltage (or current) rises faster than the critical rate-of-rise value, the device turns on (with or without a trigger) even though the actual anode voltage does not exceed the rated breakdown voltage. This critical rate-of-rise characteristic is especially important where a pulse-type signal, rather than a sine-wave voltage, is applied across the anode.

5-3 BASIC CONTROL RECTIFIER AND THYRISTOR TESTS

As in the case of diodes and transistors, control rectifiers are subjected to many tests during manufacture. Few of these tests need be duplicated in the field. One of the simplest and most comprehensive tests for a control rectifier is to operate the device in a circuit that simulates actual circuit conditions (typically, alternating current and an appropriate load at the anode; alternating current or a pulse signal at the gate), and then measure the resulting conduction angle on a dual-trace oscilloscope.

With such a test procedure, the trigger and anode voltages, as well as the load current, can be adjusted to normal (or abnormal) dynamic operating conditions and the results noted. For example, the trigger voltage can be adjusted over the supposed minimum and maximum trigger levels. Or the trigger can be removed and the anode voltage raised to the actual breakover. The conduction angle method should test all important characteristics of a control rectifier, except for turn-on, turn-off, and rate of rise (which are also discussed in this section).

5-3.1 Conduction-Angle Test

A dual-trace oscilloscope can be used to measure the conduction angle of a control rectifier or thyristor. As shown in Fig. 5-17, one trace of the oscilloscope displays the anode current, while the other trace displays the trigger voltage. Both traces must be voltage-calibrated. The anode load current is measured through a 1-Ω noninductive resistor. The voltage developed across this resistor is equal to the current. For example, if a 3-V indication is obtained on the oscilloscope trace, a 3-A current is flowing in the anode circuit. The trigger voltage is read out directly on the other oscilloscope trace. Note that a diode is shown in the trigger circuit to provide a pulsating d-c trigger. This can be removed if desired. Since the trigger is synchronized with anode current (both are obtained from the same source), the portion of the trigger cycle in which anode current flows is the conduction angle.

The test procedure is as follows:

1. Connect the equipment as shown in Fig. 5-17.
2. Place the oscilloscope in operation as described in the applicable handbook. Switch on the internal recurrent sweep. Set the sweep selector and sync selector to internal.
3. Apply power to the control rectifier. Adjust the trigger voltage, anode voltage, and anode current to the desired levels. Anode voltage can be measured by temporarily moving the oscilloscope probe (normally connected to measure gate voltage) to the anode.
4. Adjust the oscilloscope sweep frequency and sync controls to produce two or three stationary cycles of each wave on the screen.
5. On the basis of one conduction pulse equaling 180°, determine the angle of anode current flow, by reference to the trigger voltage trace. For example, in the display of Fig. 5-17, anode current starts to flow at 90° and stops at 180°, giving a conduction angle of 90°.

NOTE: If the unit under test is a triac (or similar device such as an SBS), there is a conduction display on both half-cycles.

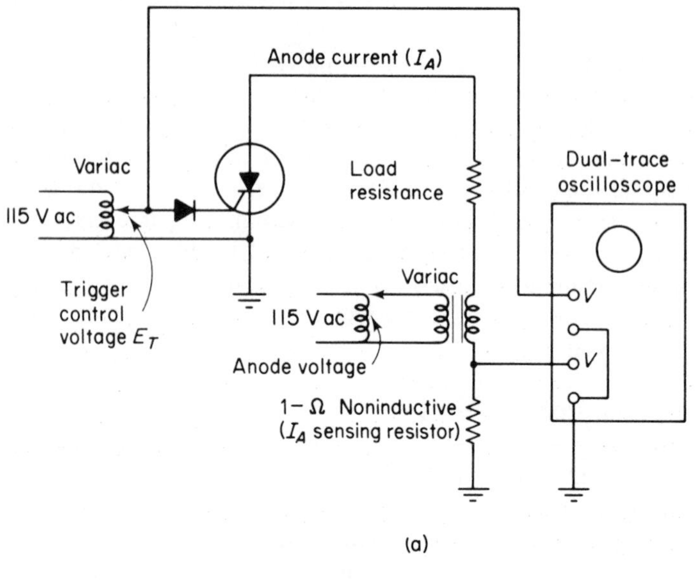

(a)

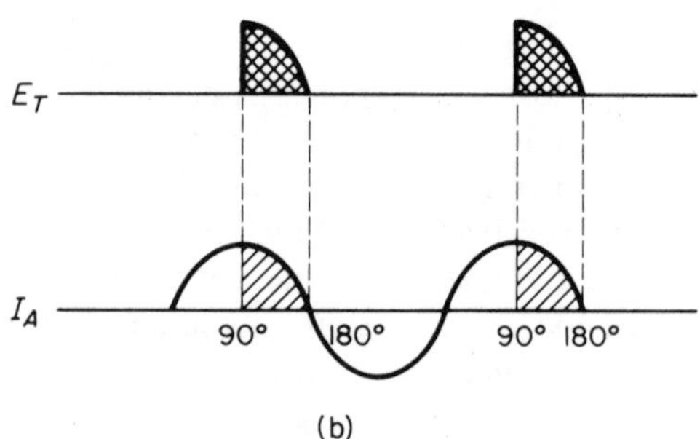

(b)

FIGURE 5-17 Basic test circuit for measurement of SCR or thyristor conduction angle.

6. To find the minimum or maximum required trigger level,vary the trigger voltage from zero across the supposed operating range, and note the level of trigger voltage when anode conduction starts.
7. To find the breakdown voltage, remove the trigger voltage and move the oscilloscope probe to the anode. Increase the anode voltage until conduction starts, and note the anode voltage level.

5-3.2 Rate-of-Rise Tests

Various manufacturers have developed a number of circuits for rate-of-rise tests. All these circuits are based on a method known as the *exponential waveform method*. The following is a description of the basic circuit and techniques for this method.

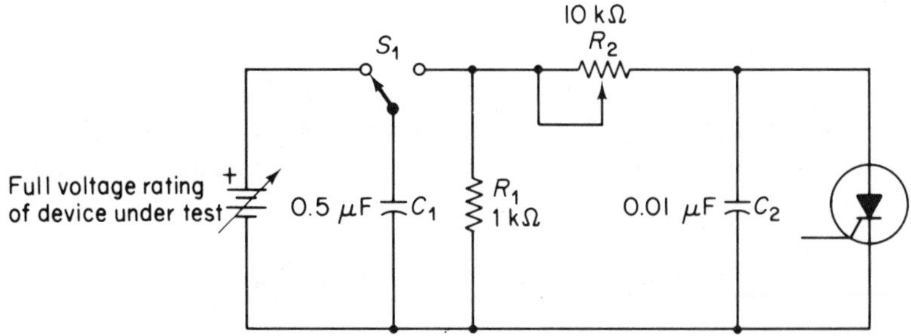

FIGURE 5-18 Basic test circuit for measurement of rate of rise (*dV/dT*).

The basic rate-of-rise test circuit is shown in Fig. 5-18. In operation, a large capacitor C_1 is charged to the *full voltage rating* of the device under test. Capacitor C_1 is then discharged through a variable time-constant network (R_2 and C_2). This is repeated with smaller time constants (higher dV/dT) until the device under test is turned on by the fast dV/dT).

The critical rate, which causes firing, is defined as

$$\frac{dV}{dT} = \frac{0.632 \times \text{anode voltage}}{R_2 \times C_2}$$

This equation describes the average slope of the essentially linear rise portion of the applied voltage, as shown in Fig. 5-19. In a practical test circuit, there are two major conditions that determine the value of the circuit components:

First, capacitor C_1 should be large enough to serve as a constant-voltage source during the discharging of C_1 and the charging of C_2. Second, capacitor C_2 should be much larger than the intrinsic cathode-to-anode capacitance of the device under test, plus any stray device and device-test wiring capacitances. Control rectifier junction capacitance has been found to be in the order of 800 pF for zero applied voltage.

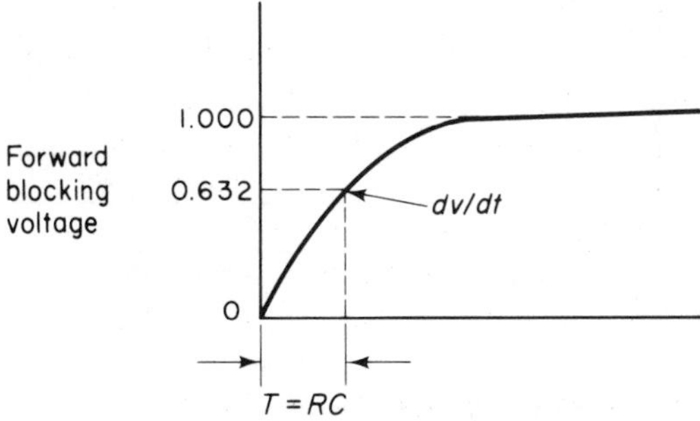

FIGURE 5-19 Exponential of applied forward voltage and definition of *dV/dT*.

As guidelines, a typical 70-A device will have a junction capacitance of about 800 pF. Also, 0.5 μF for C_1 and 0.01 μF for C_2 are practical values for the test of most control rectifiers. It should be noted that the stray inductance and capacitance of the test circuit should be minimized. This is especially true for measurement of high dV/dT values.

5-3.3 Turn-on- and Turn-off (Recovery)-Time Tests

Figure 5-20 shows a circuit capable of measuring both turn-on and turn-off (recovery) time. The circuit inductances must be kept to a minimum by using short connections, thick wires, and closely spaced return loops or wiring on a grounded chassis. External pulse sources must be provided for the circuit. These pulses are applied to transformers T_1 and T_2, and serve to turn the device under test on and off. The pulses can come from any source, but should be of the amplitude, duration, and repetition rate that correspond to the normal operating conditions of the device under test.

When a suitable gate pulse is applied to transformer T_1, the device under test is turned on. Load current can be set by resistor R_L. A predetermined time later, the turn-off control rectifier is turned on by a pulse applied to T_2. This places capacitor C_2 across the device under test, applying a reverse bias, and turning the device under test off.

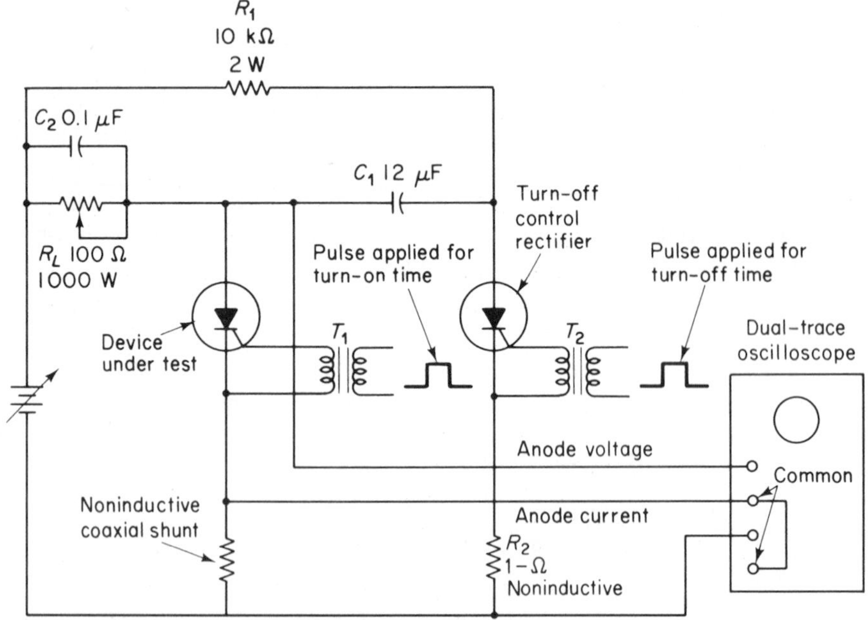

FIGURE 5-20 Test circuit for measurement of turn-on and turn-off (recovery) times.

Any oscilloscope capable of a 10-μs sweep can be used for viewing both the turn-on and turn-off action. The oscilloscope is connected with the vertical input across the device under test. Turn-on time is displayed when the oscilloscope is triggered with the gate pulse applied to the device under test. Turn-off time is displayed when the oscilloscope is triggered with the gate pulse applied to the turn-off control rectifier.

The actual spacing between the turn-on and turn-off pulses is usually not critical. However, a greater spacing causes increased conduction and heats the junction. Since operation of control rectifiers is temperature-dependent, the rise in junction temperature must be taken into account for accurate test results. (Both turn-on and turn-off times increase with an increase in junction temperature.)

Figure 5-21 shows turn-on action. Turn-on time is equal to delay time (t_d) plus rise time (t_r). Following the beginning of the gate pulse, there is a short delay before appreciable load current flows. Delay time is the time from the leading edge of the gate-current pulse (beginning of oscilloscope sweep) to the point of 10% load-current flow. (Delay time can be decreased by overdriving the gate.)

Rise time (t_r) is the time the load current increases from 10% to 90% of the total value. Rise time depends upon load inductance, load-current amplitude, junction temperature, and, to a lesser degree, upon anode voltage. The higher the inductance and load current, the longer the rise time. An increase in anode voltage tends to decrease the rise time. Capacitor C_2 (Fig. 5-20) tends to counter the load inductance, thus lessening the rise time.

By triggering the oscilloscope with the gate pulse applied to the device under test, the sweep starts at the gate pulse leading edge. Thus, the oscilloscope presentation shows the anode voltage from this point on. In noninductive circuits, when the anode voltage decreases to 90% of initial

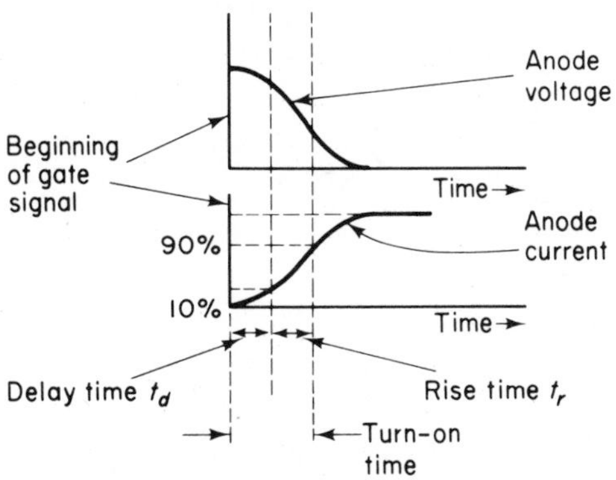

FIGURE 5-21 Definition of turn-on time (delay time plus rise time).

value, this time is equal to 10% of the load current, and is thus equal to the delay time.

With the oscilloscope set at 10-μs sweep, each square represents 1 μs. The delay time is read directly by counting the number of divisions on the oscilloscope screen. If the circuit is noninductive, the decrease from 90% to 10% of the anode voltage is approximately equal to the load current increase from 10% to 90%. The time this takes is equal to the rise time. The total time from zero time (start of oscilloscope sweep) to this 10% of the anode voltage is equal to the turn-on time. Therefore, turn-on time is determined by counting the number of divisions from the start of the oscilloscope sweep to the 90% anode-load current.

Figure 5–22 shows the reverse current and reverse recovery (turn-off) action of the device under test. Turn-off time is the time necessary for the device under test to turn off and *recover* its forward blocking ability. The reverse recovery time (t_h) is the length of the interval between the time the forward current falls to zero when going reverse and the time the current returns back to zero from the reverse direction.

In Fig. 5–20, the time available for turn-off action is determined by the value of capacitor C_1 and resistor R_2. Decreasing the value of C_1 decreases the time the device under test is reverse-biased. Resistor R_2 limits the magnitude of the reverse current. The shape of the reverse voltage and current pulses are determined by capacitor–resistor discharge. At the end of the reverse pulse, forward voltage is reapplied. Having turned off, the device under test blocks forward voltage, and no current can flow.

Figure 5–22 shows the reverse current pulse. With the oscilloscope set for a 20-μs sweep, the value of reverse recovery time (turn-off time) can be measured by counting off the divisions from the zero point on the leading

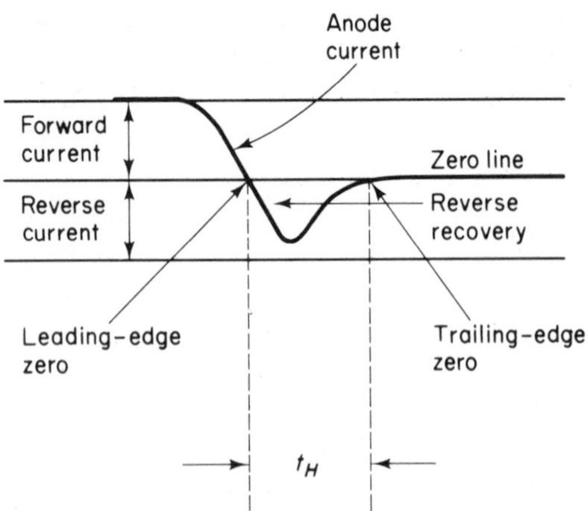

FIGURE 5–22 Definition of reverse recovery (turn-off) time.

edge of the reverse current pulse, to the zero point on the trailing edge, as shown.

5-4 BLOCKING VOLTAGE AND LEAKAGE CURRENT TESTS

This section describes test circuits and procedures for both forward and reverse blocking voltage, as well as leakage current measurements. In these circuits, the voltages and currents are measured on meters rather than on an oscilloscope. However, one circuit provides for a simultaneous oscilloscope display of the characteristics.

5-4.1 Test Circuit for Thyristors above 2 A

Figure 5-23 shows a circuit for test of the forward and reverse blocking characteristics of low-, medium-, and high-current SCRs, as well as triacs. The circuit is not recommended for very low current devices (those that operate at currents below 2 A). Although there will be no damage, the test results may be ambiguous. The test circuit operates at 60 Hz, and stray capacitance charging currents flowing in the equipment are comparable in magnitude to the very low leakage currents of small thyristors.

The circuit of Fig. 5-23 consists of a variable-voltage current-limited power supply that develops a half-wave voltage across the thyristor under test to minimize junction heating. Instrumentation consists of suitable meters to indicate blocking voltage and leakage currents. If desired, an oscilloscope can be connected to provide a visual display of the forward and reverse voltage–current characteristics of the device under test.

To measure forward breakdown voltage (generally listed as $V_{BR(FO)}$ for an SCR, and V_{FOM} for a triac) and forward leakage current, set S_1 to FWD, and raise the voltage by means of the adjustable transformer T_1 until the meter V reads rated $V_{BR(FO)}$ or V_{FOM}. Read the full-cycle average leakage current from meter I, or the peak leakage current from the voltage–current trace on the oscilloscope. To measure the actual forward breakover voltage, increase the applied voltage by means of T_1 until the oscilloscope trace indicates that the device under test is breaking over (the current will rise sharply just prior to breakdown). Be careful that the applied voltage does not exceed the peak forward voltage rating (PFV) of the device.

Since a triac conducts in both directions, it is often desirable to measure breakdown voltage and leakage current in both forward and reverse directions. A triac can be tested in the reverse direction (sometimes called the *third quadrant direction or mode*) by setting switch S_1 to *REV*, and repeating the test (read V_{FOM} on meter V, and current on meter I).

To measure reverse voltage (V_{ROM}) and current (I_{RO}) of an SCR, set S_1 to

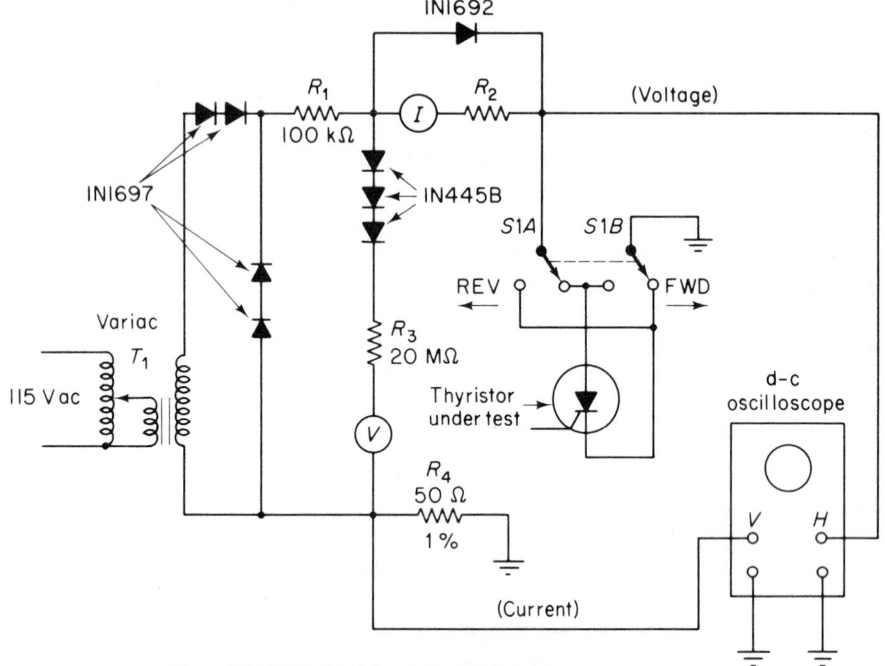

FIGURE 5-23 Test circuit for forward and reverse blocking characteristics of low-, medium-, and high-current SCRs and triacs.

REV, and raise the voltage by means of T_1 until meter V reads rated V_{ROM}. Read the full cycle average reverse leakage current $I_{RO(AV)}$ on meter I, or peak leakage current on the oscilloscope display.

5-4.2 Test Circuit for Thyristors below 2 A

The circuit of Fig. 5-24 provides a simple and inexpensive means for checking the instantaneous leakage characteristics and blocking voltage capabilities of low-current SCRs. Note that the push button switch S_1 minimizes junction heating and should not be omitted from the circuit. Test may be conducted at elevated temperatures by placing the test SCR in an oven.

To use the circuit, set switch S_2 to FWD, depress S_1 and increase the input by means of R_1 so that meter V reads the rated blocking voltage. Read

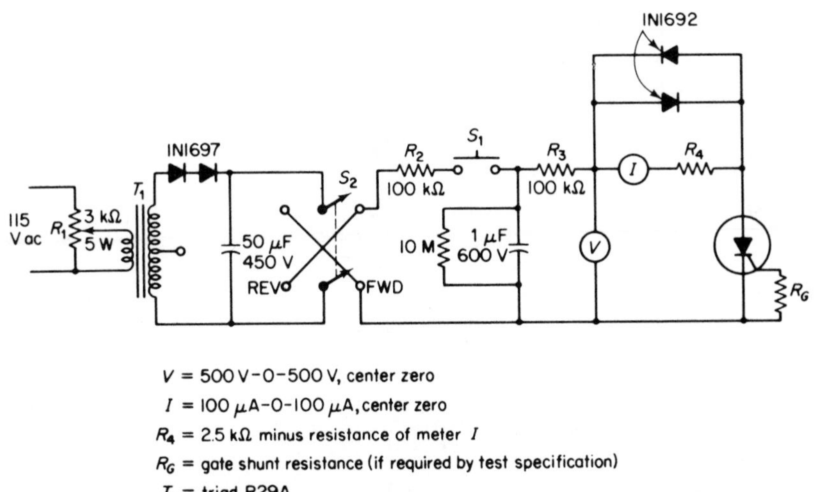

V = 500 V–0–500 V, center zero
I = 100 μA–0–100 μA, center zero
R_4 = 2.5 kΩ minus resistance of meter I
R_G = gate shunt resistance (if required by test specification)
T_1 = triad R29A

FIGURE 5-24 Test circuit for forward and reverse blocking characteristics of low (2 A or below) SCRs.

leakage current, if any, on meter I. To measure actual forward breakdown voltage, adjust R_1 until the current reading on I increases sharply and the voltage reading on V decreases. The reading on meter V just prior to this point is the forward breakover voltage of the device under test. Reverse blocking voltage and leakage current measurements are made with switch S_2 in the *REV* position.

5-5 GATE TRIGGER VOLTAGE AND CURRENT TESTS

This section describes test circuits and procedures to determine, under stated conditions, the magnitude of the gate trigger voltage and current necessary to switch a thyristor from forward blocking to the on-state. One circuit provides a pulse trigger source for SCRs and triacs, while a second circuit provides a direct-current trigger. A third circuit provides for gate test of low-current SCRs.

5-5.1 *Pulse Test Circuit for Thyristors above 2 A*

The circuit of Fig. 5-25 provides for pulse testing of thyristors with operating currents greater than 2 A. In the circuit of Fig. 5-25, the blocking voltage waveform applied to the device under test consists of a clipped half-wave with peak magnitude of 6 V or 12 V, depending on the setting of S_1. The correct

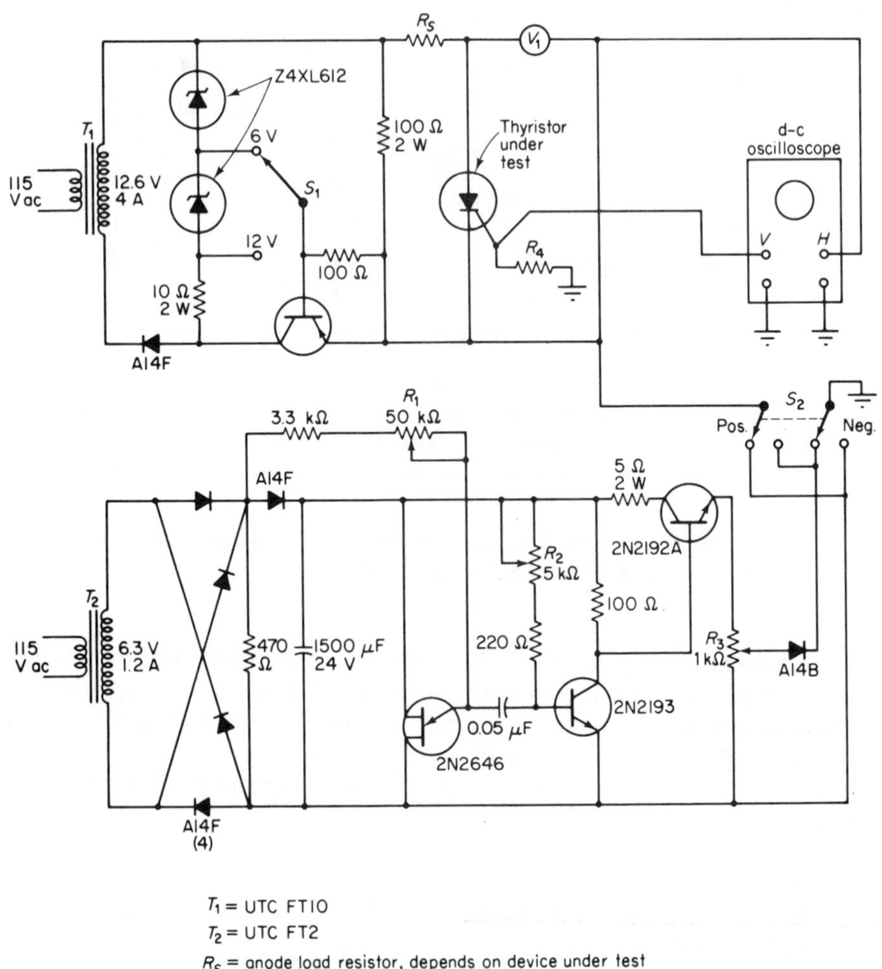

T_1 = UTC FT10

T_2 = UTC FT2

R_S = anode load resistor, depends on device under test

R_4 = current sensing resistor; select to give approximate 0.5-V drop at maximum gate current

FIGURE 5-25 Test circuit for gate trigger voltage and current characteristics of thyristors above 2 A.

setting of S_1 is determined from the specification sheet of the thyristor under test, as is the value of the anode load resistor R_S. Gate–source voltage consists of a square-wave pulse, the magnitude of which can be varied from zero to 6 V. The pulse width can also be adjusted from about 5 μs to greater than 100 μs. Gate voltage can be switched either positive or negative for testing triacs. Instrumentation consists of a gate current "looking" resistor R_4, an oscilloscope with separate vertical and horizontal amplifiers, and a d-c voltmeter V_1 to monitor when the device under test triggers.

A typical test procedure using the circuit of Fig. 5-25 is as follows:

1. Adjust R_1 so that a gate pulse occurs only once during each half-cycle of applied anode blocking voltage. Time the pulse (by adjustment of R_1) to occur approximately 4 ms after the start of the half-cycle.
2. Adjust R_2 to give the desired width of pulse. Note that gate pulse widths in excess of 100 μs result in measured values of trigger voltage and current that are equivalent to continuous d-c measurements.
3. Initially, set R_3 for zero volts output. Then gradually increase the setting of R_3 until the device under test triggers. Triggering is indicated by a sudden drop in the reading of V_1, or by a sudden step in the gate voltage–current trace observed on the oscilloscope. Because the gate impedance may change when triggering occurs, the readings of gate voltage and current must be made *just prior to triggering*.

5-5.2 Direct-Current Test Circuit for Thyristors above 2 A

The test circuit of Fig. 5-26 is essentially a d-c version of the pulse test described in Section 5-5.1. Anode supply circuitry is identical, while the pulse generator is replaced with a simple adjustable d-c power source. Instead of monitoring trigger voltage and current with an oscilloscope, d-c meters are used. As before, R_3 is turned up until the test thyristor triggers. Meters I and V_2 are read off just prior to the trigger point.

5-5.3 Gate Trigger Test Circuit for SCRs below 2 A

The measurement of low-current SCR triggering voltage and current is complicated by the fact that the gate impedance changes drastically when the SCR triggers. Also, the gate trigger voltage and current values depend on the source impedance of the test set. Thus, source impedance must be specified when making tests. The circuit of Fig. 5-27 is designed specifically for testing SCRs with current ratings below 2 A. In this circuit, a variable half-wave voltage is applied to the gate (from a controlled impedance source), and the gate voltage–current characteristics are monitored on an oscilloscope. The triggering point is detected by the sudden change in gate impedance that occurs when the SCR switches on.

Figure 5-27(b) shows a typical oscilloscope presentation of gate voltage–current characteristics during test. The trace is shown dashed beyond the trigger point. In an actual oscilloscope display, the trace suddenly jumps

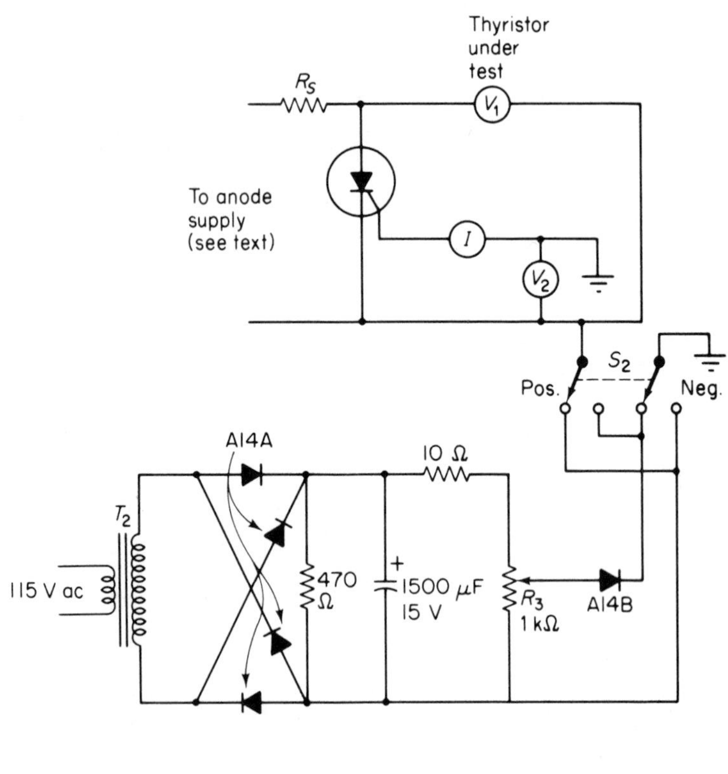

T_2 = UTC F2

V_1 = 10 V dc full scale

V_2 = 12 V dc full scale

I = select full scale to suit gate current range of interest

FIGURE 5-26 Direct-current test circuit for gate trigger voltage and current characteristics of thyristors above 2 A.

at the triggering point due to the change in gate impedance. The portion of the trace beyond the triggering point becomes somewhat reduced in intensity.

The gate trigger voltage is that value of voltage that will just cause the device to switch on. As shown in Fig. 5-27(b), the gate trigger voltage is read just prior to the switching point. Unlike trigger voltage, the gate trigger current value must be read at the point where current is maximum; although this is not necessarily the actual triggering point, the trace must first pass through this maximum, and the firing circuit design must take this into account.

Many low-current SCRs show triggering with a negative value of gate current. Figure 5-27(c) shows a typical gate voltage-current trace for this type of SCR. Note that the gate trigger voltage is always positive. On such SCRs, only the gate trigger voltage is of interest.

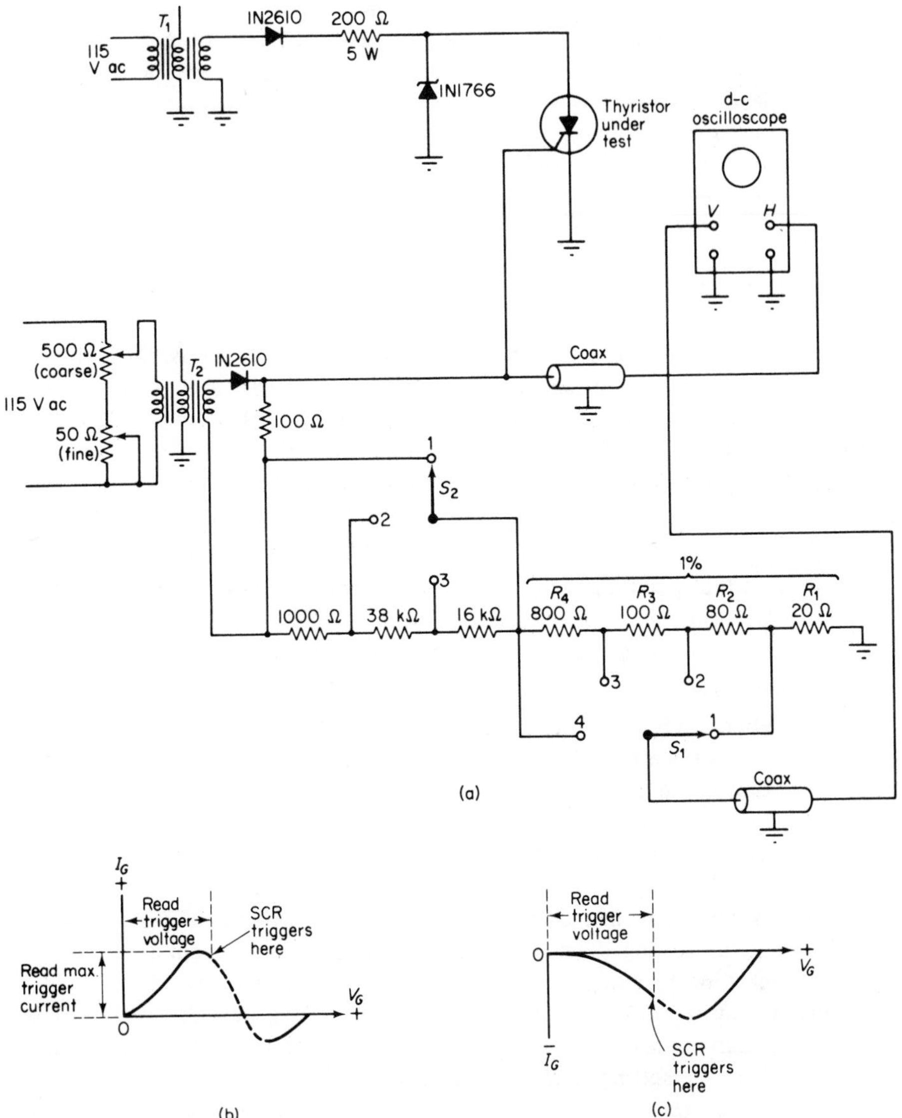

FIGURE 5-27 Gate trigger test circuit for SCRs below 2 A.

5-6 LATCHING AND HOLDING CURRENT TESTS

This section describes test circuits and procedures to measure latching and holding currents for control rectifiers and thyristors. Because latching and holding current values depend on gate conditions and anode supply voltage, these parameters must be specified as test conditions. A thyristor can show more than one value of on-voltage at a given forward current level, especially if the maximum value of anode current never rises much above holding level. This can lead to several values of holding current for the same device. To properly perform a holding-current test, turn the device on initially with a high current pulse, and then reduce the current down to the holding level. The test circuit of Fig. 5–28 fulfills this requirement, and may also be used for latching current tests.

5-6.1 Latching-Current Measurement

For latching-current measurements (switch S_3 in the LATCH position), the circuit of Fig. 5–28 consists essentially of the test thyristor in series with a current-adjusting resistor R_2, milliammeter I_1, and the 24-V supply. S_1 and R_3 provide trigger signals for the device under test. R_3 is selected to provide specified trigger current. The operating procedure is as follows:

1. Set R_2 to the maximum resistance value.
2. Gradually reduce the resistance of R_2 while pressing and releasing S_1. Each time S_1 is depressed, I_1 should deflect and then drop back to zero when S_1 is released, as long as the anode current flowing through the test thyristor is less than the latching current.
3. When latching finally occurs, I_1 should deflect and remain deflected as S_1 is released. The value of current indicated by I_1 at the transition point is the latching current of the test thyristor.

5-6.2 Holding-Current Measurement

For holding-current measurements (switch S_3 in the HOLD position), R_2 is initially left at the setting determined during the latching-current test (Section 5-6.1). Each time S_1 is depressed to trigger the test thyristor (as described for latching current test), SCR_1 also is triggered and passes an additional pulse of current through the test thyristor. The magnitude of this initial current pulse is determined by the setting of R_1 and is specified for each thyristor type. The value of the current is monitored by the 1-Ω "looking" resistor R_4 and displayed on the oscilloscope. Pulse width is fixed by C_1 and T_1. The operating procedure is as follows:

1. Set R_2 to the value determined for latching current (Section 5–6.1).
2. Press and release S_1, while gradually increasing R_2 in value until I_1 drops suddenly to zero. The reading on I_1 just prior to this point is the holding current of the device under test.

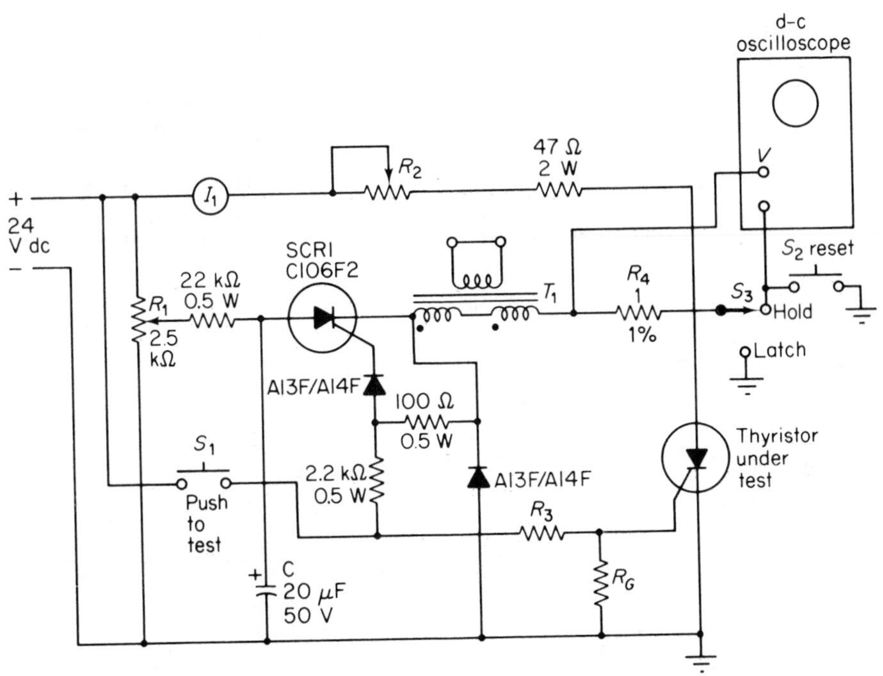

T_1 = UTC FT10, connect secondary windings in series and short the primary winding

I_1 = 500-mA dc movement

R_2 = 10 kΩ potentiometer (or 50 kΩ for testing low-current SCRs)

R_3 = to suit thyristor under test

R_G = gate shunt resistance (as required by test specification)

FIGURE 5-28 Test circuit for holding and latching current measurements.

5-7 AVERAGE FORWARD VOLTAGE TEST

The circuit of Fig. 5-29 can be used to test the forward voltage drop of a thyristor in the conducting state. This forward voltage drop is often listed as $V_{F(AVG)}$. In this test, the thyristor is subjected to a direct current as read by meter M_1, and the voltage drop is measured by meter M_2. To make the measurement, close S_1 with T_1 in a position approximately midway between the end positions. Adjust T_1 to the level of current desired for test, as read on meter M_1. Depress S_2 and read M_2.

These readings (M_1 and M_2) should be less than the maximum rated current and voltage for the device under test. A test current somewhere near the

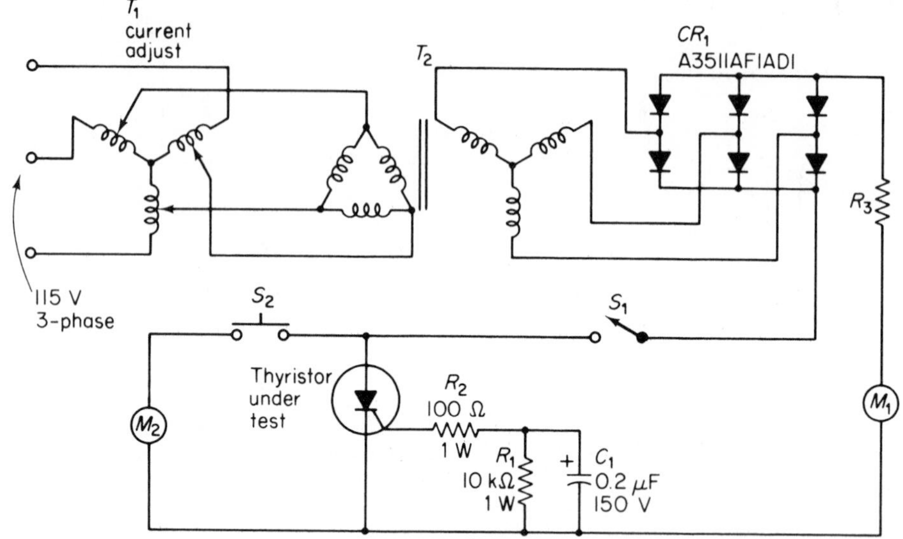

T_1 = 3-phase adjustable transformer, 115 V, 3 kW

T_2 = (3) 115/12-V, 1-kW transformers

R_3 = as required for SCR under test, typically 0.2 to 4 Ω, 1 kW

M_1 = 5-, 10-, or 100-A ammeter (as required for SCR under test

M_2 = 5-V dc voltmeter

FIGURE 5-29 Test circuit for measurement of average forward voltage.

continuous duty current rating of the thyristor is recommended. Under these conditions, the device under test should be mounted on a heat sink during test. If a heat sink is not used, the readings should be completed within 2 or 3 s to prevent overheating of the device under test. These tests can be conducted at any ambient temperature within the operating range of the thyristor.

5-8 CONTROL RECTIFIER AND THYRISTOR TESTS USING A CURVE TRACER

A curve tracer can be used to test all important characteristics of control rectifiers and thyristors. This section describes a series of tests for SCRs, triacs, and diacs.

5-8.1 SCR Tests Using a Curve Tracer

The following characteristics of an SCR can be displayed and measured with a curve tracer:

Forward blocking voltage

Reverse blocking voltage

Leakage current

Holding current

Forward voltage drop for various forward currents

Gate trigger voltage for various forward voltages

5-8.2 SCR Test Connections

For all measurements except gate trigger voltage, the SCR should be connected to the curve tracer as follows (Fig. 5-30):

SCR cathode to emitter jack, or emitter pin of socket

SCR anode to collector jack, or collector pin of socket

SCR gate to base jack, or base pin of socket

5-8.3 Forward Blocking Voltage

To measure forward blocking voltage, set the curve-tracer STEP SELECTOR to the I_{CES} position. This shorts the gate and cathode to satisfy the zero gate current requirement. Set the curve-tracer POLARITY switch to NPN. Increase the

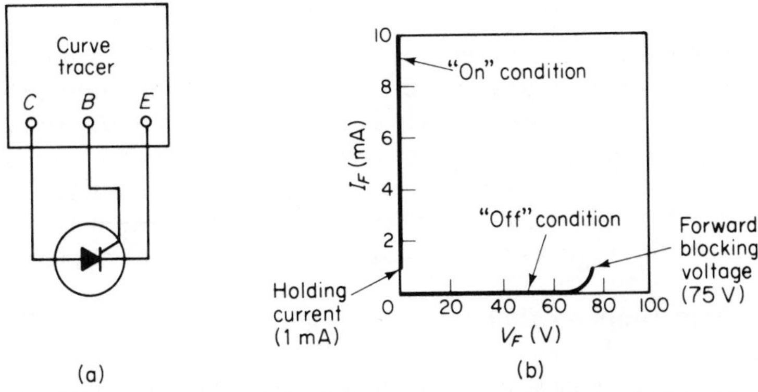

FIGURE 5-30 Typical forward blocking voltage and holding-current curve tracer displays.

curve-tracer SWEEP VOLTAGE until the SCR fires or triggers. When this occurs, the anode current suddenly increases, and the anode voltage drops to near zero, as shown in Fig. 5–30. In the "off" condition, the trace is horizontal until the forward blocking voltage is reached. In the "on" condition, the trace is vertical.

Read the highest anode voltage point in the display (75 V in Fig. 5–30). This is the maximum forward blocking voltage. Any anode current at anode voltage below the "firing" point is forward leakage current, and can be read directly from the display. (If the horizontal "off"-condition trace is above zero, it is a result of forward leakage current.)

5-8.4 Reverse Blocking Voltage

The procedure for measuring reverse blocking voltage is the same as for measuring forward blocking voltage, except that the POLARITY switch is set to PNP. The voltage at which voltage breakdown occurs, which is a sudden increase in anode current, is the reverse blocking voltage value. Any anode current at voltages below breakdown is reverse leakage current and can be read directly from the display.

5-8.5 Holding Current

Using the same procedure as described for forward blocking voltage test (Fig. 5–30), note the lowest current displayed for the "on" condition (1 mA in Fig. 5–30). This is the holding current. The measurement can also be made with the STEP SELECTOR in one of the "current per step" positions so that less sweep voltage is required to place the SCR in the "on" condition.

5-8.6 Forward Voltage Drop

The forward voltage drop during the "on" condition at various forward current levels may be measured by increasing the horizontal sensitivity of the oscilloscope, and displaying a low-voltage portion of the forward voltage. Increase the STEP SELECTOR "current per step" setting so that sweep voltage may be reduced. The vertical sensitivity may also be reduced to a typical 10 mA division for a greater range of voltage versus current. Read the forward voltage drop (for a given forward current) directly from the display.

5-8.7 Gate Trigger Voltage

The turn-on point of an SCR is dependent upon the forward voltage and gate voltage. As gate voltage is increased, less forward voltage is required to switch on the SCR. Conversely, as forward voltage is increased, less gate voltage is required to switch on the SCR. Gate trigger voltage values can be measured by connecting a d-c bias supply to the gate terminal of the SCR as shown in Fig. 5-31. The bias supply reference must be connected to the emitter jack, or the cathode of the SCR. Otherwise, the curve tracer is set up as for forward blocking voltage measurement. Two types of measurements can be made:

1. Set the sweep voltage to a specified forward anode-cathode voltage, and increase the d-c bias supply until the SCR switches on. Measure the value of gate voltage (on meter M_1) at which switching occurs.
2. Set the d-c bias supply to a specified gate voltage, and increase the sweep voltage until the SCR switches on. Read the peak value of sweep voltage (on the oscilloscope display) which is required to produce triggering.

5-8.8 Triac Tests Using a Curve Tracer

Triacs may be tested on a curve tracer exactly as for SCRs, except that forward tests should be repeated for both directions. Also, there is no reverse blocking voltage measurement, since a triac normally conducts in both directions.

5-8.9 Diac Tests Using a Curve Tracer

Breakdown voltage, leakage current, and holding current of a diac can be tested using a curve tracer. The procedures are the same as for an SCR. The diac is connected between the collector and emitter terminals of the curve tracer, and the SWEEP VOLTAGE is increased until breakdown occurs. The

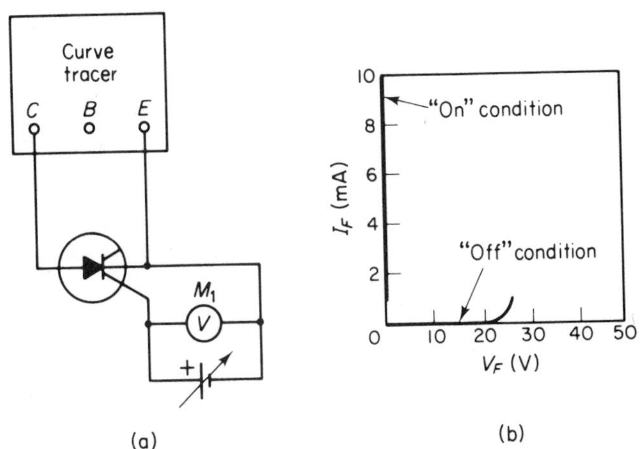

FIGURE 5-31 Testing turn-on point of an SCR using a curve.

breakdown voltage holding current and leakage can be read from the oscilloscope display. The breakdown voltage is the highest voltage in the "off" condition. Leakage current is any deflection of the horizontal "off"-condition trace from zero. Holding current is the lowest current in the "on" condition.

6

Audio-Circuit Tests

This chapter is devoted entirely to test procedures for audio circuits. These procedures can be applied to complete audio equipment (such as a stereo system), or to specific circuits (such as the audio circuits of a radio transmitter or receiver). Also, the procedures can be applied to audio circuits at any time during design or experimentation. As a minimum, the tests should be made when the circuit is first completed in experimental form. If the test results are not as desired, the component values can be changed as necessary to obtain the desired results. The circuit can then be retested in final form (with all components soldered in place). This shows if there is any change in circuit characteristics due to the physical relocation of components.

The test procedures include a series of notes regarding changes in component values on test results. This information is summarized at the end of the chapter and is of particular interest to hobbyists and experimenters. The following sections do not describe every possible test to which audio circuits can be subjected. However, they do include all basic tests necessary for "typical" audio-circuit operation.

6-1 FREQUENCY RESPONSE OF AUDIO CIRCUITS

The frequency response of an audio amplifier, or filter, can be measured with an audio signal generator and a meter or oscilloscope. When a meter is used,

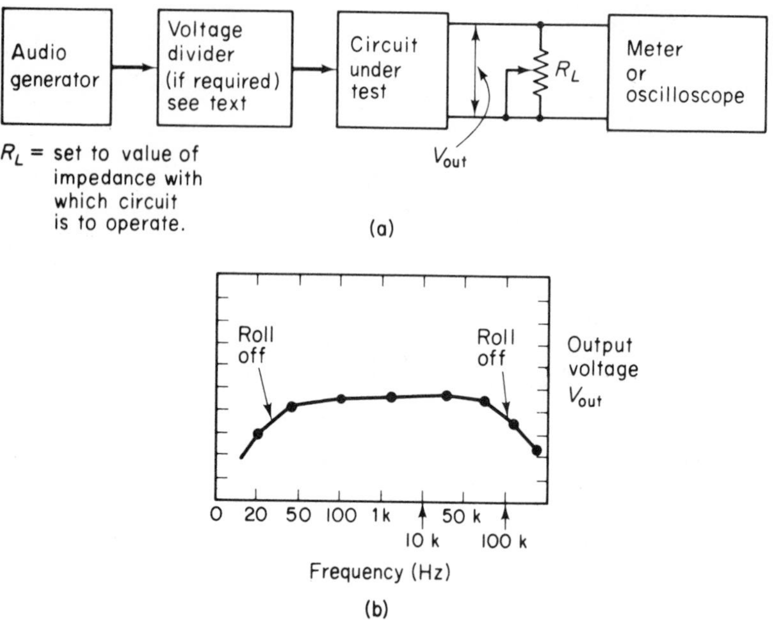

FIGURE 6-1 Frequency-response test connections and typical response curve.

the signal generator is tuned to various frequencies, and the resultant circuit output response is measured at each frequency. The results are then plotted in the form of a graph or response curve, such as shown in Fig. 6–1. The procedure is essentially the same when an oscilloscope is used to measure audio-circuit frequency response. However, an oscilloscope gives the added benefit of visual distortion analysis, as discussed in Sections 6–10 and 6–11.

6-2 BASIC FREQUENCY-RESPONSE TESTS

The basic frequency-response test or measurement procedure (with either meter or oscilloscope) is to apply a *constant-amplitude* signal to the circuit input while monitoring the circuit output. The input signal is varied in frequency (but not amplitude) across the entire operating range of the circuit. Any well-designed audio circuit should have a constant response from about 20 Hz to 20 kHz. With direct-coupled amplifiers, the response can be extended from a few hertz up to 100 kHz (and higher). The voltage output at various frequencies across the range is plotted on a graph similar to that shown in Fig. 6–1.

The basic frequency-response test procedure is as follows:

1. Connect the equipment as shown in Fig. 6–1. Set the generator, meter, and oscilloscope controls as necessary. It is assumed that

the audio generator is provided with controls to vary the output in both frequency and amplitude. If the audio generator amplitude cannot be varied in amplitude, the output can be applied to the test circuit input through a precision voltage divider, such as shown in Fig. 6-2. This divider provides test voltage in precision steps (1 mV, 10 mV, 100 mV, and 1 V). In use, the audio generator is adjusted for 10-V output, and each of the voltage divider outputs or steps is checked with a precision voltmeter. Accuracy of the step voltage divider depends on accuracy of the resistance values. Precision resistors with 1% accuracy should be sufficient for the voltage-divider circuit. It is also assumed that the audio generator is provided with an output amplitude meter. If not, the generator output can be monitored with an external meter.

2. Initially, set the generator output frequency to the low end of the frequency range. Then set the generator output amplitude to the desired input level. For example, most audio amplifiers are rated as to some specific input voltage for a given output (1-V input for full output, 1-V input for 10-W output, etc.).
3. In the absence of a realistic test input voltage, set the generator output to an arbitrary value. A simple method of finding a satisfactory input level is to monitor the circuit output (with the meter or oscilloscope) and increase the generator output at the circuit center frequency (or at 1 kHz) until the circuit is overdriven. This point is indicated when further increases in generator output do not cause

Basic Design Rules

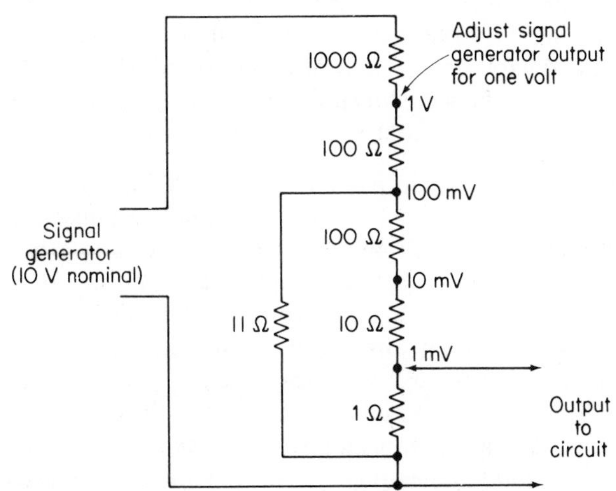

FIGURE 6-2 Voltage divider with low output impedance.

further increases in meter reading (or the output waveform peaks begin to flatten on the oscilloscope display). Set the generator output *just below* this point. Then return the meter or oscilloscope to monitor the generator voltage (at the test circuit input) and measure the voltage. Keep the generator at this voltage throughout the test.

4. If the circuit is provided with any operating or adjustment controls (volume, loudness, gain, treble, bass, etc.), set these controls to some arbitrary point when making the initial frequency-response measurement. The response measurements can then be repeated at different control settings if desired.

5. Record the circuit output voltage on the graph. Without changing the generator output amplitude, increase the generator frequency by some fixed amount, and record the new circuit output voltage. The amount of frequency increase between each measurement is an arbitrary matter. Use an increase of 10 Hz at the low end and high end (where rolloff occurs), and an increase of 100 Hz at the middle frequencies.

6. Repeat this process, checking and recording the circuit output voltage at each of the check points in order to obtain a frequency-response curve. With a typical audio-amplifier circuit, the curve resembles that of Fig. 6–1, with a flat portion across the middle frequencies and a rolloff at each end. A bandpass filter has a similar response curve. High-pass and low-pass filters produce curves with rolloff at one end only. (High-pass has a rolloff at the low end, and vice versa.)

7. After the initial frequency-response check, the effect of operating or adjustment controls should be checked. Volume, loudness, and gain controls should have the same effect all across the frequency range. Treble and bass controls may also have some effect at all frequencies. However, a treble control should have the greatest effect at the high end, whereas a bass control affects the low end most.

8. Note that generator output may vary with changes in frequency, a fact often overlooked in making a frequency-response test of any circuit (not just audio circuits). Even precision laboratory generators can vary in amplitude output with changes in frequency, thus resulting in considerable error. It is recommended that the generator output be monitored after each change in frequency (some audio generators have a built-in output meter). Then, if necessary, the generator output amplitude can be reset to the correct value. Within extremes, it is more important that the generator output amplitude *remain constant* rather than at some specific value when making a frequency-response check.

6-3 VOLTAGE-GAIN TESTS

Voltage-gain test of an audio circuit is made in the same way as frequency response. The ratio of output voltage to input voltage (at any given frequency, or across the entire frequency range) is the voltage gain. Since the input voltage (generator output) must be held constant for a frequency-response test, a voltage-gain curve should be identical to a frequency-response curve.

6-4 POWER OUTPUT AND GAIN TESTS

The *power output* of an audio circuit is found by noting the output voltage V_{OUT} across load resistance R_L (Fig. 6-1), at any frequency, or across the entire frequency range. Power output is found by $(V_{OUT})^2/R_L$.

To find *power gain* of an audio circuit, it is necessary to find both the input and output power. Input power is found in the same way as output power, except that the impedance at the input must be known (or calculated). This is not always practical in some audio circuits, especially in designs where input impedance is dependent upon transistor gain. With input power known (or estimated), the power gain is the ratio of output power to input power.

An *input sensitivity* specification is often used in place of power gain for some audio circuits and amplifiers. Input sensitivity specifications require a minimum power output with a given voltage input (such as 100-W output with 1-V input).

6-5 POWER-BANDWIDTH TESTS

Many audio circuit or amplifier specifications include a power-bandwidth factor. Such specifications require that the audio circuit deliver a given power output across a given frequency range. For example, a certain audio amplifier circuit produces full power output up to 20 kHz (as shown in Fig. 6-3), even

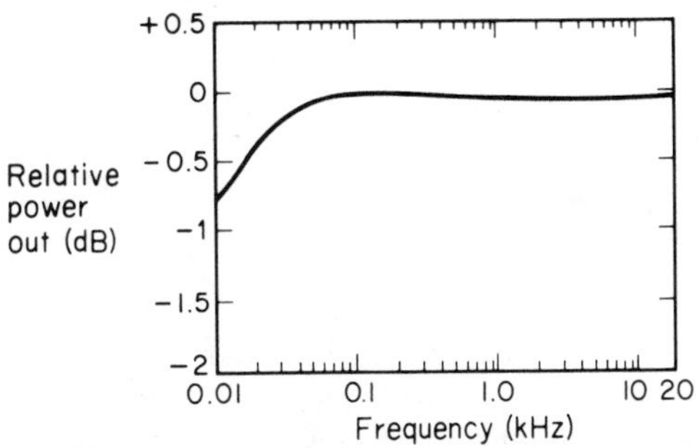

FIGURE 6-3 Typical power-bandwidth graph.

though the frequency response is flat up to 100 kHz. That is, voltage (without a load) remains constant up to 100 kHz, while power output (across a normal load) remains constant up to 20 kHz.

6-6 LOAD-SENSITIVITY TESTS

An audio circuit is sensitive to changes in load. This is especially true of audio power amplifiers but can also be the case with voltage amplifiers. An amplifier produces maximum power when the output impedance of the amplifier circuit is the same as the load impedance. This is shown by the curve of Fig. 6-4 (the load sensitivity for a typical audio amplifier circuit). If the load is twice the output circuit impedance (ratio of 2.0), the output is reduced to approximately 50%. If the load is 40% of the output impedance (ratio of 0.4), the output power is reduced to approximately 25%. Generally, a power-amplifier circuit should be checked for load sensitivity during some stage of design. Such a test often shows defects in design that are not easily found with the usual frequency-response and power-output tests.

The circuit for load-sensitivity test is the same as for frequency response (Fig. 6-1), except that the load resistance R_L must be variable. (Never use a wire-wound load resistance. The reactance can result in considerable error. If a non-wire-wound variable resistance of sufficient wattage rating is not available, use several fixed carbon or composition resistances arranged to produce the desired resistance values.)

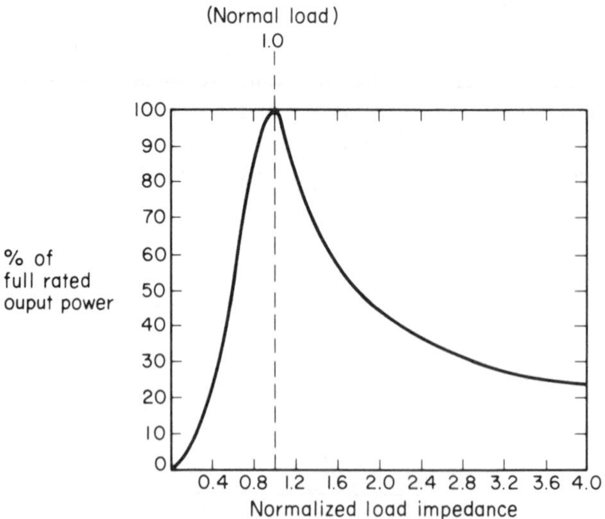

FIGURE 6-4 Typical output power versus load impedance (load sensitivity) graph.

Measure the power output at various load impedance/output impedance ratios. To make a comprehensive test of an audio circuit under design, repeat the load-sensitivity test across the entire frequency range.

6-7 DYNAMIC OUTPUT IMPEDANCE TESTS

The load-sensitivity test described in Section 6–6 can be reversed to find the dynamic output impedance of an audio circuit. The connections and procedure (Fig. 6–1) are the same, except that the load resistance R_L is varied until maximum output power is found. Power is removed and R_L is disconnected from the circuit. The d-c resistance of R_L (measured with an ohmmeter) is equal to the dynamic output impedance. Of course, the value applies only at the frequency of measurement. The test should be repeated across the entire frequency range of the circuit.

6-8 DYNAMIC INPUT IMPEDANCE TESTS

To find the dynamic input impedance of an audio circuit, use the circuit of Fig. 6–5. The test conditions should be identical to those for frequency response, power output, and so on. That is, the same audio generator, operating load, meter or oscilloscope, and frequencies should be used.

The signal source is adjusted to the frequency (or frequencies) at which the circuit is operated. Switch *S* is moved back and forth between position *A* and *B*, while resistance *R* is adjusted until the voltage reading is the same in both positions of the switch. Resistor *R* is then disconnected from the circuit, and the d-c resistance of *R* is measured with an ohmmeter. The d-c resistance of *R* is then equal to the dynamic impedance at the circuit input.

Accuracy of the impedance measurement depends on the accuracy with which the d-c resistance is measured. A noninductive resistance must be used. The impedance found by this method applies only to the frequency used during the test.

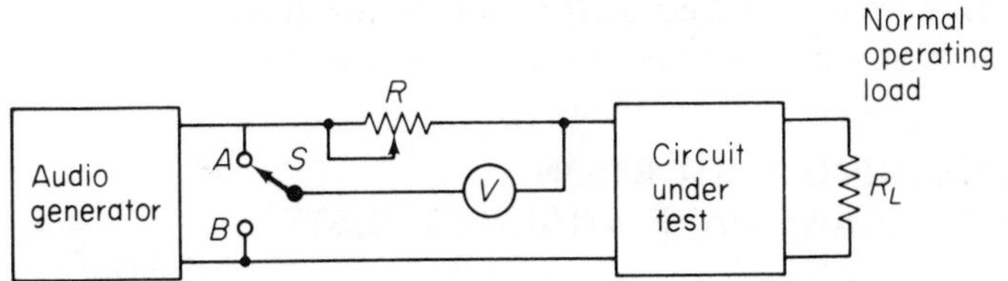

FIGURE 6-5 Test circuit for measurement of dynamic input impedance.

6-9 AUDIO-CIRCUIT SIGNAL-TRACING TESTS

An oscilloscope is the most logical instrument for tracing and testing signals throughout audio circuits, whether the circuits are complete audio-amplifier systems or a single stage. The oscilloscope duplicates every function of an electronic voltmeter in troubleshooting, signal tracing, and performance testing. In addition, the oscilloscope offers the advantage of a visual display for such common audio-circuit conditions as distortion, hum, noise, ripple, and oscillation.

An oscilloscope is used in a manner similar to that of an electronic voltmeter when signal-tracing audio circuits. A signal is introduced into the input by the signal generator. The amplitude and waveform of the input signal are measured on the oscilloscope. The oscilloscope probe is then moved to the input and output of each stage, in turn, until the final output is reached. The gain of each stage is measured as a voltage on the oscilloscope. In addition, it is possible to observe any change in waveform from that applied to the input. Thus, stage gain and distortion (if any) are established quickly with an oscilloscope.

6-10 CHECKING DISTORTION BY SINE-WAVE ANALYSIS TESTS

The connections for audio-circuit signal-tracing tests with sine waves are shown in Fig. 6-6. The procedure for checking audio-circuit distortion by means of sine waves is essentially the same as that described in Section 6-9. The primary concern, however, is *deviation* of the amplifier (or stage) output waveform from the input waveform. If there is no change (except in amplitude), there is no distortion. If there is a change in the waveform, the nature of the change often reveals the cause of distortion. For example, the presence of second or third harmonics distorts the fundamental signal.

In practice, analyzing sine waves to pinpoint audio-circuit problems that produce distortion is a difficult job, requiring considerable experience. Unless the distortion is severe, it may pass unnoticed. Sine waves are best used where *harmonic distortion* or *intermodulation distortion* meters are combined with oscilloscopes for distortion analysis. If an oscilloscope is to be used alone, *square waves* provide the best basis for distortion analysis. (The reverse is true for frequency-response and power measurements.)

6-11 CHECKING DISTORTION BY SQUARE-WAVE ANALYSIS TESTS

The procedure for checking distortion by means of square waves is essentially the same as for sine waves. Distortion analysis is more effective with square

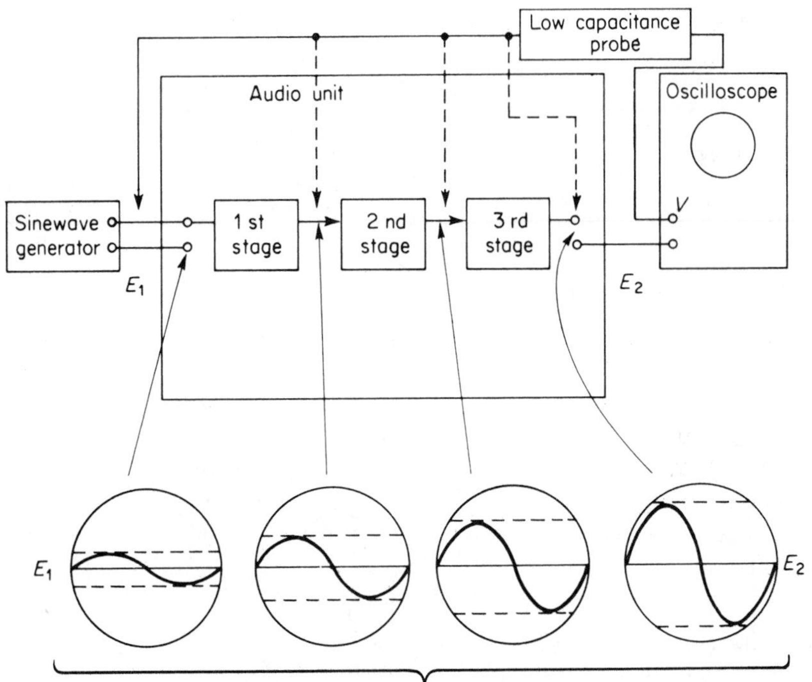

FIGURE 6-6 Basic audio signal tracing with an oscilloscope.

waves because of their high odd-harmonic content, and because it is easier to see a deviation from a straight line with sharp corners than from a curving line.

As in the case of sine-wave distortion testing, square waves are introduced into the circuit input, while the output is monitored on an oscilloscope, as shown in Fig. 6–7. The primary concern is deviation of the amplifier (or stage) output waveform from the input waveform (which is also monitored on the oscilloscope). If the oscilloscope has the dual-trace feature, the input and output can be monitored simultaneously. If there is a change in waveform, the nature of the change often reveals the cause of distortion. For example, a comparison of the square-wave response shown in Fig. 6–8 against the "typical" patterns of Fig. 6–7 indicates possible low-frequency phase shift. In any event, such a comparison indicates some problem in the low-frequency response.

The third, fifth, seventh, and ninth harmonics of a clean square wave are emphasized. Thus, if an audio circuit passes a signal of some given frequency, and produces a clean square-wave output, it is safe to assume that the

Oscilloscope
Squarewave generator
Amplifier In Out
V
V
Connections for dual–trace input

Reduced H–F response
Poor H–F response
H–F emphasis
H–F overshoot
Narrow band emphasis
Ringing (oscillation)
Poor overall response
L–F phase shift (leading)
L–F phase shift (lagging)
Narrow band attenuation
Reduced L–F gain
Emphasized L–F gain
Reduced L–F response
Poor L–F response

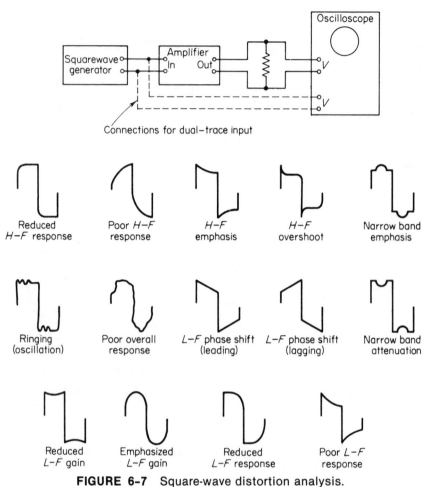

FIGURE 6-7 Square-wave distortion analysis.

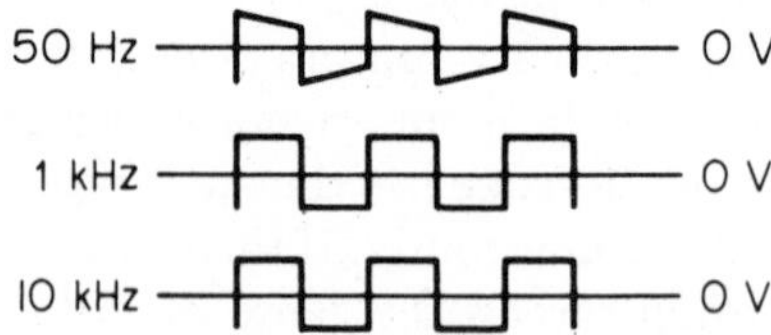

FIGURE 6-8 Square-wave response of audio amplifier.

frequency response is good up to at least nine times the fundamental frequency. For example, if the response at 10 kHz is as shown in Fig. 6–8, it is reasonable to assume that the response is the same at 90 to 100 kHz. This is convenient since not all audio generators provide an output up to 100 kHz (required by many audio-circuit design specifications).

6-12 HARMONIC DISTORTION TESTS

No matter what audio circuit is used or how well the circuit is designed, there is always the possibility of odd or even harmonics being present with the fundamental. These harmonics combine with the fundamental and produce distortion, as is the case when any two signals are combined. The effects of second and third harmonic distortion are shown in Fig. 6-9. As shown, when

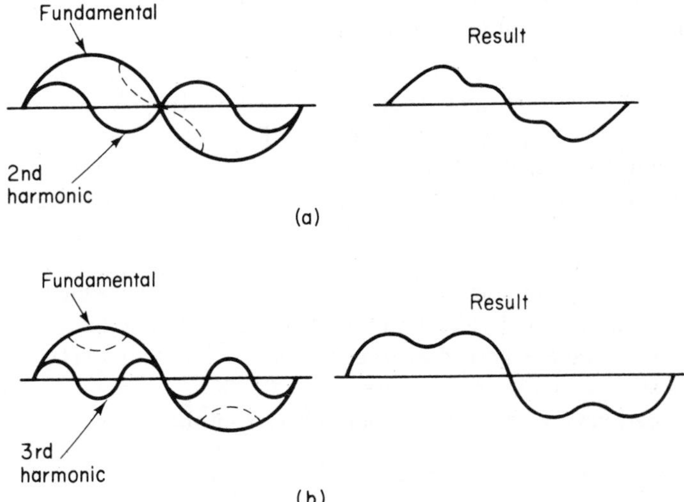

FIGURE 6-9 Effects of second and third harmonic distortion.

harmonic signals swing negative simultaneously with a positive swing of the fundamental, or vice versa, the fundamental signal is distorted by the combination.

Commercial harmonic distortion meters operate on the *fundamental suppression* principle. As shown in Fig. 6-10, a sine wave is applied to the circuit input, and the circuit output is measured on the oscilloscope. The output is then applied through a filter that suppresses the fundamental frequency. Any output from the filter is then the result of harmonics. This output can also be displayed on the oscilloscope. (Some commercial harmonic distortion meters use a built-in meter instead of, or in addition to, an external oscilloscope.) When the oscilloscope is used, the frequency of the filter output signal can be checked to determine harmonic content. For example, if the input is 1 kHz and the output (after filtering) is 3 kHz, third harmonic distortion is indicated.

The percentage of harmonic distortion can also be determined by this

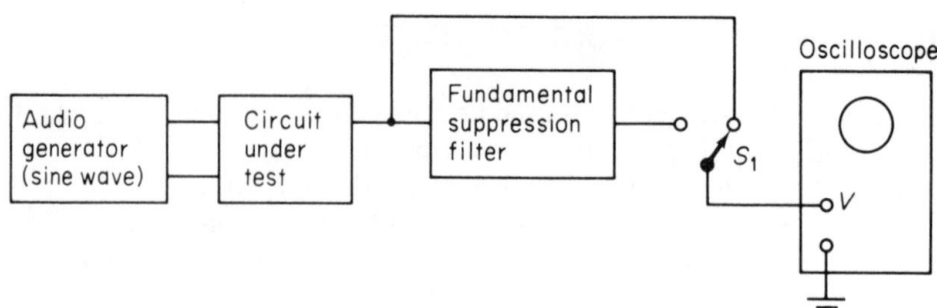

FIGURE 6-10 Basic harmonic distortion meter circuit using fundamental suppression principle.

method. For example, if the output without filter is 100 mV, and with filter is 3 mV, a 3% harmonic distortion is indicated.

In some commercial harmonic distortion meters, the filter is tunable so that the amplifier can be tested over a wide range of fundamental frequencies. In other harmonic distortion meters, the filter is fixed in frequency but can be detuned slightly to produce a sharp null.

When a design circuit is tested over a wide range of frequencies for harmonic distortion, and the results plotted on a graph similar to that of Fig. 6–11, the percentage is known as *total harmonic distortion* (THD). Note that the THD shown in Fig. 6–11 is less than 0.2%. Also note that harmonic distortion can vary with frequency and power output.

6-13 INTERMODULATION DISTORTION TESTS

When two signals of different frequency are mixed in any circuit, there is a possibility of the lower-frequency signal amplitude-modulating the higher-frequency signal. This produces a form of distortion known as *intermodulation distortion.*

Commercial intermodulation distortion meters consist of a signal generator and high-pass filter as shown in Fig. 6–12. The signal-generator portion of the meter produces a high-frequency signal (usually about 7 kHz) that is modulated by a low-frequency signal (usually 60 Hz). The mixed signals are applied to the circuit input. The circuit output is connected through a high-pass filter to the oscilloscope vertical channel. The high-pass filter removes the low-frequency (60-Hz) signal. Thus, the only signal appearing on the oscilloscope vertical channel should be the high-frequency (7-kHz) signal. If any 60-Hz signal is present on the display, it is being passed through as modulation on the 7-kHz signal.

Figure 6–13 shows an intermodulation test circuit that can be fabricated in the shop or laboratory. Note that the high-pass filter is designed to pass

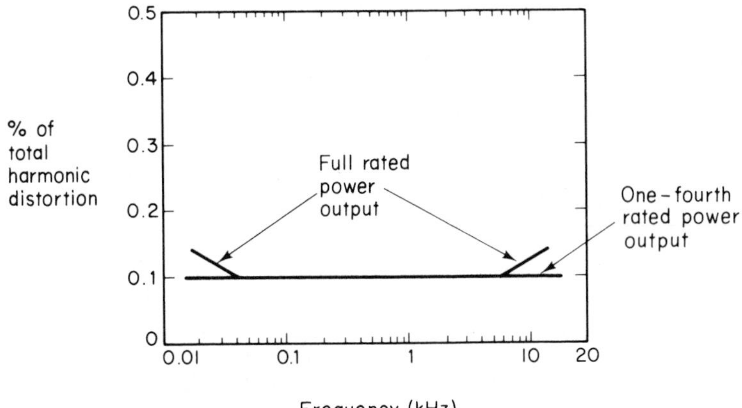

FIGURE 6-11 Typical total harmonic distortion (TMD) graph.

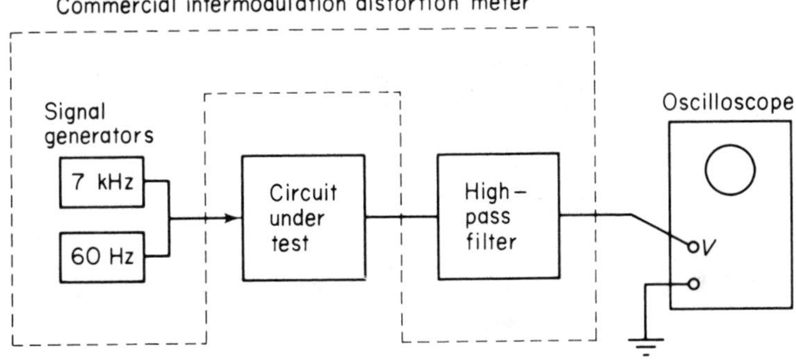

FIGURE 6-12 Basic intermodulation distortion meter circuit.

signals above 200 Hz. The purpose of the 39-kΩ and 10-kΩ resistors is to set the 60-Hz signals at four times the 7-kHz signal. Most audio generators provide for a line-frequency output (60 Hz) that can be used as the low-frequency modulation source.

If the laboratory circuit of Fig. 6–13 is used instead of a commercial meter, set the generator line-frequency output to some fixed value (1 V, 2 V, etc.). Then set the generator audio output (7 kHz) to the same value. If the line-frequency output is not adjustable, measure the actual value of the line-frequency output, and then set the generator audio output to the same value.

The percentage of intermodulation distortion can be calculated using the equation of Fig. 6–13.

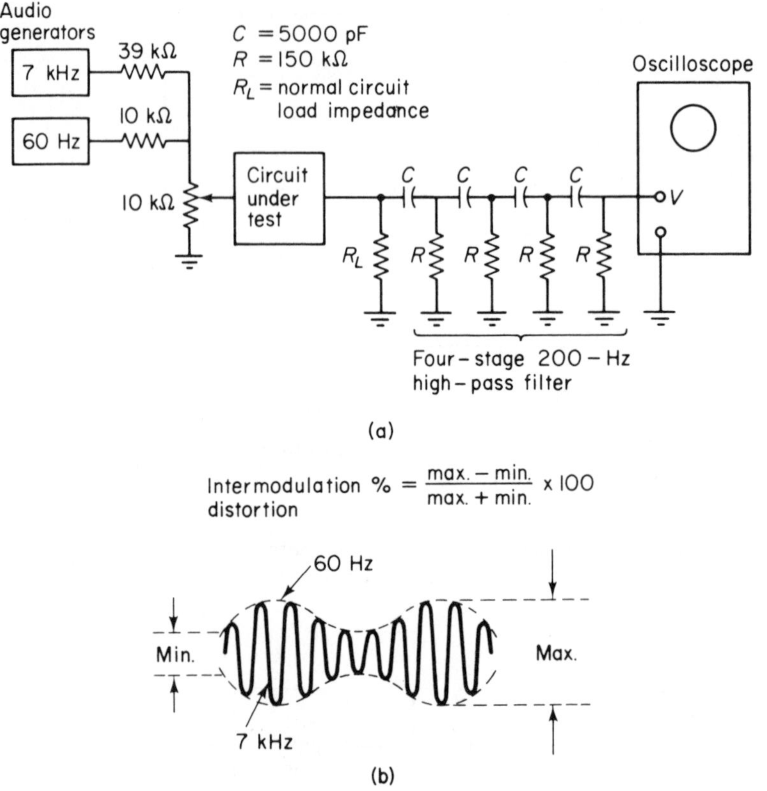

FIGURE 6-13 Test circuit for measurement of intermodulation distortion percentage.

6-14 BACKGROUND NOISE TESTS

If the vertical channel of an oscilloscope is sufficiently sensitive, an oscilloscope can be used to test and measure the background noise level of an audio circuit, as well as to test for the presence of hum, oscillation, and so on. The oscilloscope vertical channel should be capable of a measurable deflection with about 1 mV (or less), since this is the background noise level of many amplifiers.

The basic procedure consists of measuring circuit output with the volume or gain control (if any) at maximum, but without an input signal. The oscilloscope is superior to a voltmeter for noise-level measurement since the frequency and nature of the noise (or other signal) are displayed visually.

The basic connections for background-noise-level tests are shown in Fig. 6-14. The oscilloscope gain or sensitivity control is increased until there is a noise or "hash" indication. It is possible that a noise indication could be

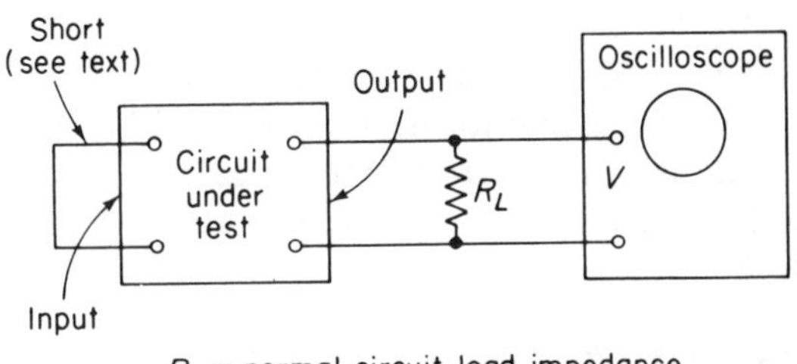

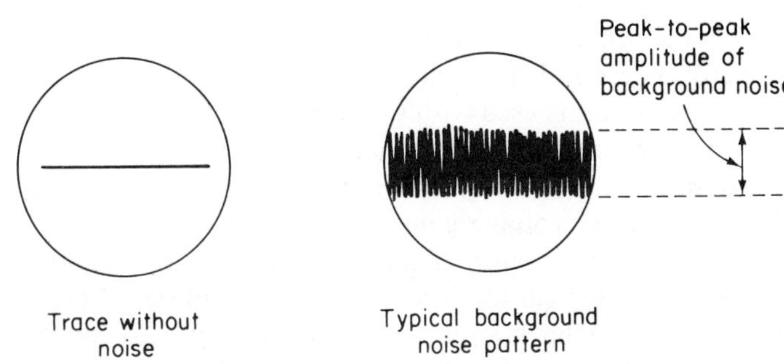

FIGURE 6-14 Measuring circuit background noise and hum.

caused by pickup in the test connections. If in doubt, disconnect the test leads from the circuit, but not from the oscilloscope. If the noise indications are removed when the test leads are disconnected, the noise is being produced in the circuit under test.

If you suspect that there is 60-Hz line hum present in the circuit output (picked up from the power supply or any other source), set the oscilloscope "sync" control to line. If a stationary pattern appears, it is due to line hum.

If a signal appears that is not at the line frequency, the signal can be due to oscillation in the circuit or to stray pickup. Short the circuit input terminals. If the signal remains on the oscilloscope display, it is probably oscillation of the circuit under test.

6-15 FEEDBACK AUDIO-AMPLIFIER TESTS

In all of the tests described thus far in this chapter, it is assumed that the audio circuit or amplifier has no feedback. Testing circuits without feedback is a relatively simple procedure. When the circuit has feedback, the task is more difficult. Problems such as the measurement of gain (voltage or power) and frequency response can be of particular concern.

For example, if you try opening the loop to make gain measurements, you usually find so much gain that the amplifier saturates and the measurements are meaningless. On the other hand, if you start making waveform measurements on a working closed-loop amplifier circuit, you often find that the input and output signals are normal (or near normal), although many of the waveforms are distorted inside the loop. For this reason, feedback loops, especially internal circuit feedback loops, require special attention during test.

6-15.1 Typical Feedback Audio-Amplifier Circuits

Figure 6-15 is the schematic of a basic feedback amplifier circuit. Note the various waveforms around the circuit. These waveforms are similar to those that appear if the circuit is used with sine waves. Note that there is an approximate 15% distortion inside the feedback loop (between Q_1 and Q_2) but only 0.5% distortion at the output. This is only slightly greater distortion than the input 0.3%. Open-loop voltage gain for this circuit is approximately 4300; closed-loop voltage gain is approximately 1000. The gain ratio (open-loop to closed-loop) of 4:1 is typical for feedback audio-amplifier circuits.

Transistors in feedback amplifier circuits behave just like transistors in any other circuits. That is, the transistors respond to all the same rules for gain and input/output impedance. Specifically, each transistor amplifies the signal appearing between emitter and base. It is here that the greatest dif-

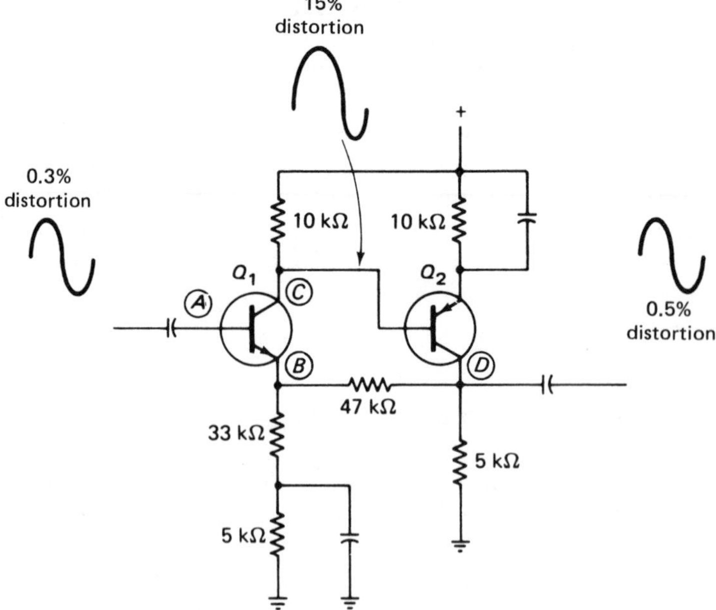

FIGURE 6-15 Basic feedback amplifier circuit.

ference between gain in feedback amplifier circuits and gain in nonfeedback (open-loop) circuits occurs.

Transistor Q_1 in Fig. 6–16 has a varying signal on both the emitter and the base, rather than on only one element. In a nonfeedback amplifier circuit, the signal usually varies at only one element, either the emitter or the base. Because most feedback circuits use negative feedback, the signals at both the base and the emitter are in-phase. The resultant gain is much less than when one of these elements is fixed (no feedback, open loop).

This accounts for the circuit's great gain increase when the loop is opened. Either the base or the emitter of the transistor stops changing (no signal), and a much larger effective input signal appears at the base–emitter control element. Assume that a perfect input signal is applied to the input (point A of Fig. 6–16). If the circuit is perfect (produces no distortion), the signal returning to B is also undistorted. Because the circuit uses negative feedback, the signal that travels around the loop a second time is undistorted as well. If the circuit is not perfect (assume an extreme case of clipping distortion), the returning signal will show that effect of distortion, as illustrated in Fig. 6–16.

To simplify the explanation, assume that clipping is introduced in Q_1 and that Q_2 is perfect. Now, the signals applied at the control point of Q_1 are not identical. The resultant applied signal at the control point of Q_1 is quite distorted. In effect, the distortion is a mirror image of the distortion introduced by Q_1. Transistor Q_1 then amplifies this distortion and its own counter-distortion. The result, after many trips around the feedback loop, is that there can be distortion inside the loop, but that this is counterbalanced by the feedback. The final output from Q_2 is undistorted, or relatively free of circuit-induced distortion. The higher the amplification and the greater the feedback,

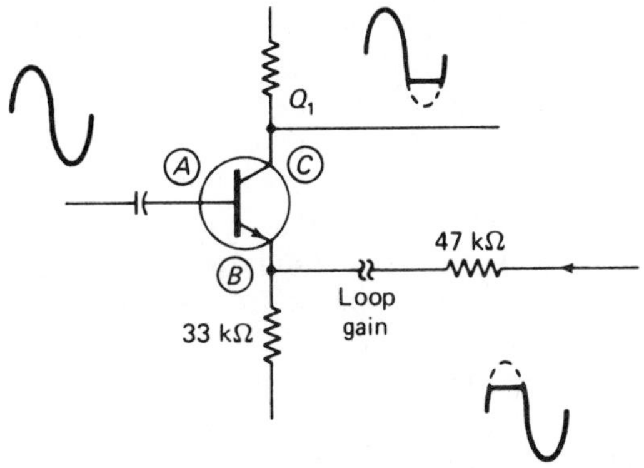

FIGURE 6–16 Amplifier-induced distortion in signal returning to point *B*.

the more effective this cancellation becomes, and the lower the output distortion becomes.

This last fact marks the basic difference in testing a feedback amplifier circuit. In any amplifier circuit, there are three basic causes of distortion: *overdriving,* operating the transistors at the *wrong bias point,* and the *inherent* nonlinearity of any solid-state device.

Overdriving can be the result of many causes (too much input signal, too much gain in the previous stage, etc.). The net result is that the output signal is clipped on one peak as a consequence of the transistor being driven into saturation, and on the other peak by driving the transistor below cutoff.

Operating at the *wrong bias point* can also produce clipping, but of only one peak. For example, if the input signal is 1 V and the transistor is biased as 1 V, the input swings from 0.5 to 1.5 V. Assume that the transistor saturates at any point where the base goes above 1.6 V and is cut off when the base goes below 0.5 V. No problem occurs when the bias is correct at 1 V. But now assume that the bias point is shifted (because of component aging, transistor leakage, etc.) to 1.3 V. When the 1-V input signal is applied, the base swings from 0.8 to 1.8 V, and the transistor saturates when one peak goes from 1.6 to 1.8 V. If, on the other hand, the bias point is shifted down to 0.7 V, the base swings from 0.2 V to 1.2 V, and the opposite peak is clipped as the transistor goes into cutoff.

Even if the transistor is not overdriven, it is still possible to operate a transistor on a nonlinear portion of its gain curve because of wrong bias. Some portion of the input/output gain curve of all transistors is more linear than other portions. That is, the output increases (or decreases) directly in proportion to input. An increase of 10% at the input produces an increase of 10% at the output. Ideally, transistors are operated at the center of this linear curve. If the bias point is changed, the transistor can operate on a portion of the curve that is less linear than the desired point.

The *inherent nonlinearity* of any solid-state device (diode, transistor, etc.) can produce distortion even if a stage is not overdriven and is properly biased. That is, the output does not increase (or decrease) directly in proportion to the input. For example, an increase of 10% at the input can produce an increase of 13% at the output. This is one of the main reasons for feedback in amplifier circuits where low distortion is required.

In summary, a negative feedback loop operates to minimize distortion, in addition to stabilizing gain. The feedback takeoff point has the least distortion of any point within the loop. From a practical testing standpoint, if the final output distortion and the overall gain are within limits, all circuit stages within the loop can be considered to be operating properly. Even if there is some abnormal gain in one or more of the stages, the overall feedback system has compensated for the problem. Of course, if the overall gain and/or distortion are not within limits, the individual stages must be tested.

6-15.2 *Feedback Amplifier Test Techniques*

Give special attention to the following notes when testing any feedback amplifier circuit.

Opening the Loop. Some test literature recommends that the loop be opened and the circuits tested under no-feedback conditions. In some cases, this can cause circuit damage. Even if there is no damage, the technique is rarely effective. Open-loop gain is usually so great that some circuit stage will block or distort badly. If the technique is used, as it must be for some circuits, keep in mind that the distortion is *increased* when the loop is opened. That is, a normally closed-loop circuit can show considerable distortion when operated as an open-loop device, even though the circuit is good.

Measuring Stage Gain. Care should be taken when measuring the gain of amplifier stages in a feedback amplifier. For example, in Fig. 6–15, if you measure the signal at the base of Q_1, the base-to-ground voltage is not the same as the input voltage. To get the correct value, connect the low side of the measuring device (a-c voltmeter or oscilloscope) to the emitter and the other lead (high side) to the base, as shown in Fig. 6–17. In effect, measure the signal that appears *across* the base–emitter junction. This measurement includes the effect of the feedback signal.

As a general safety precaution, never connect the ground lead of a voltmeter or oscilloscope to the base of a transistor unless the lead connects back to an isolated inner chassis on the meter or oscilloscope. The reason for this precaution is that large a-c ground-loop currents (between the measuring

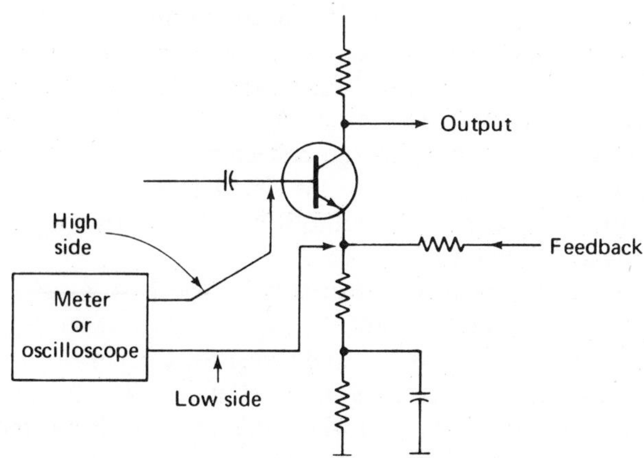

FIGURE 6–17 Measuring input-signal voltage or waveforms.

device and the equipment being tested) can flow through the base–emitter junction and possibly burn out the transistor.

Low-Gain Problems. As we have noted, low gain in a feedback amplifier can also result in distortion. That is, if gain is normal in a feedback amplifier, some distortion can be overcome. With low gain, the feedback may not be able to bring the distortion within limits. Keep in mind that most feedback amplifiers have a very high open-loop gain that is set to some specific value by the ratio of resistor values (feedback-resistor value to load-resistor value). If the closed-loop gain is low, it usually means that the open-loop gain has fallen far enough so that the resistors no longer determine the gain. For example, if the a-c beta (Chapter 1) of Q_2 shown in Fig. 6–15 is lowered, the open-loop gain is lowered. Also, the lower beta lowers the input impedance of Q_2 which, in turn, reduces the effective value of the load resistor for Q_1. This also has the effect of lower overall gain.

In testing such a situation, if measurements indicate low gain but transistor voltages are normal, the transistors are suspect. Do not overlook the possibility of open or badly leaking emitter-bypass capacitors. If the capacitors are open or leaking (acting as a resistance in parallel with the emitter resistor), there will be considerable negative feedback and little a-c gain. Of course, a completely shorted emitter-bypass capacitor produces an abnormal d-c voltage indication at the transistor emitter. Capacitor testing is discussed further in Section 6–16.1.

Distortion Problems. As we have discussed, distortion can be caused by improper bias, overdriving (too much gain), or underdriving (too little gain, preventing the feedback signal from countering the distortion). One problem often overlooked in a feedback amplifier with a pattern of distortion is overdriving that results from transistor leakage. (The problem of transistor leakage is discussed further in Section 6–16.2.)

Generally, it is assumed that the collector–base leakage reduces gain because the leakage is in opposition to the signal current flow. Although this is true in the case of a single-stage circuit, it may not be true when more than one feedback stage is involved.

Whenever there is collector–base leakage, the base assumes a voltage nearer to that of the collector (nearer than is the case without leakage). This increases both transistor forward bias and transistor current flow. An increase in the transistor current causes a lower input resistance, which may or may not cause a gain reduction (depending on where the transistor is located in the circuit).

If the feedback amplifier circuit is *direct-coupled,* the effects of feedback are increased. This is because the operating point (base bias) of the following stage is changed, possibly resulting in distortion. For example, in

Fig. 6-15, the collector of Q_1 is connected directly to the base of Q_2. If Q_1 starts to leak (or if the collector-base leakage increases with age), the base of Q_2 (as well as the collector of Q_1) shifts its operating point.

6-16 ANALYZING EXPERIMENTAL AND DESIGN PROBLEMS WITH TEST RESULTS

The following notes apply primarily to solving design problems (poor frequency response, lack of gain, etc.) in experimental audio circuits. However, many of the procedures can be applied to other circuits.

When experimental circuits fail to perform properly (or as hoped they would perform) a planned procedure for isolating the problem is very helpful. Keep in mind that experimental circuit test and troubleshooting is difficult at best. This is especially true when the circuit involves more than one stage, since the stages are interdependent. A special problem arises in analyzing the failure of experimental circuits during test. The first requirement in logical troubleshooting is a thorough knowledge of the circuit's performance when operating normally. However, a failure in a trial circuit just designed can be the result of component failure or improper experimental values for components. For example, an existing circuit may show low gain, based on past performance. A newly designed circuit may show the same results, simply because it is the best gain possible with the selected experimental components.

Start all analysis of experimental circuits using a basic troubleshooting procedure. Try to isolate trouble on a stage-by-stage basis. For example, if the circuit has two or more stages, and the gain is low for the overall circuit, measure the gain for each stage. With trouble isolated to a particular stage, try to determine which half of the stage is at fault. Any transistor stage has two halves, input and output. Generally, the input is base-emitter, with the emitter-collector acting as the output. Keep in mind that a defect in one half affects the other half. An obvious example of this is where low input current (base) produces low output current (collector). Sometimes less obvious is the case where output affects input. For example, in a circuit with an emitter resistor (for feedback stabilization) an open collector appears to reduce the input impedance.

Circuits with loop feedback present a particular problem during test, as discussed. A closed feedback loop causes all stages to respond as a unit, making it difficult to know which stage is at fault. This problem can be solved by opening the feedback loop. To do so, however, creates another problem, since the operating point of one (or possibly all) stages is disturbed.

Look for any transistor that is *full-on* (collector voltage very low in comparison to the supply voltage), or *full-off* (collector voltage very high, probably at or near the supply voltage). Either of these conditions in a linear amplifier (audio or otherwise) is the result of component failure or improper design. Design faults that appear under no-signal conditions (such as improper operating point) are generally the result of improper bias relationships. Faults that appear only when a signal is applied can also be caused by poor bias relationships, but can also result from wrong component values.

A high-gain circuit is generally more difficult to bias than is a low-gain circuit. Use the following procedure when it is difficult to find a good bias point (or operating point) for a high-gain amplifier. Increase the input signal until the output waveform appears as a square wave (that is, overdrive the circuit as shown in Fig. 6–18). Then keep reducing the input signal, while adjusting the bias, until both positive and negative peaks are *clipped by the same amount*. If it is impossible to find any bias point where the signal peaks can be clipped symmetrically, a defect in design or components can be suspected. Unsymmetrical clipping can be caused by operating the transistor on a nonlinear portion of its curve, or by the fact that the transistor does not have a linear curve (or a very short linear curve). Try a different transistor. If the results are the same, change the circuit trial values (collector, emitter, and base resistances).

If it is impossible to obtain any operating point, except very near full-on or full-off, this can be caused by excessive *positive feedback*. While this condi-

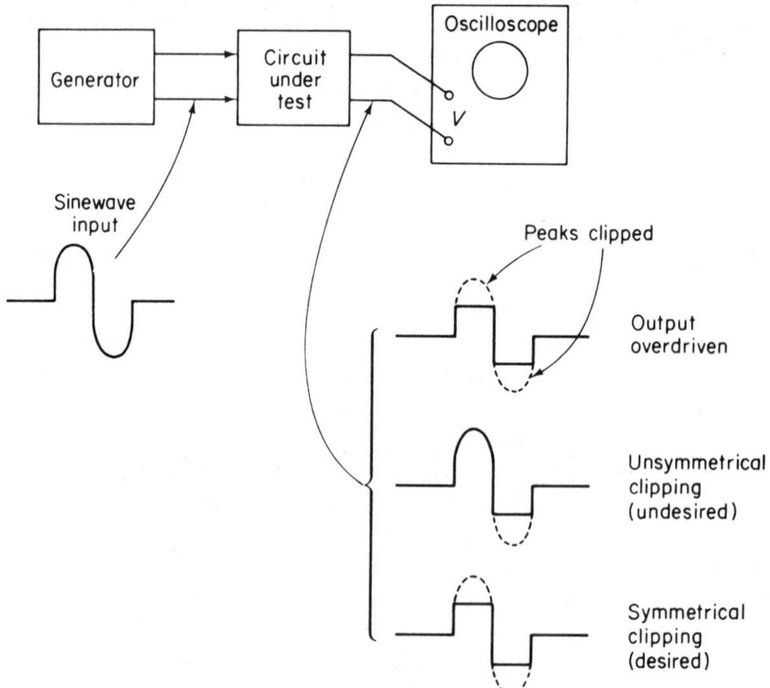

FIGURE 6–18 Determining proper bias (*Q* point) by means of oscilloscope displays.

tion is desirable in a multivibrator or oscillator, it must be avoided in the design of linear amplifiers. Positive feedback can cause a linear amplifier to act like a flip-flop or oscillator.

High-gain amplifier circuits should be carefully checked to find any tendency to oscillate or to exhibit abnormal noise (Section 6–14). High-gain circuits may also be very sensitive to supply-voltage changes. When a design circuit has been completed, it is often helpful to repeat all the basic test procedures with various supply voltages. Unless otherwise specified by design requirements, any circuit should perform equally well with a ±10% supply-voltage variation.

If oscillation occurs in any amplifier circuit (high or low gain) during test, try moving the input and output leads. An experimental circuit may oscillate because input and output leads are close together. It may be necessary to physically relocate parts, or to shield parts, of a circuit to prevent feedback that results in oscillation. Low-frequency oscillation is often the result of poor supply-voltage filter, or too many stages connected to the same supply-voltage point. Try isolating the stages with separate supply-voltage-filter capacitors.

The effects of capacitance are the most common causes of poor frequency response in experimental circuits. The most common cause of *poor low-frequency response* is low capacitor values. The effects of capacitors in circuits during test are described further in Section 6–16.1. The most common cause of *poor high-frequency response* is the input capacitance of transistors. As frequency increases, transistor input capacitance decreases, changing the input impedance. This change in input impedance usually results in decreased gain, all other factors remaining equal. Generally, poor high-frequency response is not a problem over the audio range (up to about 20 kHz), but can be a problem beyond about 100 kHz. The only practical solutions are to reduce stage gain, or change transistors.

6–16.1 Effects of Capacitors during a Circuit Test

During the testing process of an experimental circuit, suspected capacitors can be removed from the circuit and tested on bridge-type capacitor checkers. This will establish that the capacitor value is correct. If the checker shows the value to be correct, it is reasonable to assume that the capacitor is not open, shorted, or leaking.

From another standpoint, if a capacitor shows no shorts, opens, or leakage, it is reasonable to assume that the capacitor is good. Thus, for practical purposes, a simple test procedure that shows the possibility of shorts, opens, or leakage is usually sufficient.

There are two basic methods for a quick test of capacitors. One method involves using the circuit voltages. The other technique requires an ohmmeter.

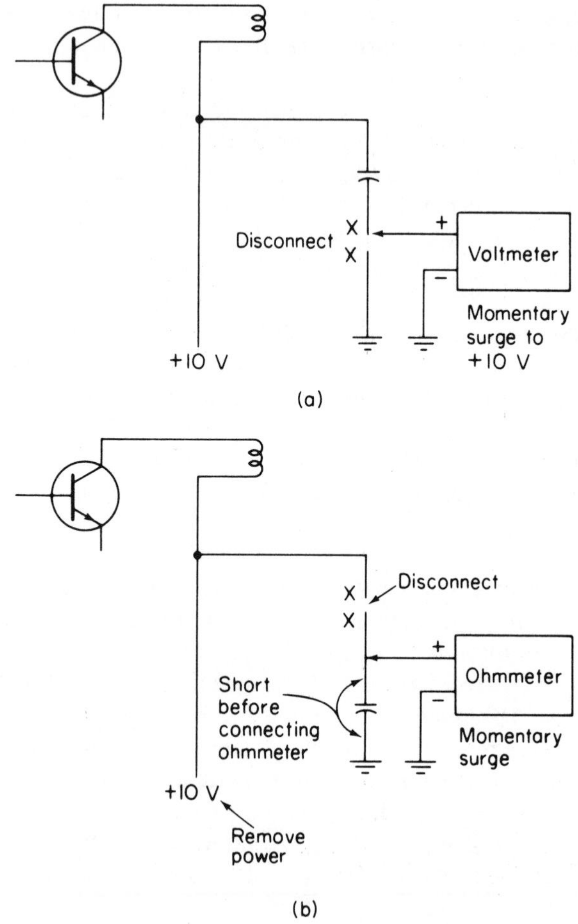

FIGURE 6-19 Testing capacitors with circuit voltages (power applied) and with ohmmeter (power removed).

Testing Capacitors with Circuit Voltages. As shown in Fig. 6–19(a), this method involves disconnecting one lead of the capacitor (the ground or cold lead) and connecting a voltmeter between the disconnected lead and ground. In a good capacitor, there should be a momentary voltage indication (or surge) as the capacitor charges up to the voltage at the hot end.

If the voltage indication remains high, the capacitor is probably shorted.

If the voltage indication is steady but not necessarily high, the capacitor is probably leaking.

If there is no voltage indication whatsoever, the capacitor is probably open.

Testing Capacitors with an Ohmmeter. As shown in Fig. 6–19(b), this method involves disconnecting one lead of the capacitor (usually the hot end) and connecting an ohmmeter across the capacitor. Make certain that all power is removed from the circuit. As a precaution, after the power is removed, short across the capacitor to make sure that no charge is retained.

In a good capacitor, there should be a momentary resistance indication (or surge) as the capacitor charges up to the voltage of the ohmmeter battery.

If the resistance indication is near zero and remains so, the capacitor is probably shorted.

If the resistance indication is steady at some high value, the capacitor is probably leaking.

If there is no resistance indication whatsoever, the capacitor is probably open.

Functions of Capacitors in Solid-State Circuits. The most common functions of capacitors in solid-state circuits are emitter bypass, coupling, and decoupling (or bypass).

An *emitter-bypass* capacitor permits high gain, but with stability, in a circuit. The emitter resistor in a circuit (such as R_4 in Fig. 6–20) is used to stabilize the transistor d-c gain and prevent thermal runaway. With an emitter resistor in the circuit, any increase in collector current produces a greater drop in voltage across the resistor. When all other factors remain the same, the change in emitter voltage reduces the base–emitter forward-bias differential, thus tending to reduce collector current flow. When circuit stability is more important than gain, the emitter resistor is not bypassed. When signal gain must be high, the emitter resistance is bypassed with a capacitor to permit passage of the signal. If the emitter-bypass capacitor is open, stage gain is reduced drastically, although the transistor d-c voltages remain substantially the same.

If there is a low-gain symptom in any circuit with an emitter bypass and the voltages appear normal, check the bypass capacitor with an ohmmeter or voltmeter, as described. As an alternative, shunt the emitter-bypass capacitor with a known good capacitor of the same value, and retest the circuit. As a precaution, shut off the power before connecting the shunt capacitor. Then reapply power and repeat the test procedure. (Shutting off the power prevents damage to the transistor caused by large current surges.)

The function of *coupling capacitor* C_1 in Fig. 6–20 is to pass signals from the previous stage to the base of Q_1. If C_1 is shorted or leaking badly, the voltage from the previous stage is applied to Q_1. This forward-biases Q_1, causing heavy current flow and possible burnout of the transistor. In any event, Q_1 is driven into saturation, and stage gain is reduced.

If C_1 is open, there is little or no change in the voltage at Q_1, but the test signal from the previous stage does not appear at the base of Q_1. From a test

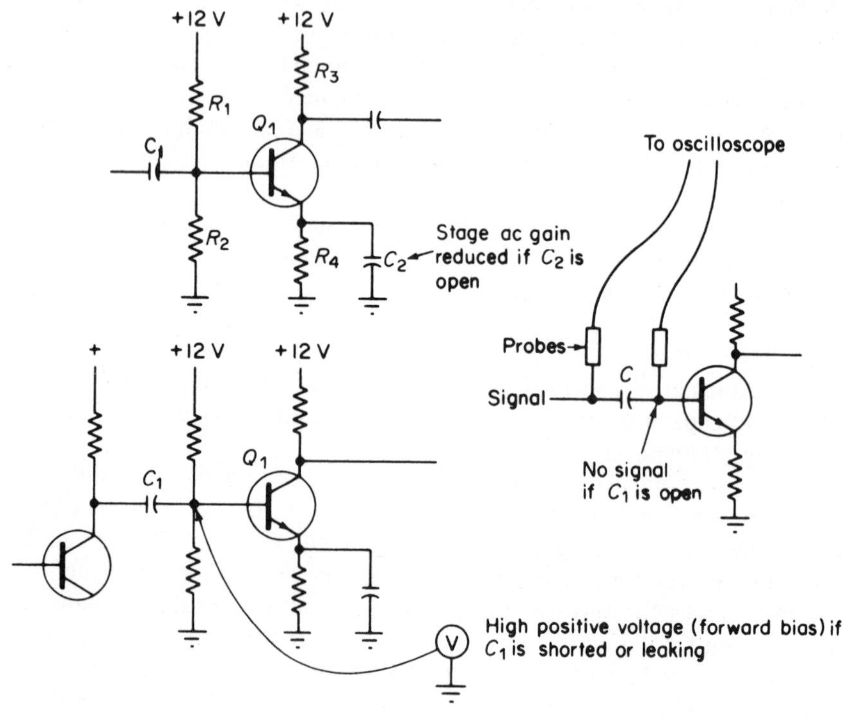

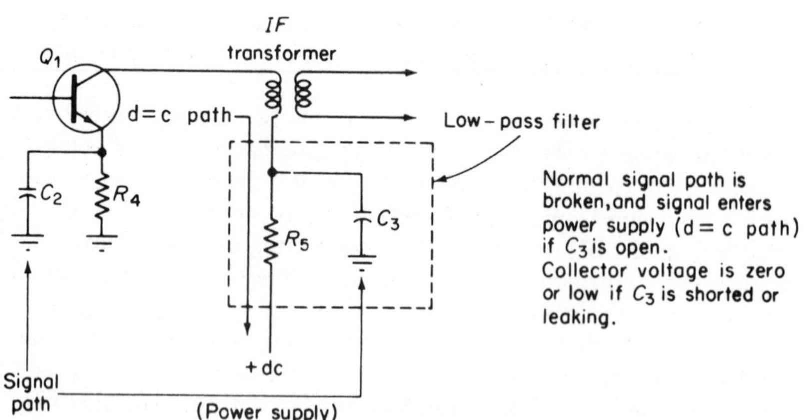

FIGURE 6-20 Effects of capacitor failure in solid-state circuits.

standpoint, a shorted or leaking C_1 will probably show up as distortion of the signal waveform. If C_1 is suspected of being shorted or leaky, replace C_1. An open C_1 will show up as a lack of test signal at the base of Q_1, with a normal signal at the previous stage. If an open C_1 is suspected, replace C_1 or try shunting with a known-good capacitor, whichever is convenient.

The function of the *decoupling* or *bypass capacitor* in Fig. 6–20 is to pass operating-signal frequencies to ground (to provide a return path) and to prevent signals from entering the power-supply line or other circuits connected to the line. In effect, C_3 and C_5 form a low-pass filter that passes direct-current and very low frequency signals (well below the operating frequency of the circuit) through the power-supply line. Higher-frequency signals are passed to ground and do not enter the power-supply line.

If C_3 is shorted or leaking badly, the power-supply voltage will be shorted to ground or greatly reduced. This reduction of collector voltage will make the stage totally inoperative or will reduce the output, depending on the amount of leakage in C_3.

If C_3 is open, there will be little or no change in the voltages at Q_1. However, the signals will appear in the power-supply line. Also, signal gain will be reduced, and the signal waveform will be distorted. In some cases, at higher signal frequencies, the signal simply cannot pass through the power-supply circuits. Because there is no path through an open C_3, the signal will not appear on the collector circuit. From a practical test standpoint, the results of an open C_3 depend on the values of R_5 (and the power-supply components) as well as on the signal frequency involved.

6–16.2 Effects of Transistor Leakage during a Circuit Test

When there is considerable leakage in a solid-state amplifier, the gain is reduced to zero, and/or the signal waveform is drastically distorted. Such a condition also produces abnormal waveforms and transistor voltages. These indications make the trouble easy to locate during test. The analysis problem becomes really difficult when there is just enough leakage to reduce amplifier gain but not enough leakage to distort the waveform seriously or to produce transistor voltages that are way off.

Collector–base leakage is the most common form of transistor leakage and produces a classic condition of low gain (in a single stage). When there is any collector–base leakage, the transistor is forward-biased, or the forward bias is increased, as shown in Fig. 6–21.

Collector–base leakage has the same effect as a resistance between the collector and base. The base assumes the same polarity as the collector (although at a lower value), and the transistor is forward-biased. If leakage is sufficient, the forward bias can be enough to drive the transistor into or near

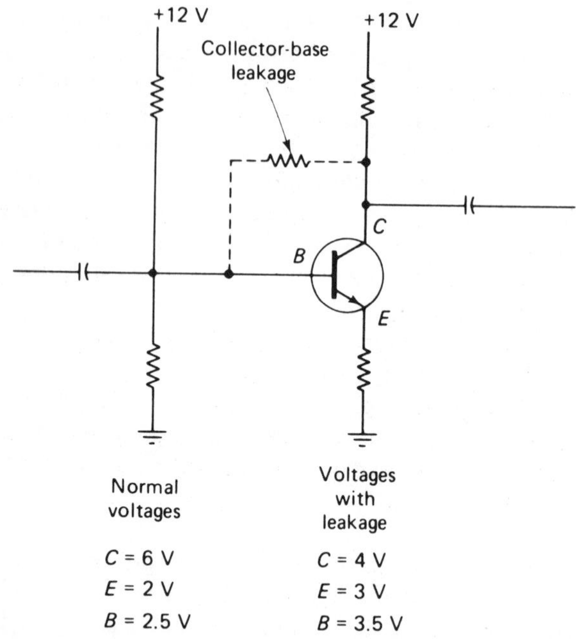

FIGURE 6-21 Effect of collector-base leakage on transistor element voltage.

saturation. When a transistor is operated at or near the saturation point, the gain is reduced (for a single stage), as shown in Fig. 6-22.

If the normal transistor-element voltages are known, excessive transistor leakage can be spotted easily because all the transistor voltages will be off. For example, in Fig. 6-21, the base and emitter will be high, and the collector will be low (when measured in reference to ground).

If the normal operating voltages are not known, the transistor can appear to be good because all the voltage relationships are normal. That is, the collector-base junction is reverse-biased (collector more positive than base for an NPN), and the emitter-base junction is forward-biased (emitter less positive than base for an NPN).

A simple way to test transistor leakage is shown in Fig. 6-23. Measure the collector voltage to ground. Then short the base to the emitter, and remeasure the collector voltage. If the transistor is not leaking, the base-emitter short will turn the transistor off, and the collector voltage will rise to the same value as the supply. If there is any leakage, a current path will remain (through the emitter resistor, emitter-base short, collector-base leakage path, and collector resistor). There will be some voltage drop across the collector resistor, and the collector will have a voltage at some value lower than the supply.

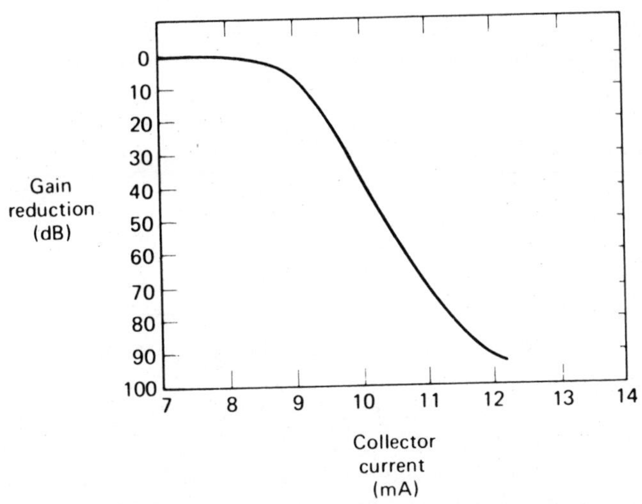

FIGURE 6-22 Relative gain of solid-state amplifier at various average collector-current levels.

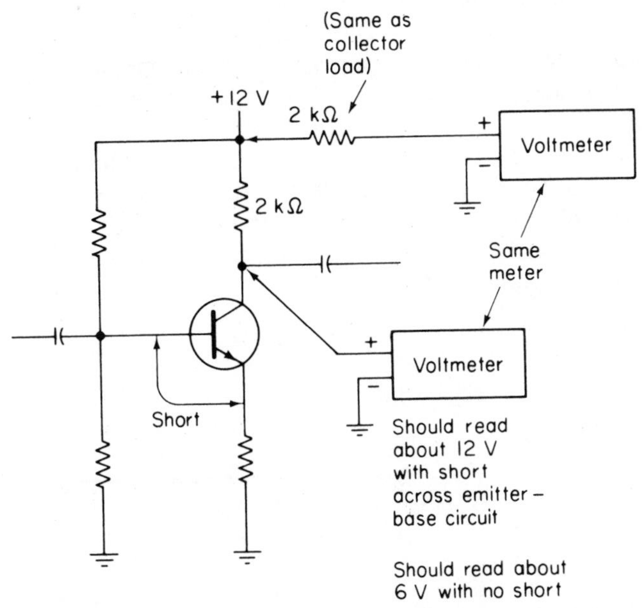

FIGURE 6-23 Checking for transistor leakage in amplifier circuits.

Note that most meters draw current, and this current passes through the collector resistor. This can lead to some confusion during test, particularly if the meter draws heavy current (has a low ohms-per-volt rating). To eliminate any doubt, connect the meter to the supply through a resistor with the same value as the collector resistor. The drop, if any, should be the same as it is when the transistor is measured to ground. If the drop is much different (lower) than when the collector is measured, the transistor is leaking.

For example, assume that in the circuit of Fig. 6–23 the supply is 12 V, the collector resistance is 2 kΩ, and the collector measures 4 V with respect to ground. This means that there is an 8-V drop across the collector resistor and a collector current of 4 mA (8/2000 = 4 mA). Normally, the collector is operated at about one-half the supply voltage (in this case, 6 V). However, simply because the collector is at 4 V instead of 6 V does not make the circuit faulty. Some circuits are designed that way.

In any event, the transistor should be checked for leakage with the emitter–base short test shown in Fig. 6–23. Now, assume that the collector voltage rises to 10.5 V when the base and emitter are shorted. This indicates that the transistor is cutting off but that there is still some current flow through the collector resistor, about 1 mA (2/2000 = 1 mA).

A current flow of 1 mA is high for a meter. However, to confirm a leaking transistor, connect the same meter through a 2-kΩ resistor (same as the collector load) to the 12-V supply (preferably at the same point where the collector resistor connects the power supply). Now, assume that the indication is 11.7 V through the external resistor. This indicates that there is some transistor leakage.

The amount of transistor leakage can be estimated as follows: 11.7 − 10.5 = 1.2-V drop and 1/2/2000 = 0.6 mA. However, from a practical test standpoint, the presence of any current flow with the transistor supposedly cut off is sufficient cause to replace the transistor.

7

Power-Supply-Circuit Tests

This chapter is devoted entirely to test procedures for power-supply circuits. The basic function of a power supply is to convert alternating current into direct current. In the case of a dc–dc converter, direct current is converted to direct current at a different voltage (usually higher). Either way, the power-supply function can be checked quite simply by measuring the output voltage. However, for a thorough test of a power supply, the output voltage should be measured without a load, with a full load, and possibly with a partial load.

If a power supply delivers the full-rated output voltage into a full-rated load, the basic power-supply function is met. In addition, it is often helpful to measure the regulating effect of a power supply, the power-supply internal resistance, and the amplitude of any ripple at the power-supply output. Also, since operation of most power-supply circuits is dependent upon characteristics of a transformer, it is convenient (particularly in the experimental stage) to test the transformer as a separate component.

7-1 POWER-SUPPLY-OUTPUT TESTS

Figure 7–1 is the schematic of a basic power-supply test circuit. This arrangement permits the power supply to be tested at no-load, half-load, and full-load, depending upon the position of load selector switch S_1. With S_1 in position 1, there is no load on the power supply. At positions 2 and 3, there is half-load and full-load, respectively. The load resistors R_1 and R_2 should be noninductive so that a reactance is not placed on the power-supply output. (Generally, the output load is considered as pure resistance for test purposes.)

As shown by the equations in Fig. 7–1, the values of R_1 and R_2 are chosen on the basis of output voltage and load current (maximum or half-load). For example, if the power supply is designed for an output of 25 V at 500-mA full-load, the value of R_2 should be 25/0.5, or 50 Ω. The value of R_1 should be 25/0.25, or 100 Ω. For practical purposes, where more than one power supply is to be tested, R_1 and R_2 should be variable, and adjusted to the correct resistance value before test (using an ohmmeter with power removed.). If only one supply is to be tested, or as a temporary test setup, fixed resistors can be used. Either way, the resistors should be composition (not wire-wound), and must be capable of dissipating the rated power without overheating. For example, using the previous values for R_1 and R_2, the power dissipation of R_1 is 25 × 0.5, or 12.5 W (use at least 15 W), and the dissipation for R_2 is 25 × 0.25, or 6.25 W (use at least 10 W).

To measure power-supply output, use the following procedure:

1. Connect the equipment as shown in Fig. 7–1.

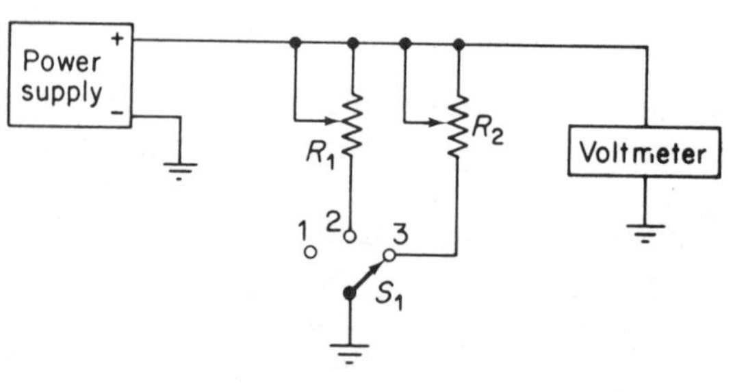

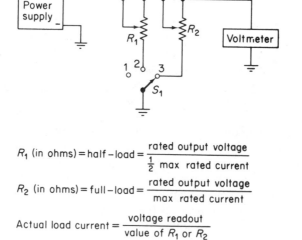

$$R_1 \text{ (in ohms)} = \text{half-load} = \frac{\text{rated output voltage}}{\frac{1}{2} \text{ max rated current}}$$

$$R_2 \text{ (in ohms)} = \text{full-load} = \frac{\text{rated output voltage}}{\text{max rated current}}$$

$$\text{Actual load current} = \frac{\text{voltage readout}}{\text{value of } R_1 \text{ or } R_2}$$

$$\text{\% regulation} = \frac{\text{no-load voltage} - \text{full-load voltage}}{\text{full-load voltage}} \times 100$$

$$\text{Power supply internal resistance} = \frac{\text{no-load voltage} - \text{full-load voltage}}{\text{current (amperes)}}$$

FIGURE 7–1 Basic power-supply test circuit.

2. If R_1 and R_2 are adjustable, set them to the correct values, using the equations.
3. Apply power. Measure the output voltage at each position of S_1.
4. Calculate the current at positions 2 and 3 of S_1, using the equation for actual load current. For example, assume that R_1 is 100 Ω, and the meter indicates 22 V at position 2 of S_1. Then, actual load current is 22/100, or 220 mA. If the power-supply output is 25 V at position 1, and dropped to 22 V at position 2, this means that the power supply does not produce full output with a load. This is an indication of poor regulation (possibly resulting from poor design).

7-2 POWER-SUPPLY-REGULATION TESTS

Power-supply regulation is usually expressed as a percentage and is determined with the following equation:

$$\% \text{ regulation} = \frac{(\text{no-load voltage}) - (\text{full-load voltage})}{\text{full-load voltage}} \times 100$$

A low percentage-of-regulation figure is desired since it indicates that the output voltage changes very little with load. The following steps are used when measuring power-supply regulation. The same circuit can be used for power supplies with or without a regulation circuit.

1. Connect the equipment as shown in Fig. 7-1.
2. If R_2 is adjustable, set it to the correct value, using the equation.
3. Apply power. Measure the output voltage at position 1 (no-load) and position 2 (full-load).
4. Using the equation, calculate the percentage of regulation. For example, if the no-load voltage is 25 V and the full-load voltage is 20 V, the percentage of regulation is (25–20)/20, or 25% (very poor regulation).
5. Note that power-supply regulation is usually poor (high percentage) when the internal resistance is high.

7-3 POWER-SUPPLY-INTERNAL-RESISTANCE TEST

Power-supply internal resistance is determined with the following equation:

$$\text{internal resistance} = \frac{(\text{no-load voltage}) - (\text{full-load voltage})}{\text{current}} \times 100$$

A low internal resistance is most desirable since it indicates that the output voltage changes very little with load. To measure power-supply internal resistance, use the following procedure:

1. Connect the equipment as shown in Fig. 7-1.
2. If R_2 is adjustable, set it to the correct value, using the equation.
3. Apply power. Measure the output voltage at position 1 (no-load) and position 3 (full-load).
4. Calculate the actual load current at position 3 (full-load). For example, if R_2 is adjusted to 50 Ω and the output voltage at position 3 is 20 V (as in the previous example), the actual load current is 20/50, or 400 mA.
5. With the no-load voltage, full-load voltage, and actual current established, find the internal resistance using the equation given. For example, with no-load voltage of 25 V, a full-load voltage of 20 V, and a current of 400 mA (0.4 A), the internal resistance is (25–20)/0.4, or 12.5 Ω.

7-4 POWER-SUPPLY-RIPPLE TESTS

Any power supply, no matter how well regulated or filtered, has some ripple. This ripple can be measured with a meter or oscilloscope. Usually, the factor of most concern is the ratio between ripple and full d-c output voltage. For example, if 3 V of ripple is measured, together with a 100-V d-c output, the ratio is 3:100 (or 3% ripple).

7-4.1 Measuring Power-Supply Ripple with a Meter

1. Connect the equipment as shown in Fig. 7-1.
2. If R_2 is adjustable, set it to the correct value, using the equation. Ripple is usually measured under full-load power.
3. Apply power. Measure the d-c output voltage at position 3 (full-load).
4. Set the meter to measure alternating current. Any voltage measured under these conditions is a-c ripple.
5. Find the percentage of ripple, as a ratio between the two voltages (ac/dc).
6. One problem often overlooked in measuring ripple is that any ripple voltage (single-phase, three-phase, full-wave, half-wave, etc.) is not a pure sine wave. Most meters provide accurate a-c voltage indications

only for pure sine waves. A more satisfactory method of measuring ripple is with an oscilloscope, where the peak value can be measured directly.

7-4.2 Measuring Power-Supply Ripple with an Oscilloscope

An oscilloscope will display the ripple waveform, from which the amplitude, frequency, and nature of the ripple voltage can be determined. Usually, the oscilloscope is set to the a-c mode when measuring ripple, since this blocks the d-c output of the power supply. Normally, ripple voltage is small in relation to the power-supply voltage. Thus, if the oscilloscope gain is set to display the ripple, the power-supply d-c voltage will drive the display off the screen.

1. Connect the equipment as shown in Fig. 7-2.
2. Apply power. If the test is to be made under load conditions, close

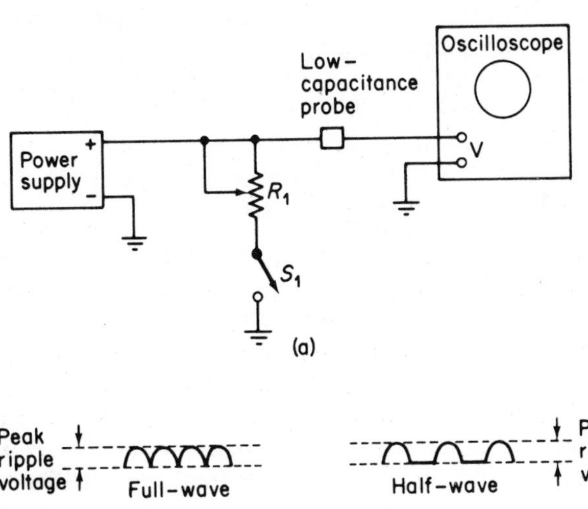

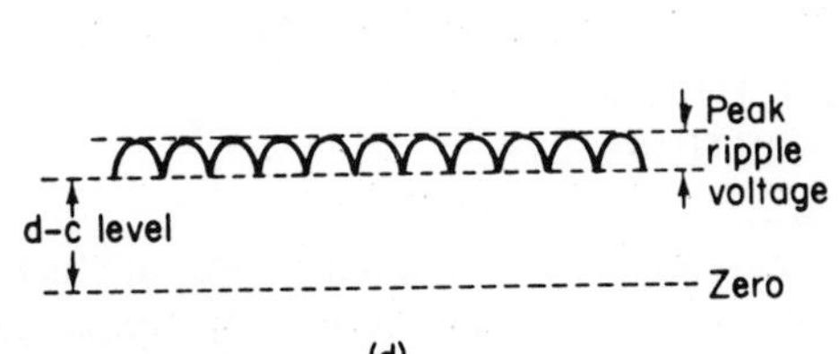

FIGURE 7-2 Measuring power-supply ripple with an oscilloscope.

switch S_1. Open switch S_1 for a no-load test. The value of R_1 must be chosen to provide the power supply with the desired load.

3. Adjust the oscilloscope sweep frequency and "sync" controls to produce two or three stationary cycles of each wave on the screen.
4. Measure the peak amplitude of the ripple on the voltage-calibrated vertical scale of the oscilloscope.
5. Measure the frequency of the ripple on the horizontal oscilloscope scale. When measuring ripple frequency, note that a full-wave produces two ripple "humps" per cycle [as shown in Fig. 7–2(b)], whereas a half-wave power supply produces one "hump" per cycle [Fig. 7–2(c)].
6. If the power-supply d-c output must be measured simultaneously with the ripple, set the oscilloscope for d-c mode. The baseline should be deflected upward to the d-c level, and the ripple should be displayed at that level [Fig. 7–2(d)]. If this drives the display off the screen, reduce the vertical gain and measure the d-c output. Then return the vertical gain to a level where the ripple can be measured, and use the vertical-position control to bring the display back onto the screen for ripple measurement. This procedure will work for most laboratory oscilloscopes. On some shop-type oscilloscopes where the vertical gain control must be set at a given "calibrate" position, the procedure is difficult, if not impossible. If this is so, it is easier to measure the d-c voltage with a meter.
7. A study of the ripple waveform can reveal defects in the power supply (due to poor design or defective components). For example, if the power supply is unbalanced (one rectifier passing more current than the others), the ripple "humps" are unequal in amplitude. If there is noise or fluctuation in the power supply (particularly in zener diodes, if used), the ripple "humps" will vary in amplitude or shape. If the ripple varies in frequency, the a-c source is varying. If a full-wave power supply produces a half-wave output, one rectifier is not passing current.

7–5 MEASURING TRANSFORMER CHARACTERISTICS

If a power-supply regulator design does not prove satisfactory, the fault may be with the transformer. Although it is usually simple to substitute other power-supply components (rectifiers, capacitors, etc.) during an experimental or design stage, it is more practical to test a transformer (unless another transformer is readily available for substitution).

The obvious test is to measure the transformer windings for opens, shorts, and the proper resistance value with an ohmmeter. In addition to basic resistance checks, it is possible to test a transformer's proper polarity markings, regulation, impedance ratio, and center-tap balance with a voltmeter.

7-5.1 Measuring Transformer Phase Relationships

When two supposedly identical transformers must be operated in parallel, and the transformers are not marked as to phase or polarity, the phase relationship of the transformers can be checked using a voltmeter and power source. The test circuit is shown in Fig. 7-3. For power transformers, the source should be line voltage (115 V). Other transformers can be tested with a lower voltage dropped from a line source or (in extreme cases) from an audio generator.

The transformers are connected in proper phase relationship if the meter reading is zero or very low. The transformers are out of phase if the secondary output voltage is double that of the normal secondary output. This condition can be corrected by reversing either the primary or secondary leads (but not both) of one transformer (but not both transformers).

If the meter indicates some secondary voltage, it is possible that one transformer has a greater output than the other. This condition results in considerable local current in the secondary winding and produces a power loss (if not actual damage to the transformer).

7-5.2 Checking Transformer Polarity Markings

Many transformers are marked as to polarity or phase. These markings may consist of dots, color-coded wires, or a similar system. Unfortunately, transformer polarity markings are not always standard. This can prove very confusing during the experimental stage.

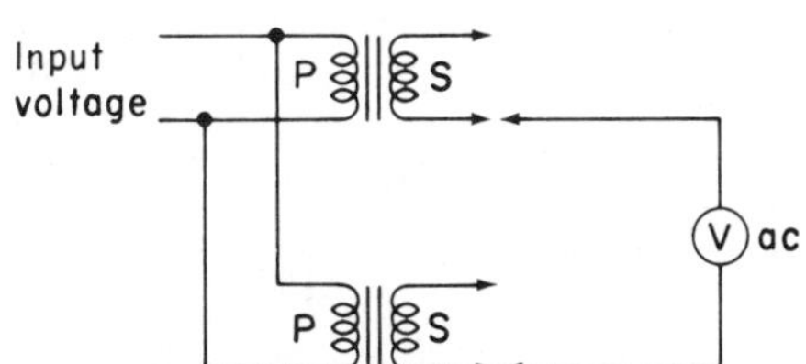

FIGURE 7-3 Measuring transformer phase relationships.

Generally, transformer polarities are indicated on schematics as dots next to the terminals. When standard markings are used, the dots mean that if electrons are flowing into the primary terminal with the dot, the electrons will flow out of the secondary terminal with the dot. Therefore, the dots have the same polarity as far as the external circuits are concerned. No matter what system is used, the dots or other markings show *relative phase,* since instantaneous polarities are changing across the transformer windings.

From a practical test standpoint, there are only two problems of concern: the relationship of the primary to the secondary, and the relationship of markings on one transformer to those on another.

The phase relationship of primary to secondary can be found using the test circuit of Fig. 7–4. First check the voltage across terminals 1 and 3, then across 1 and 4 (or 1 and 2). Assume that there is 3 V across the primary, with 7 V across the secondary. If the windings are as shown in Fig. 7–4(a), the 3 V is added to the 7 V and appears as 10 V across terminals 1 and 3. If the windings are as shown in Fig. 7–4(b), the voltages oppose each other, and appear as 4 V (7–3) across terminals 1 and 3.

The phase relationship of one transformer marking to another can be found using the test circuit of Fig. 7–5. Assume that there is a 3–V output from the secondary of transformer *A,* and a 7–V output from transformer *B.* If the markings are consistent on both transformers, the two voltages oppose, and 4 V is indicated. If the markings are not consistent, the two voltages add, resulting in a 10–V reading.

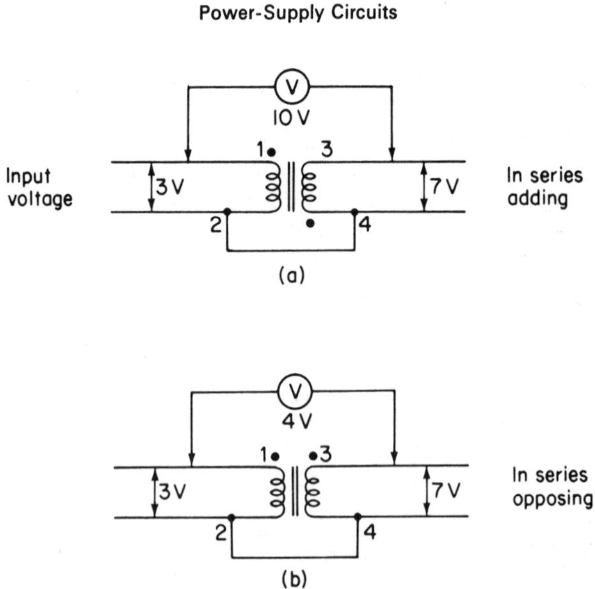

FIGURE 7–4 Checking transformer polarity markings.

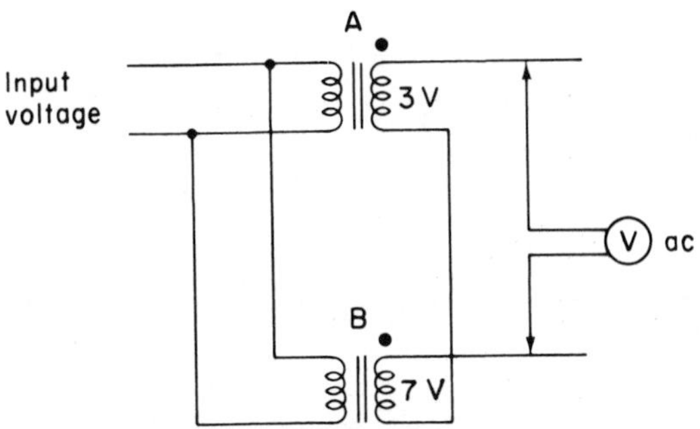

FIGURE 7-5 Testing consistency of transformer polarities or phase markings.

7-5.3 Checking Transformer Regulation

All transformers have some regulating effect, even though not used in a power supply with a regulator circuit. That is, the output voltage of a transformer tends to remain constant with changes in load. Regulation is usually expressed as:

$$\% \text{ regulation} = \frac{(\text{no-load voltage}) - (\text{full-load voltage})}{\text{full-load voltage}} \times 100$$

However, some manufacturers rate their transformers by:

$$\% \text{ regulation} = \frac{(\text{no-load voltage}) - (\text{full-load voltage})}{\text{no-load voltage}} \times 100$$

Whichever factor is used, keep in mind that all transformers do not provide good regulation, even though designed for power supply use. The regulation effect produced by a transformer is determined by core material, winding arrangement, wire size, etc. From a practical design standpoint, a transformer with very poor regulation requires extra filtering and/or regulation. Often, an inexpensive transformer will show poor regulation. Thus, the savings on the transformer are offset by the increased cost of filter and regulator components.

Transformer regulation can be tested using the circuit of Fig. 7-6. The value of R_l (load) should be selected to draw maximum rated current from the secondary. For example, if the transformer is to be used with a power supply designed to deliver 100 W, and the secondary output voltage is 25 V, the value of R_l should be such that 4 A flows, or $R_l = 25/4 = 6.25\ \Omega$.

First, measure the secondary output voltage without a load, and then with a load. Use either equation to find the percentage of regulation.

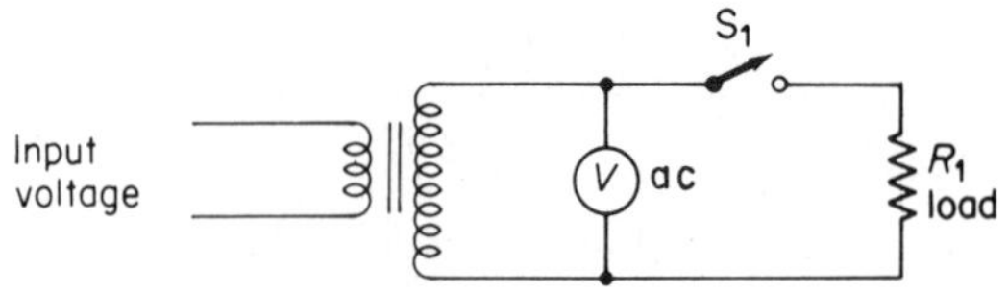

Input voltage

S_1

V ac

R_1 load

$$\% \text{ regulation} = \frac{\text{no-load } V - \text{full-load } V}{\text{full-load } V} \times 100$$

or

$$\% \text{ regulation} = \frac{\text{no-load } V - \text{full-load } V}{\text{no-load } V} \times 100$$

FIGURE 7-6 Testing transformer regulation.

7-5.4 Checking the Transformer Impedance Ratio

The impedance ratio of a transformer is the square of the winding ratio. For example, if the winding ratio of a transformer is 15:1, the impedance ratio is 225:1. Any impedance value placed across one winding is reflected onto the other winding by a value equal to the impedance ratio. Assuming an impedance ratio of 225:1 and an 1800-Ω impedance placed on the primary, the secondary then has a reflected impedance of 8 Ω. Similarly, if a 10-Ω impedance is placed on the secondary, the primary has a reflected impedance of 2250 Ω.

Impedance ratio is related to turns ratio (primary to secondary). However, turns ratio information is not always available, so the ratio must be calculated using a test circuit as shown in Fig. 7-7.

Measure both the primary and secondary voltage. Divide the larger

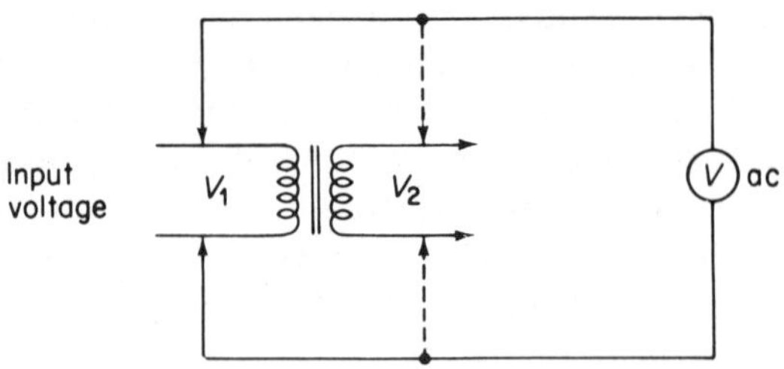

$$\text{Turns ratio} = V_1/V_2 \text{ or } V_2/V_1$$

$$\text{Impedance ratio} = (V_1/V_2)^2 \text{ or } (V_2/V_1)^2$$

FIGURE 7-7 Measuring transformer turns and impedance ratios.

voltage by the smaller, noting which is primary and which is secondary. For convenience, set either the primary or secondary to some exact voltage.

The *turns ratio* is equal to one voltage divided by the other.

The *impedance ratio* is the square of the turns ratio.

For example, assume that the primary shows 115 V, with 23 V at the secondary. This indicates a 5:1 turns ratio and a 25:1 impedance ratio.

7-5.5 Checking the Transformer-Winding Balance

There is always some imbalance in center-tapped transformers. That is, the turns ratio and impedance are not exactly the same on both sides of the center tap. A large imbalance can impair operation of any circuit, but especially a circuit such as that used in dc–dc converters.

It is possible to find a large imbalance by measuring the d-c resistance on either side of the center tap. However, a small imbalance might not show up.

It is usually more practical to measure the voltage on both sides of a center tap, as shown in Fig. 7–8. If the voltages are equal, the transformer winding is balanced. If a large imbalance is indicated by a large voltage difference, the winding should be checked with an ohmmeter for shorted turns, poor design, and so on.

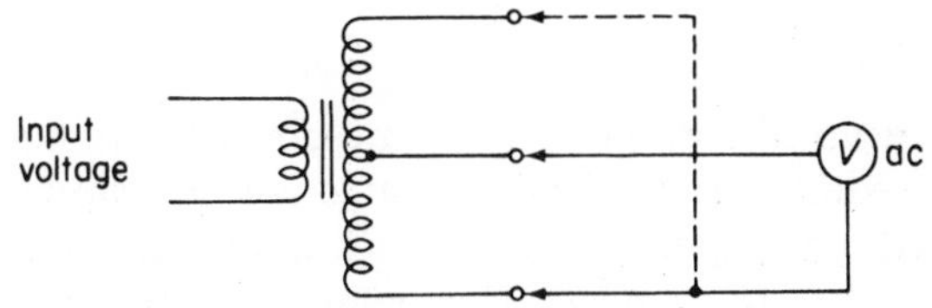

FIGURE 7-8 Testing transformer winding balance.

8

Radio-Frequency-Circuit Tests

This chapter is devoted entirely to test procedures for radio-frequency circuits. These procedures can be applied to a complete radio-frequency device (such as a CB set), or to specific circuits (such as the radio-frequency circuits of a radio transmitter or receiver). Communications equipment circuits are discussed further in Chapter 9. Also, the procedures described in this chapter can be applied to radio-frequency circuits at any time during design or experimentation. As a minimum, the test should be made when the circuit is first completed in experimental form. If the test results are not as desired, the component values should be changed as necessary to obtain the desired results.

Experimental radio-frequency (RF) circuits should always be retested in final form (with all components soldered in place). This shows if there is any change in circuit characteristics due to the physical relocation of components. Such tests are especially important at the higher radio frequencies. Often, there is capacitance or inductance between components, from components to wiring, and between wires. These stray "components" can add to the reactance and impedance of circuit components. When the physical location of parts and wiring are changed, the stray reactances change and alter circuit performance.

8-1 BASIC RF VOLTAGE MEASUREMENT

When the voltages to be measured are at radio frequencies and are beyond the frequency capabilities of the meter circuits or oscilloscope amplifiers, an RF probe is required. Such probes rectify the RF signals into a d-c output which is almost equal to the peak RF voltage. The d-c output of the probe is then applied to the meter or oscilloscope input and is displayed as a voltage readout in the normal manner.

If a probe is available as an accessory for a particular meter, that probe should be used in favor of any experimental or homemade probe. The manufacturer's probe is matched to the meter in calibration, frequency compensation, and so on. If a probe is not available for a particular meter or oscilloscope, the following notes discuss the fabrication of probes suitable for measurement and test of all RF circuits.

8-1.1 Half-Wave RF Probe

The half-wave probe of Fig. 8–1 provides an output to the meter (or oscilloscope) that is approximately equal to the peak value of the voltage being measured. Since most meters are calibrated to read in root-mean-square (RMS) values, the probe output must be reduced to 0.707 of the peak value, by means of R_1. The value of R_1 can be found by calculation. For practical purposes, a variable resistor should be used during calibration and then replaced by a fixed resistor of the correct value. The following steps describe the calibration and fabrication procedure:

1. Connect the experimental probe circuit to a signal generator and meter.

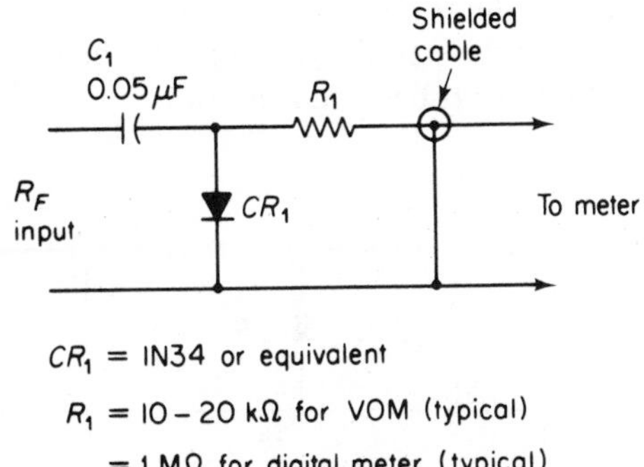

FIGURE 8–1 Basic half-wave RF probe circuit.

2. Set the meter to measure d-c voltage. Either a VOM or digital meter can be used, but best results are found with a high input impedance meter.
3. Adjust the signal generator voltage amplitude to some precise value, such as 10 V RMS, as measured on the generator's output meter.
4. Adjust the calibrating resistor R_1 until the meter indicates the same value (10 V RMS).
5. As an alternative procedure, adjust the signal generator for a 10-V peak output, then adjust R_1 for a reading of 7.07 on the meter being calibrated.
6. Remove the power, disconnect the circuit, measure the value of R_1, and replace the variable resistor with a fixed resistor of the same value.
7. Repeat the test with the fixed resistance in place. If the reading is correct, mount the circuit in a suitable package, such as within a test prod. Repeat the test with the circuit in final form. Also repeat the test over the entire frequency range of the probe. Generally, the probe provides satisfactory response up to about 250 MHz.
8. Keep in mind that the meter must be set to measure direct current, since the probe output is dc.

8-1.2 Demodulator Probe

When RF signals contain modulation, a demodulator probe (Fig. 8-2) is most effective for testing RF circuits during experimentation, or at any time. For example, an RF signal modulated by a fixed audio tone can be applied to an RF amplifier being tested. The demodulator measures the RF amplifier response in terms of both RF signal and audio signal.

The demodulator probe is similar to the half-wave probe except for the low capacitance of C_1 and the parallel resistor R_2. These two components act as a filter. The demodulator probe produces both an a-c and a d-c output. The

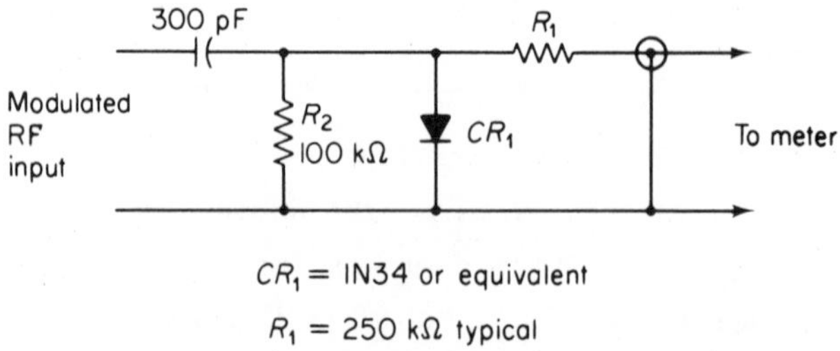

FIGURE 8-2 Basic demodulation RF probe circuit.

RF signal is converted into a d-c voltage approximately equal to the peak value. The low-frequency modulation voltage on the RF signal appears as ac at the probe output.

In use, the meter is set to dc and the RF signal is measured. Then the meter is set to ac and the modulating voltage is measured. The calibrating resistor R_1 is adjusted so that the d-c scale reads the RMS value. The procedure for calibration and fabrication of the demodulator probe is the same as for the half-wave RF probe (Section 8-1.1), except that the schematic of Fig. 8-2 should be used. Also, keep in mind that R_1 should be adjusted on the basis of the RF signal (not the modulating signal) with the meter set to dc.

8-2 MEASURING THE RESONANT FREQUENCY OF *LC* CIRCUITS

RF equipment is based on the use of resonant circuits (or "tank" circuits), consisting of a capacitor and a coil (inductance) connected in series or parallel as shown in Fig. 8-3. At the resonant frequency, the inductive and capacitive reactances are equal, and the *LC* circuit acts as a high impedance (if it is a parallel circuit) or a low impedance (if it is a series circuit). In either case, any combination of capacitance and inductance has some resonant frequency.

Either (or both) the capacitance or inductance can be variable to permit tuning of the resonant circuit over a given frequency range. When the inductance is variable, tuning is usually done by means of a metal slug inside the coil. The metal slug is screwdriver-adjusted to change the inductance (and thus the inductive reactance) as required. Typical RF circuits used in receivers (AM, FM, communications, etc.) often include two resonant circuits in the form of a transformer (RF or IF transformer, etc.). Either the capacitance or inductance can be variable. Figure 8-3 contains equations that show the relationships among capacitance, inductance, reactance, and frequency, as they relate to resonant circuits. Note that there are two sets of equations. One set of equations includes reactance (inductive and capacitive). The other set omits reactance.

A meter can be used in conjunction with an RF signal generator to find the resonant frequency of either series or parallel *LC* circuits. The generator must be capable of producing a signal at the resonant frequency of the circuit, and the meter must be capable of measuring the frequency. If the resonant frequency is beyond the normal range of the meter, an RF probe must be used. The following steps describe the measurement procedure.

1. Connect the equipment as shown in Fig. 8-4. Use the connections of Fig. 8-4(a) for parallel resonant *LC* circuits, or the connections of Fig. 8-4(b) for series resonant *LC* circuits.
2. Adjust the generator output until a convenient midscale indication is

Resonance and impedance

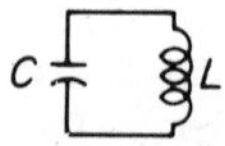

Parallel* (Infinite impedance) — Series (zero impedance)

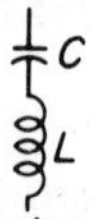

$$F = \frac{1}{6.28\sqrt{LC}}$$

$$F\,(\text{kHz}) = \frac{10^6}{6.28\sqrt{L\,(\mu\text{H}) \times C\,(\text{pF})}}$$

$$F(\text{kHz}) = \frac{159}{\sqrt{L\,(\mu\text{H}) \times C\,(\mu\text{F})}}$$

$$L\,(\mu\text{H}) = \frac{2.54 \times 10^4}{F\,(\text{kHz})^2 \times C\,(\mu\text{F})}$$

$$F(\text{MHz}) = \frac{0.159}{\sqrt{L\,(\mu\text{H}) \times C\,(\mu\text{F})}}$$

$$C\,(\mu\text{F}) = \frac{2.54 \times 10^4}{F\,(\text{kHz})^2 \times L\,(\mu\text{H})}$$

*Approximate; accurate when circuit Q is 10 or higher

Inductive reactance

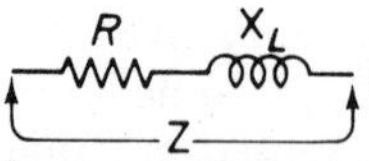

Series

$$Z = \sqrt{R^2 + X_L^2} \qquad Q = \frac{X_L}{R} \qquad L = \frac{X_L}{6.28\,F}$$

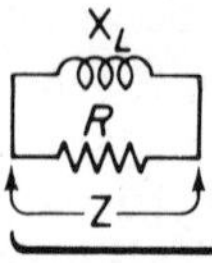

Parallel

$$Z = \frac{RX_L}{\sqrt{R^2 + X_L^2}} \qquad Q = \frac{R}{X_L} \qquad F = \frac{X_L}{6.28\,L}$$

$$X_L = 6.28 \times F(\text{Hz}) \times L(\text{H})$$

$$X_L = 6.28 \times F(\text{kHz}) \times L(\text{mH})$$

$$X_L = 6.28 \times F(\text{MHz}) \times L\,(\mu\text{H})$$

Capacitive reactance

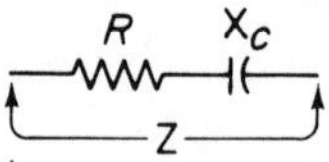

Series

$$Z = \sqrt{R^2 + X_C^2} \qquad Q = \frac{X_C}{R} \qquad F = \frac{1}{6.28\,CX_C}$$

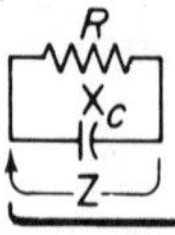

Parallel

$$Z = \frac{RX_C}{\sqrt{R^2 + X_C^2}} \qquad Q = \frac{R}{X_C} \qquad C = \frac{1}{6.28\,FX_C}$$

$$X_C = \frac{1}{6.28 \times F(\text{Hz}) \times C(\text{F})}$$

$$X_C = \frac{159}{F(\text{kHz}) \times C\,(\mu\text{F})}$$

FIGURE 8-3 Resonant circuit equations.

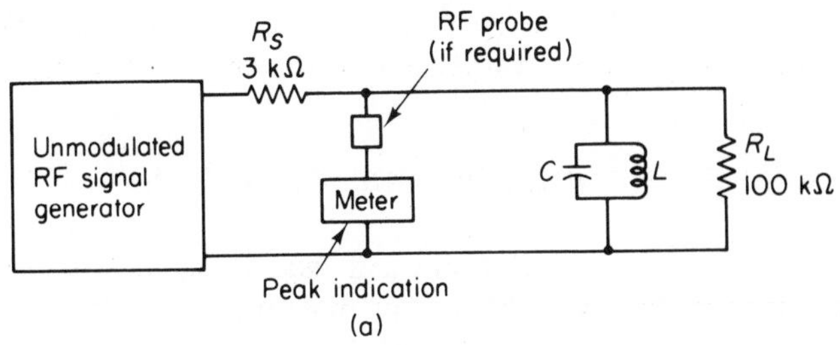

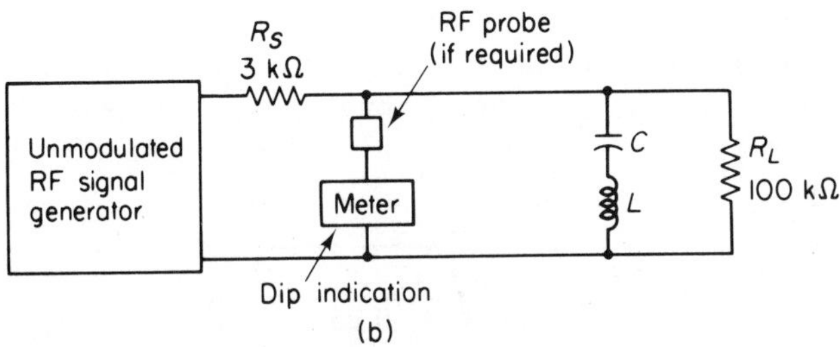

FIGURE 8-4 Measuring resonant frequency of *LC* circuits: (a) parallel; (b) series.

obtained on the meter. Use an unmodulated signal output from the generator.

3. Starting at a frequency well below the lowest possible frequency of the circuit under test, slowly increase the generator output frequency. If there is no way to judge the approximate resonant frequency, use the lowest generator frequency.
4. If the circuit being tested is parallel-resonant, watch the meter for a maximum, or peak, indication.
5. If the circuit being tested is series-resonant, watch the meter for a minimum, or dip, indication.
6. The resonant frequency of the circuit under test is the one at which there is a maximum (for parallel) or minimum (for series) indication on the meter.
7. There may be peak or dip indications at harmonics of the resonant frequency. Therefore, the test is most efficient when the approximate resonant frequency is known.
8. The value of load resistor R_L is not critical. The load is shunted across the *LC* circuit to flatten or broaden the resonant response (to lower the circuit *Q*, as discussed in Section 8-5). Thus, the voltage maximum or minimum is approached more slowly. A suitable trial

value for R_L is 100 kΩ. A lower value of R_L sharpens the resonant response, and a higher value flattens the curve.

8-3 MEASURING THE INDUCTANCE OF A COIL

The equations necessary to calculate the self-inductance of a single-layer, air-core coil are given in Fig. 8–5. Note that maximum inductance is obtained when the ratio of coil radius to coil length is 1.25, that is, when the length is 0.8 of the radius. RF coils wound for that ratio are the most efficient (maximum inductance for minimum physical size).

As an example, assume that it is desired to design a coil with 0.5 μH inductance on a 0.25-in. radius (air-core, single-layer). Using the equations of Fig. 8–5, for maximum efficiency the coil length should be 0.8R, or 0.2 inch. Then

$$N = \sqrt{\frac{17 \times 0.5}{0.25}}$$

$$= \sqrt{34}$$

$$= 5.8 \text{ turns}$$

For practical purposes, use 6 turns and spread the turns slightly. The additional part of a turn increases inductance, but the spreading decreases the inductance. After the coil is made, the inductance should be checked with an inductance bridge or as follows. A meter can be used in conjunction with an RF signal generator and a fixed capacitor, of known value and accuracy, to find the inductance of a coil. The generator must be capable of producing a signal at the resonant frequency of the test circuit, and the meter must be capable of measuring the frequency. If the resonant frequency is beyond the normal range of the meter, an RF probe must be used. The following steps describe the measurement procedure.

1. Connect the equipment as shown in Fig. 8–6. Use a capacitive value such as 10-μF, 100 pF, or some other even number to simplify the calculation.
2. Adjust the generator output until a convenient midscale indication is obtained on the meter. Use an unmodulated signal output from the generator.
3. Starting at a frequency well below the lowest possible resonant frequency of the inductance–capacitance combination under test, slowly increase the generator frequency. If there is no way to judge the approximate resonant frequency, use the lowest generator frequency.

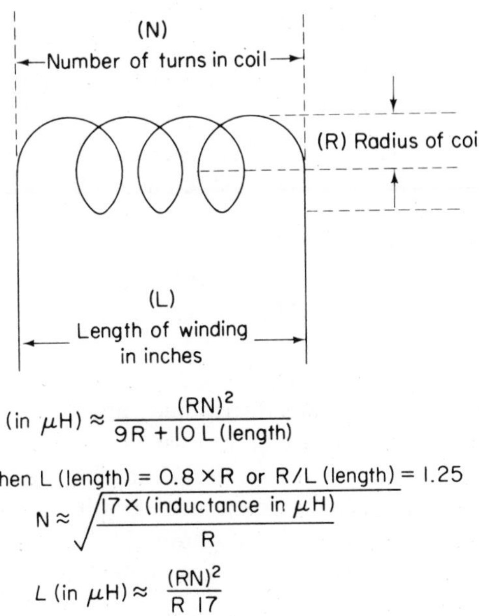

FIGURE 8-5 Calculations for self-inductance of single-layer air-core coil.

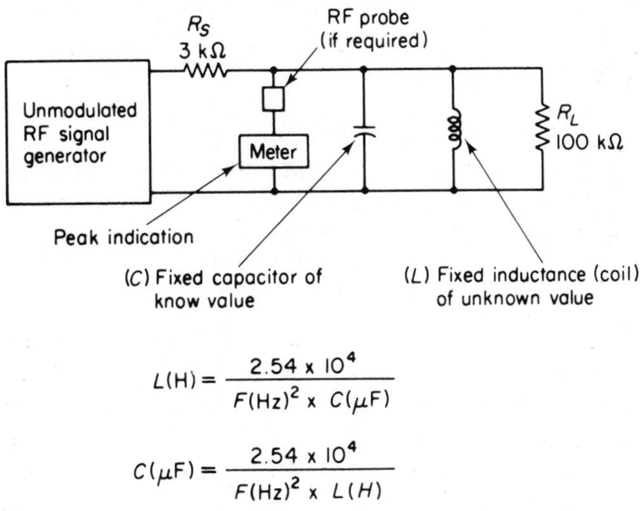

FIGURE 8-6 Measuring inductance and capacitance in *LC* circuits.

4. Watch the meter for a maximum or peak indication. Note the frequency at which the peak indication occurs. This is the resonant frequency of the circuit.
5. Using this resonant frequency, and the known capacitance value, calculate the unknown inductance using the equation of Fig. 8-6.
6. Note that the procedure can be reversed to find an unknown capacitance value, when a known inductance value is available.

8-4 MEASURING THE SELF-RESONANCE AND DISTRIBUTED CAPACITANCE OF A COIL

No matter what design or winding method is used, there is some distributed capacitance in any coil. When the distributed capacitance combines with the coil's inductance, a resonant circuit is formed. The resonant frequency is usually quite high in relation to the frequency at which the coil is used. However, since self-resonance may be at or near a harmonic of the frequency to be used, the self-resonant effect may limit the coil's usefulness in *LC* circuits. Some coils, particularly RF chokes, may have more than one self-resonant frequency.

A meter can be used in conjunction with an RF signal generator to find both the self-resonant frequency and distributed capacitance of a coil. The generator must be capable of producing a signal at the resonant frequency of the circuit, and the meter must be capable of measuring voltages at that frequency. Use an RF probe if required. The following steps describe the measurement procedure.

1. Connect the equipment as shown in Fig. 8-7.
2. Adjust the generator output until a convenient midscale indication

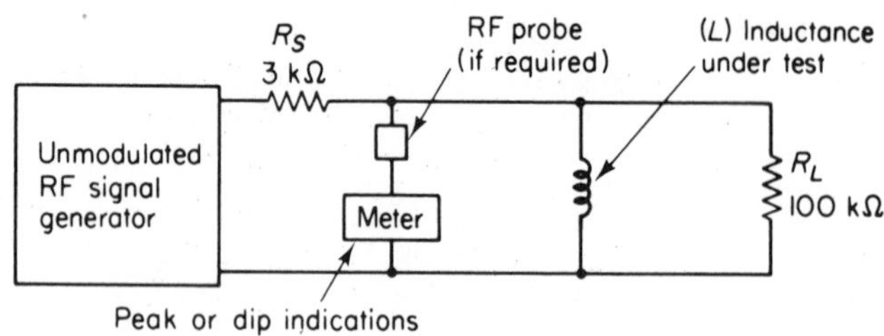

$$C(\mu F) = \frac{2.54 \times 10^4}{F(\text{Hz})^2 \times L(\mu H)}$$

FIGURE 8-7 Measuring self-resonance and distributed capacitance of coils.

is obtained on the meter. Use an unmodulated signal output from the generator.

3. Tune the signal generator over its entire frequency range, starting at the lowest frequency. Watch the meter for either peak or dip indications. Either a peak or a dip indicates that the inductance is at a self-resonant point. The generator output frequency at that point is the self-resonant frequency. Make certain that peak or dip indications are not the result of changes in generator output level. Even the best laboratory generators may not produce a flat (constant level) output over the entire frequency range.
4. Since there may be more than one self-resonant point, tune through the entire signal generator range. Try to cover a frequency range up to at least the third harmonic of the highest frequency involved in a resonant circuit design.
5. Once the resonant frequency has been found, calculate the distributed capacitance using the equation of Fig. 8–8. For example, assume that a coil with an inductance of 7 μH is found to be self-resonant at 50 MHz.

$$C \text{ (distributed capacitance)} = \frac{2.54 \times 10^4}{(50)^2 \times 7} = 1.45 \text{ pF}$$

8-5 MEASURING THE Q OF RESONANT CIRCUITS

A resonant circuit has a Q, or quality, factor. The circuit Q depends upon the ratio of reactance to resistance. If a resonant circuit has pure reactance, the Q is high (actually infinite). However, this is not practical. For example, any coil has some d-c resistance, as do the leads of the capacitor. Also, as frequency increases, the a-c resistance presented by lead increases. The sum total of these resistances is usually lumped together and is considered as a resistor in series or parallel with the circuit. The total resistance, usually termed the effective resistance, is not to be confused with the reactance.

The resonant circuit Q depends upon the individual Q factors of inductance and capacitance used in the circuit. For example, if both the inductance and capacitance have a high Q, the circuit will have a high Q, provided that a minimum of resistance is produced when the inductance and capacitance are connected to form a resonant circuit.

From a practical test standpoint, a resonant circuit that has a high Q produces a sharp resonance curve (narrow bandwidth), whereas a low Q produces a broad resonance curve (wide bandwidth). For example, a high resonant circuit provides good harmonic rejection and efficiency, in comparison

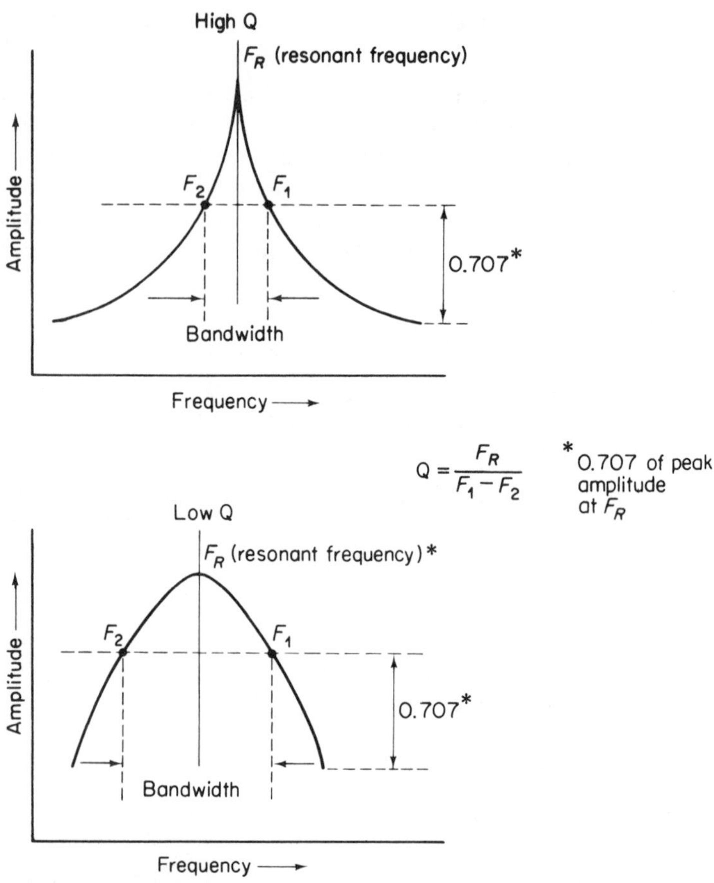

FIGURE 8-8 Relationship of bandpass characteristics to Q of resonant circuits.

with a low-Q circuit, all other factors being equal. The selectivity of a resonant circuit is thus related directly to Q. A very high Q (or high selectivity) is not always desired. Sometimes it is necessary to add resistance to a resonant circuit to broaden the response (increase the bandwidth, decrease the selectivity).

Usually, resonant-circuit Q is measured at points on either side of the resonant frequency where the signal amplitude is down 0.707 of the peak resonant value, as shown in Fig. 8–8. Note that Q must be increased for increases in resonant frequency if the same bandwidth is to be maintained. For example, if the resonant frequency is 10 MHz with a bandwidth of 2 MHz, the required circuit Q is 5. If the resonant frequency is increased to 50 MHz, with the same 2-MHz bandwidth, the required Q is 25. Also, note that Q must be decreased for increases in bandwidth if the same resonant frequency is to be maintained. For example, if the resonant frequency is 30 kHz, with a bandwidth of 2 kHz, the required Q is 15. If the bandwidth is increased to 10 kHz, with the same 30-kHz resonant frequency, the required Q is 3.

The Q of a circuit can be measured using a signal generator and a meter with an RF probe. A high-impedance digital meter generally provides the least loading effect on the circuit and thus provides the most accurate indication. Figure 8–9(a) shows the test circuit in which the signal generator is connected directly to the input of a complete stage, and Fig. 8–9(b) shows the indirect method of connecting the signal generator to the input.

When the stage or circuit has sufficient gain to provide a good reading on the meter with a nominal output from the generator, the indirect method

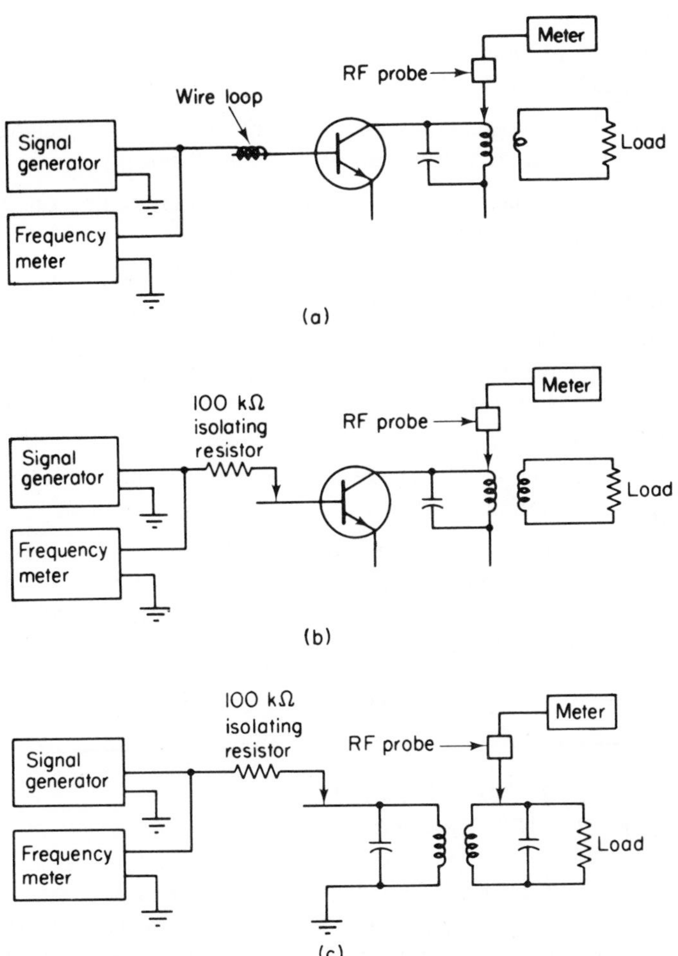

FIGURE 8–9 Measuring Q of resonant circuits.

(with isolating resistor) is preferred. Any signal generator has some output impedance (such as a 50-Ω output resistor). When this resistance is connected directly to the tuned circuit, the Q is lowered, and the response becomes broader. (In some cases, the generator output impedance seriously detunes the circuit.)

Figure 8–9(c) shows the test circuit for a single component (such as an IF transformer). The value of the isolating resistance is not critical, and is typically in the range of 100 kΩ. The procedure for determining Q using any of the circuits in Fig. 8–9 is as follows:

1. Connect the equipment as shown in Fig. 8–9(a), (b), or (c), as applicable. Note that a load is shown in Fig. 8–9(c). When a circuit is normally used with a load, the most realistic Q measurement is made with the circuit terminated in that load value. A fixed resistance can be used to simulate the load. The Q of a resonant circuit often depends on the load value.
2. Tune the signal generator to the circuit resonance frequency. Operate the generator to produce an unmodulated output.
3. Tune the generator frequency for maximum reading on the meter. Note the generator frequency.
4. Tune the generator below resonance until the meter reading is 0.707 of the maximum reading. Note the generator frequency. To make the calculation more convenient, adjust the generator output level so that the meter reading is some even value, such as 1 V or 10 V, after the generator is turned for maximum. This makes it easy to find the 0.707 mark.
5. Tune the generator above resonance until the meter reading is 0.707 of the maximum reading. Note the generator frequency.
6. Calculate the circuit Q using the equation of Fig. 8–9. For example, assume that the maximum meter indication occurs at 455 kHz (F_R), the below-resonance indication is 453 kHz (F_2), and the above-resonance indication is 458 kHz (F_1). Then,

$$458 - 453 = 5{,}455/5 = \text{a } Q \text{ of } 91.$$

8–6 MEASURING THE IMPEDANCE OF RESONANT CIRCUITS

Any resonant circuit has some impedance at the resonant frequency. The impedance changes as frequency changes. This includes transformers (tuned and untuned), tank circuits, and so on. In theory, a series-resonant circuit has zero impedance, while a parallel-resonant circuit has infinite impedance, at the resonant frequency. In practical circuits, this is impossible since there is always some resistance in the circuit.

It is often convenient to find the impedance of an experimental resonant circuit, at a given frequency. Also, it may be necessary to find the impedance of a component in an experimental circuit so that other circuit values can be designed around the impedance. For example, an IF transformer presents an impedance at both the primary and secondary windings. These values may not be specified in the transformer datasheet.

The impedance of a resonant circuit or component can be measured using a signal generator and a meter with an RF probe. A high-impedance digital meter provides the least loading effect on the circuit and therefore provides the most accurate indication.

The procedure for impedance measurement at radio frequencies is the same as for audio frequencies, as discussed in Section 6–8, except as follows. An RF signal generator must be used as the signal source. The meter must be provided with an RF probe. If the circuit or component under measurement has both an input and output (such as a transformer), the opposite side or winding must be terminated in the normal load, as shown in Fig. 8–10. A fixed, noninductive, resistance can be used to simulate the load. If the impedance of a tuned circuit is to be measured, tune the circuit to peak or dip, then measure the impedance at resonance. Once the resonant impedance is found, the signal generator can be tuned to other frequencies to find the corresponding impedance.

The RF signal generator is adjusted to the frequency (or frequencies) at which impedance is to be measured. Switch *S* is moved back and forth between positions *A* and *B*, while resistance *R* is adjusted until the voltage reading is the same in both positions of the switch. Resistor *R* is then disconnected from the circuit, and the d-c resistance of *R* is measured with an ohmmeter. The d-c resistance of *R* is then equal to the dynamic impedance at the circuit input. Accuracy of the impedance measurement depends upon the accuracy with which the d-c resistance is measured. A noninductive resistance must be used. The impedance found by this method applies only to the frequency used during the test.

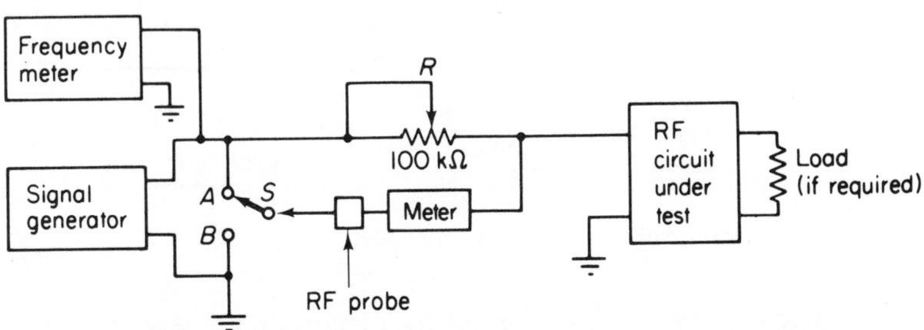

FIGURE 8–10 Measuring impedance of RF circuits.

8-7 TESTING TRANSMITTER RF CIRCUITS

It is possible to test and adjust transmitter RF circuits using a meter and an RF probe. If an RF probe is not available (or as an alternative), it is possible to use a circuit such as shown in Fig. 8-11. This circuit is essentially a pickup coil which is placed near the RF circuit inductance, and a rectifier that converts the RF into a d-c voltage or measurement on a meter. The basic procedures are as follows:

1. Connect the equipment as shown in Fig. 8-12. If the circuit being measured is an amplifier, without an oscillator, a drive signal must be supplied by means of a signal generator. Use an unmodulated signal, at the correct operating frequency.
2. In turn, connect the meter (through an RF probe or the special circuit of Fig. 8-11) to each stage of the RF circuit. Start with the first stage (this is the oscillator if the circuit under test is a complete transmitter), and work toward the final (or output) stage.
3. A voltage indication should be obtained at each stage. Usually, the voltage indication increases with each amplifier stage, as you proceed from oscillator to the final amplifier. However, some stages may be frequency multipliers and provide no voltage amplification.
4. If a particular stage is to be tuned, adjust the tuning control for a maximum reading on the meter. If the stage is to be operated with a

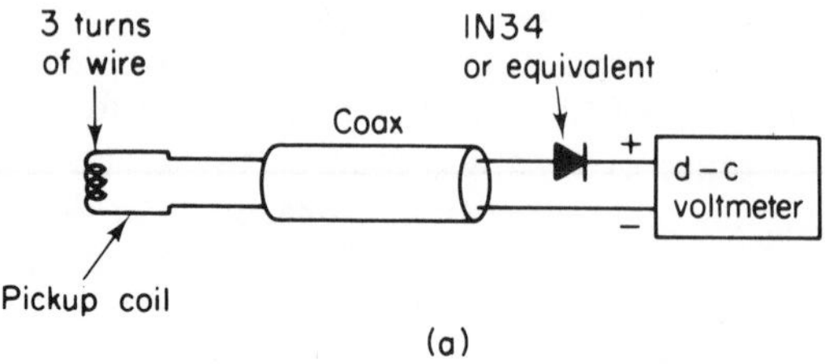

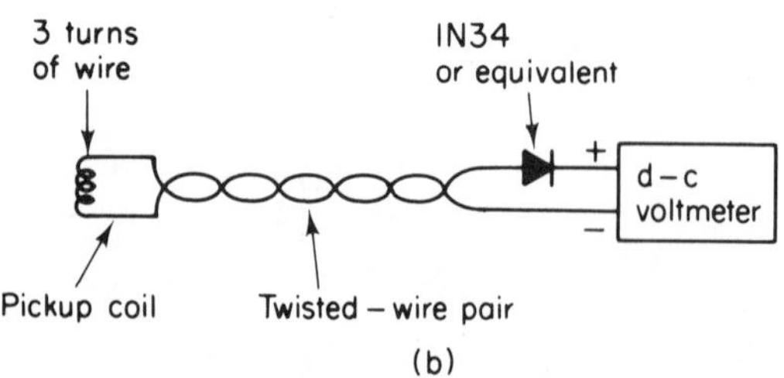

FIGURE 8-11 (a) Test circuit for pickup and measurement of RF signals. (b) Alternative.

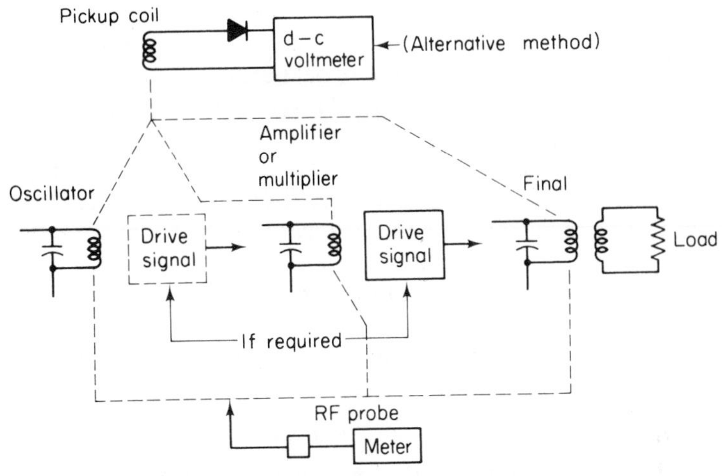

FIGURE 8-12 Testing RF transmitter circuits.

load (such as the final amplifier into an antenna), the load should be connected, or a simulated load should be used. A fixed, noninductive resistance provides a good simulated load at frequencies up to about 250 MHz.

5. It should be noted that this tuning method or measurement technique does not guarantee each stage is at the desired operating frequency. It is possible to get maximum readings on harmonics. However, it is conventional to design RF transmitter circuits so they do not tune to both the desired operating frequency and a harmonic. Generally, RF amplifier tank circuits tune on either side of the desired frequency, but not to a harmonic (unless the circuit is seriously detuned, or the design calculations are hopelessly inaccurate).

8-8 TESTING RECEIVER RF CIRCUITS WITH A METER AND SIGNAL GENERATOR

The procedures for receiver circuit test using a meter and signal generator are as follows. Both AM and FM receivers require alignment of the IF and RF stages. An FM receiver also requires the alignment of the detector stage (discriminator or ratio detector). The normal sequence for alignment in a complete FM receiver is (1) detector, (2) IF amplifier and limiter stages, and (3) RF and local oscillator (mixer/converter). The alignment sequence for an AM receiver is (1) IF stages, and (2) RF and local oscillator. The following procedures can be applied to a complete receiver, or to individual stages, at any point during design.

If a complete receiver is being tested, and the receiver includes an AVC-AGC circuit, the AGC must be disabled. This is best accomplished by placing a fixed bias, of opposite polarity to the signal normally produced by the detector, on the AGC line. The fixed bias should be of sufficient amplitude to overcome the bias signal produced by the detector (usually on the order of 1 or 2 V). When such bias is applied, the stage gain is altered from the normal condition. Once alignment is complete, the bias should be removed. If individual circuits are to be tested, the precautions regarding AGC can be ignored.

8-8.1 FM Detector Alignment and Test

1. Connect the equipment as shown in Fig. 8–13 (for a discriminator) or Fig. 8–14 (for a ratio detector).
2. Set the meter to measure d-c voltage.
3. Adjust the signal generator frequency to the intermediate frequency (usually 10.7 MHz for a broadcast FM receiver). Use an unmodulated output from the signal generator.
4. Adjust the secondary winding (either capacitor or tuning slug) of the discriminator transformer for zero reading on the meter. Adjust the transformer slightly each way and make sure the meter moves smoothly above and below the exact zero mark. (A meter with a zero-center scale is most helpful when adjusting FM detectors.)
5. Adjust the signal generator to some point below the intermediate

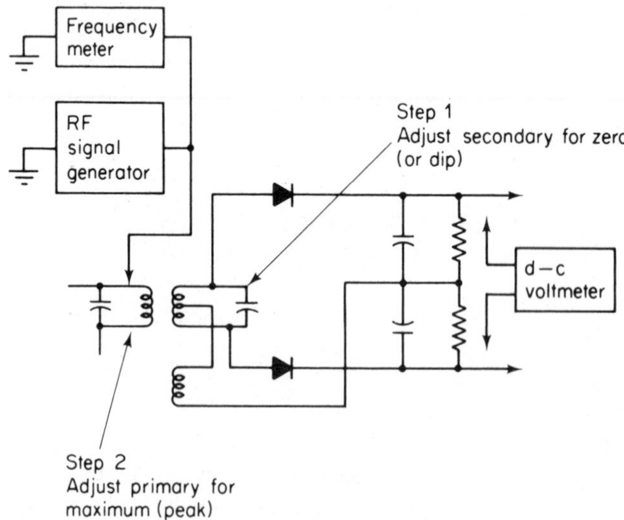

FIGURE 8–13 Test connections for FM discriminator alignment.

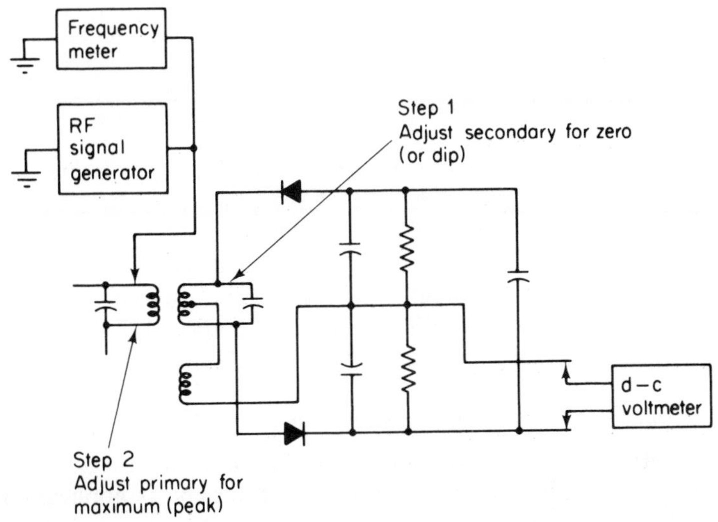

FIGURE 8-14 Test connections for FM ratio detector alignment.

frequency (to 10.625 MHz for an FM detector with a 10.7-MHz IF). Note the meter reading. If the meter reading goes downscale against the pin, reverse the meter polarity or test leads (the RF probe is not used for FM detector alignment).

6. Adjust the signal generator to some point above the intermediate frequency exactly equivalent to the amount set below the IF in step 5. For example, if the generator is set to 0.075 MHz below the IF (10.7 − 0.075 = 10.625), then set the generator to 10.775 (10.7 + 0.075 = 10.775).
7. The meter should read approximately the same in both steps 5 and 6, except that the polarity is reversed. For example, if the meter reads seven scale divisions below zero for step 5, and seven scale divisions above zero for step 6, the detector is balanced. If a detector circuit under test cannot be balanced under these conditions, the fault is usually a serious mismatch of diodes or other components.
8. Return the generator output to the intermediate frequency (10.7 MHz).
9. Adjust the primary winding of the detector transformer (either capacitor or tuning slug) for a maximum reading on the meter. This sets the primary winding at the correct resonant frequency of the IF amplifier.
10. Repeat steps 4 through 8 to make sure that adjustment of the trans-

former primary has not disturbed the secondary setting. Invariably, two settings interact.

8-8.2 AM and FM Alignment and Test

The alignment procedure for the IF stages of an AM receiver are essentially the same as those for an FM receiver. However, the meter must be connected at different points in the corresponding detector, as shown in Fig. 8-15. In either case the meter is set to measure direct current, and the RF probe is not used. In those cases where IF stages are being tested without a detector (such as during design), an RF probe is required. As shown in Fig. 8-15, the RF probe is connected to the secondary of the final IF output transformer.

1. Connect the equipment as shown in Fig. 8-15.
2. Set the meter to measure direct current, and connect it to the appropriate test point (with or without an RF probe as applicable).
3. Place the signal generator in operation and adjust the generator fre-

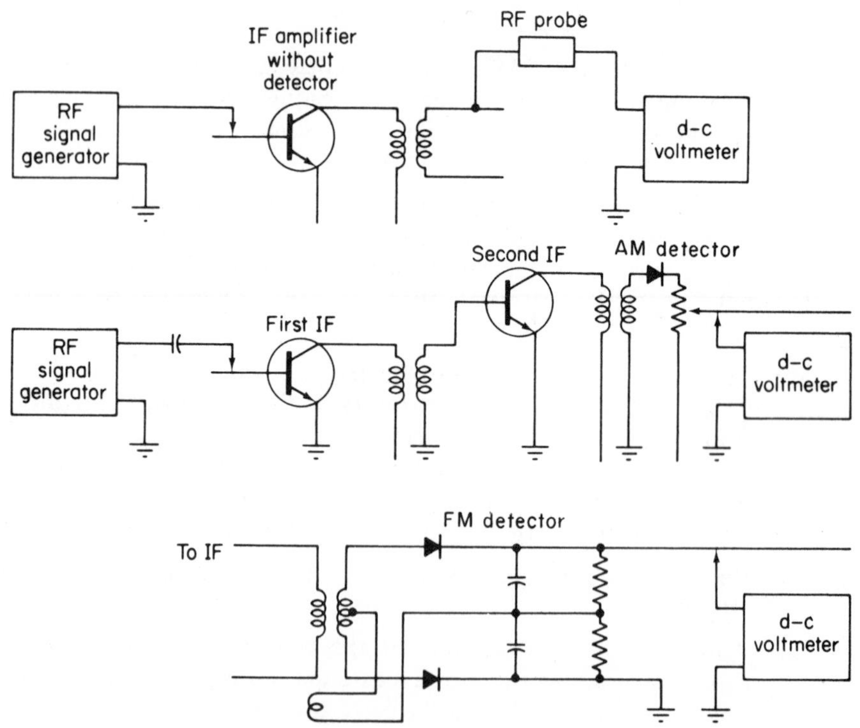

FIGURE 8-15 IF alignment for AM and FM resonant circuits.

quency to the receiver intermediate frequency (typically 10.7 MHz for FM and 455 kHz for AM). Use an unmodulated output from the signal generator.

4. Adjust the windings of the IF transformers (capacitor or tuning slug) in turn, starting with the last stage and working toward the first stage. Adjust each winding for maximum reading.
5. Repeat the procedure to make sure the tuning on one transformer has no effect on the remaining adjustments. Often, the adjustments interact.

8-8.3 AM and FM RF Alignment and Test

The alignment procedures for the RF stages (RF amplifier, local oscillator, mixer/converter) of an AM receiver are essentially the same as for an FM receiver. Again, it is a matter of connecting the meter to the appropriate test point. The same test points used for IF alignment can be used for aligning the RF stages as shown in Fig. 8-16. However, if an individual RF stage is to be aligned, the meter must be connected to the secondary winding of the RF-stage output transformer, through an RF probe. The procedure is as follows:

1. Connect the equipment as shown in Fig. 8-16.

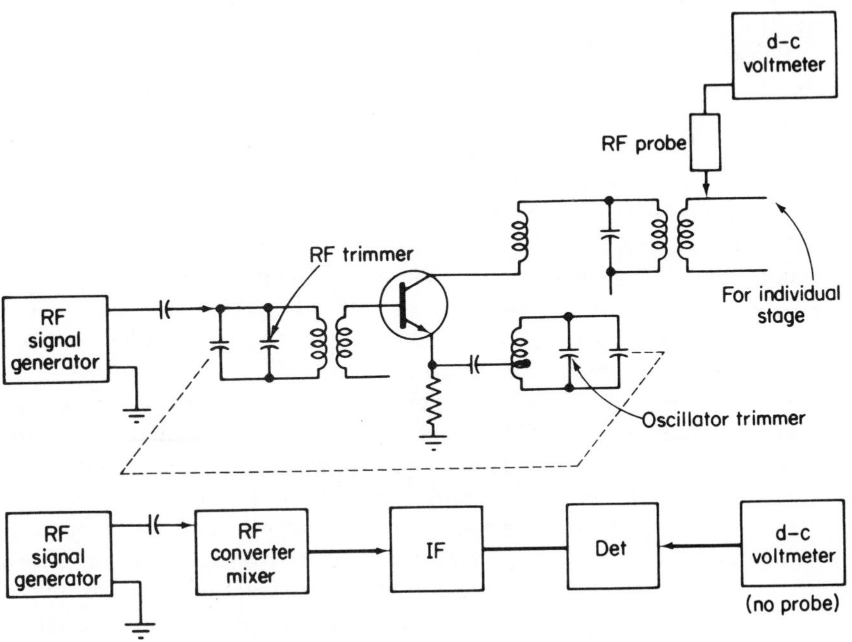

FIGURE 8-16 Alignment of RF amplifier and local oscillator (converter/mixer).

2. Set the meter to measure direct current, and connect it to the appropriate test point (with or without an RF probe as applicable).
3. Adjust the generator frequency to some point near the high end of the receiver operating frequency (typically, 107 MHz for a broadcast-FM receiver, and 1400 kHz for an AM-broadcast receiver). Use an unmodulated output from the signal generator.
4. Adjust the RF-stage trimmer for maximum reading on the meter.
5. Adjust the generator frequency to the low end of the receiver operating frequency (typically 90 MHz for FM and 600 kHz for AM).
6. Adjust the oscillator-stage trimmer for maximum reading on the meter.
7. Repeat the procedure to make sure the resonant circuits "track" across the entire tuning range.

8-9 TESTING RECEIVER RF CIRCUITS WITH A SWEEP GENERATOR/OSCILLOSCOPE

The response characteristics of AM and FM receiver RF circuits can be checked, or the IF stages aligned, using a sweep generator/oscilloscope combination. The sweep generator must be capable of sweeping over the entire IF range. If maximum accuracy is desired, a marker generator must also be used.

Before going into the specific test/alignment procedures, let us review the basic sweep generator/oscilloscope test technique, as shown in Fig. 8–17. The primary function of a sweep generator is the sweep-frequency alignment of TV and FM receivers. In this application, sweep generators are used with oscilloscopes to display the bandpass characteristics of the receiver under test. A sweep generator is an FM generator. When a sweep generator is set to a given frequency, this is center frequency. In essence, a sweep generator is a frequency-modulated RF generator. The usual frequency modulation rate is 60 Hz for most TV and FM sweep generators. Other sweep rates can be used, but since power lines usually have a 60-Hz frequency, this frequency is both convenient and economical for the sweep rate.

Some sweep generators have a built-in marker generator. Marker signals are necessary to pinpoint frequencies when making sweep frequency alignments and tests. Although sweep generators are accurate in both center frequency and sweep width, it is almost impossible to pick out a particular frequency along the spectrum of frequencies being swept. Thus, fixed-frequency "marker" signals are injected into the circuit together with the sweep-frequency-generator output. On sweep generators without a built-in marker generator, markers can be added by means of an absorption-type marker adder. These marker adders can be built in or they are available as accessories.

Where a marker adder would provide too limited a number of fixed fre-

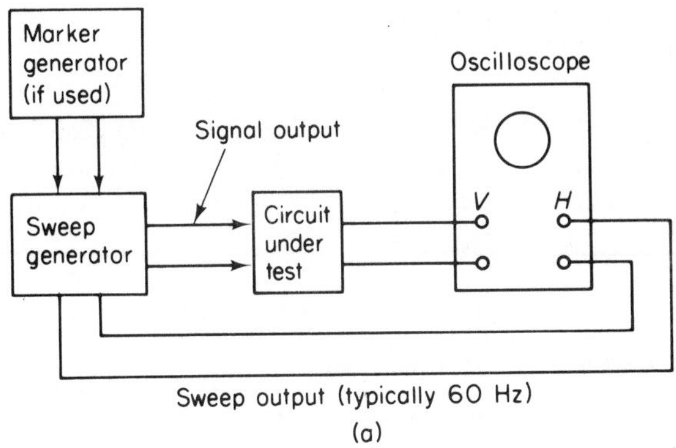

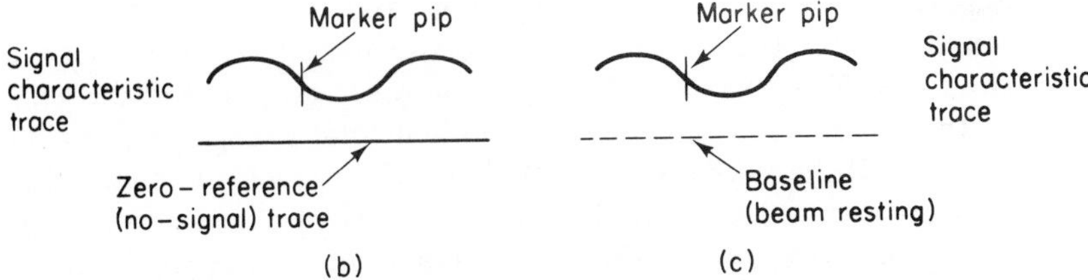

FIGURE 8-17 (a) Basic sweep generator/oscilloscope test connections. (b) Oscilloscope display with sweep signal blanking. (c) Oscilloscope display without sweep signal blanking.

quency points, a marker generator can be used in conjunction with the sweep generator and oscilloscope. Basically, a marker generator is an RF signal generator which has highly accurate dial markings and which can be calibrated precisely against internal or external signals. The sweep generator is tuned to sweep the band of frequencies passed by the wideband circuits (tuner, IF, video, filter, etc.) and a trace representing the response characteristics of the circuits is displayed on the oscilloscope, as shown in Fig. 8-17. The marker generator is used to provide calibrated markers along the response curve. When the marker signal from the marker generator is coupled into the test circuit, a vertical "pip" or marker appears on the curve as shown. When the marker generator is tuned to a frequency within the passband accepted by the equipment under test, the marker indicates the position of that frequency on the sweep trace.

Another feature found on some sweep generators is a blanking circuit. When the sweep generator output is swept across its spectrum, the frequencies go from low to high, then return from high to low. With the blanking circuit actuated, the return or retrace is blanked off. This makes it possible to view a zero-reference line on the oscilloscope during the retrace period.

8-9.1 Basic Sweep Generator/Oscilloscope Test Procedure

The following steps describe the basic procedure for using a sweep generator with an oscilloscope. The remaining paragraphs of this section describe procedures for using the sweep generator/oscilloscope combination to test the RF and IF circuits of radio receivers.

1. Connect the equipment as shown in Fig. 8-17.
2. Place the oscilloscope in operation.
3. Place the sweep generator in operation.
4. Switch off the oscilloscope internal recurrent sweep.
5. Set the oscilloscope sweep selector and sync selector to external. Under these conditions, the oscilloscope horizontal sweep should be obtained from the generator sweep output, and the length of the horizontal sweep should represent total sweep spectrum. For example, if the sweep is from 10 to 20 kHz, the left-hand end of the horizontal trace represents 10 kHz and the right-hand end represents 20 kHz. Any point along the horizontal trace represents a corresponding frequency. For example, the midpoint on the trace represents 15 kHz. If a rough approximation of frequency is desired, the horizontal gain control can be adjusted until the trace occupies an exact number of scale divisions, such as 10 cm for the 10–20 kHz sweep signal. Each centimeter division then represents 1 kHz.
6. If a more accurate frequency measurement is desired, the marker generator must be used. The marker generator output frequency is adjusted until the marker pip is aligned at the desired point on the trace. The frequency is then read from the marker-generator frequency dial.
7. The response curve (oscilloscope trace) depends upon the device under test. If the device has a passband (as do most receiver circuits), and the sweep generator is set so that the sweep is wider than the passband, the trace will start low at the left, rise toward the middle, and then drop off at the right, as shown in Fig. 8-17. The sweep generator/oscilloscope method tells at a glance the overall passband characteristics of the device (sharp response, flat response, irregular response at certain frequencies, etc.). The exact frequency limits of the passband can be measured with the marker generator pip.
8. Switch the sweep generator blanking control on or off as desired. Some sweep generators do not have a blanking function. With the blanking function in effect, there is a zero-reference line on the

trace. With the blanking function off, the horizontal baseline does not appear. The sweep generator blanking function is not to be confused with the oscilloscope blanking (which is bypassed when the sweep signal is applied directly to the horizontal amplifier).

8-9.2 Alignment of AM and FM IF Amplifiers with an Oscilloscope

Test the IF amplifiers with sweep generator and oscilloscope as follows:

1. Connect the equipment as shown in Fig. 8–18.
2. Place the oscilloscope in operation. Switch off the internal recurrent sweep. Set the oscilloscope sweep selector and sync selector to external.
3. Place the sweep generator in operation. Switch sweep generator blanking control on or off as desired. Adjust the sweep generator to cover the complete IF range. Usually, AM IF center frequency is 455 kHz and requires a sweep about 30 kHz wide. FM IF center frequency is 10.7 MHz and requires a sweep of about 300 kHz.
4. Check the IF response curve appearing on the oscilloscope against those of Fig. 8–18, or against the receiver specifications.
5. If it is desired to determine the exact frequencies at which IF response occurs, adjust the marker generator until the marker pip is aligned at the point of interest. Read the frequency, or band of frequencies, from the marker generator frequency dial.
6. The amplitude of any point on the response curve can be measured directly on the oscilloscope (assuming that the vertical display is voltage-calibrated (as it is on most oscilloscopes).
7. Adjust the IF alignment controls to produce the desired response curve, as specified in the receiver service data.

8-9.3 Alignment of AM and FM Front Ends with an Oscilloscope

The response characteristics of AM and FM receiver RF stages (RF amplifier, mixer or first detector, oscillator) or "front end" can be checked, or aligned, using a sweep generator/oscilloscope combination. The procedure is essentially the same as for IF alignment (Section 8–9.2) except that the sweep generator output is connected to the antenna input of the receiver, whereas the input to the first detector or mixer is applied to the oscilloscope vertical chan-

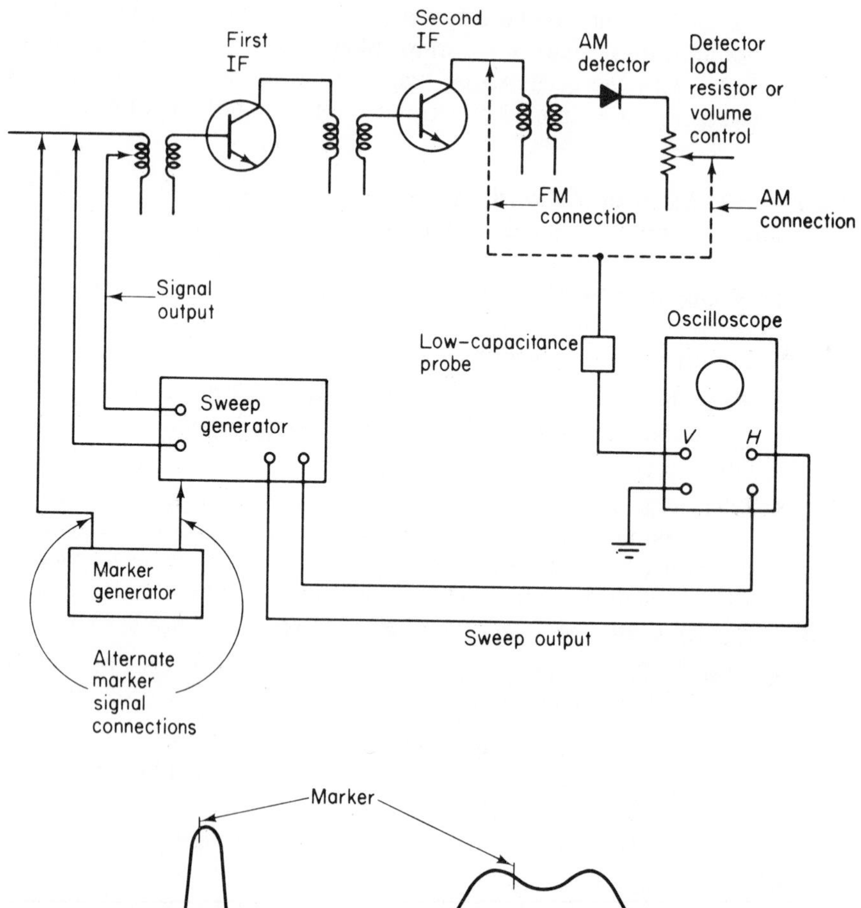

FIGURE 8-18 Alignment of IF amplifiers with an oscilloscope.

nel. The sweep generator must be capable of sweeping over the entire RF range. If maximum accuracy is desired, a marker generator must also be used.

1. Connect the equipment as shown in Fig. 8-19.
2. Place the oscilloscope in operation. Switch off the internal recurrent sweep. Set the oscilloscope sweep selector and sync selector to external.

3. Place the sweep generator in operation. Switch the sweep generator blanking control on or off as desired. Adjust the sweep generator to cover the complete RF range. The center frequency depends on the receiver. Usually, an AM receiver requires a 30-kHz sweep width, and an FM receiver needs about 300 kHz.
4. Check the RF response curve appearing on the oscilloscope against those of Fig. 8–19, or against the receiver specifications.
5. If it is desired to determine the exact frequency at which RF response occurs, the marker generator can be adjusted until the marker pip is aligned at the point of interest. The frequency, or band of frequencies, can be read from the marker generator frequency dial.
6. The amplitude of any point of the response curve can be measured directly on the oscilloscope (assuming that the vertical system is voltage-calibrated).
7. Adjust the RF alignment controls to produce the desired response curve, as specified in the receiver service data. Usually, the RF response of an AM receiver is similar to that of Fig. 8–19(b), whereas an FM receiver has a broad response similar to that of Fig. 8–19(c).

8-9.4 Alignment of an FM Detector with an Oscilloscope

The detector of an FM receiver (either discriminator or ratio detector) can be aligned using the sweep generator/oscilloscope combination. The test connections are similar to front-end and IF alignment. The sweep generator output is connected to the last IF stage input, whereas the oscilloscope vertical channel is connected across the FM detector load resistor. The sweep generator must be capable of sweeping over the entire IF range. If maximum accuracy is desired, a marker generator must also be used.

1. Connect the equipment as shown in Fig. 8–20(a).
2. Place the oscilloscope in operation. Switch off the internal recurrent sweep. Set the oscilloscope sweep selector and sync selector to external.
3. Place the sweep generator in operation. Switch the sweep generator blanking control on or off as desired. Set the sweep generator frequency to the receiver IF (usually 10.7 MHz). Adjust the sweep width to about 300 kHz.
4. Check the detector response curve appearing on the oscilloscope against that of Fig. 8–20(b), or against the receiver specifications.
5. Adjust the last IF stage and detector alignment controls so that peaks 2 and 4 of the response curve are equal in amplitude above and below

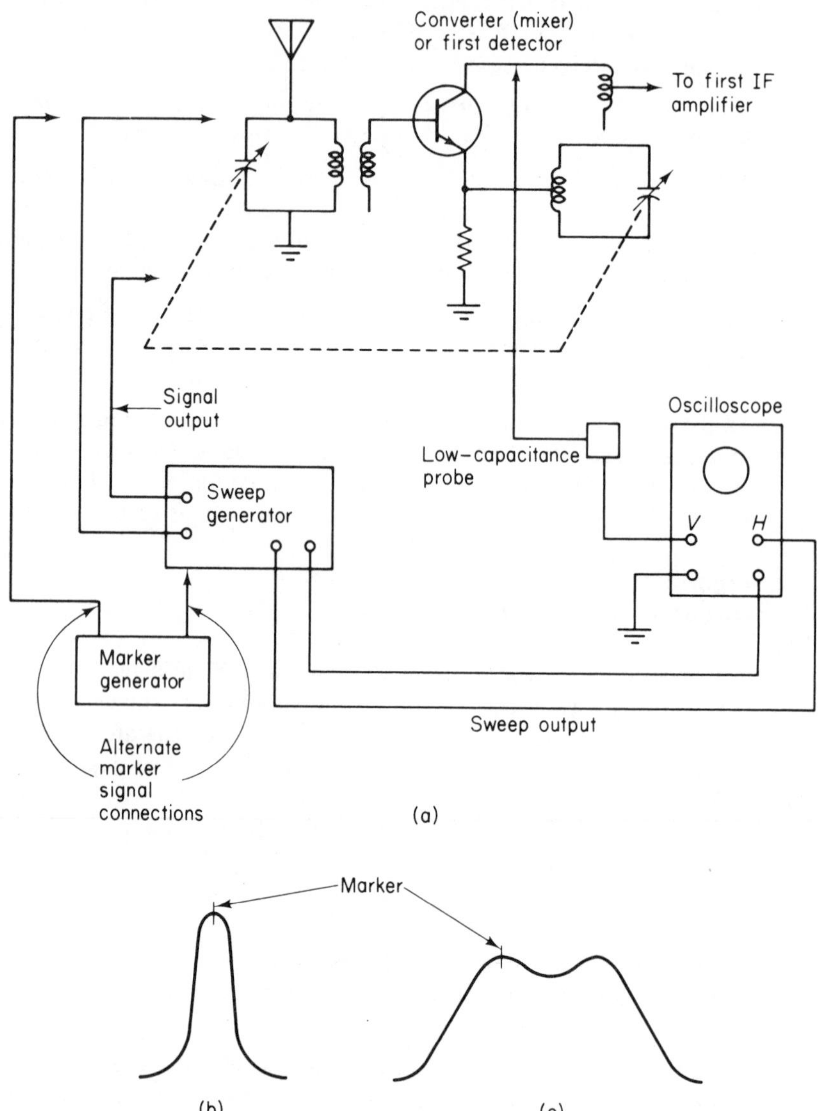

FIGURE 8-19 (a) Alignment of RF (front end) stages with an oscilloscope. (b) Typical AM response (sharp). (c) Typical FM response (broad).

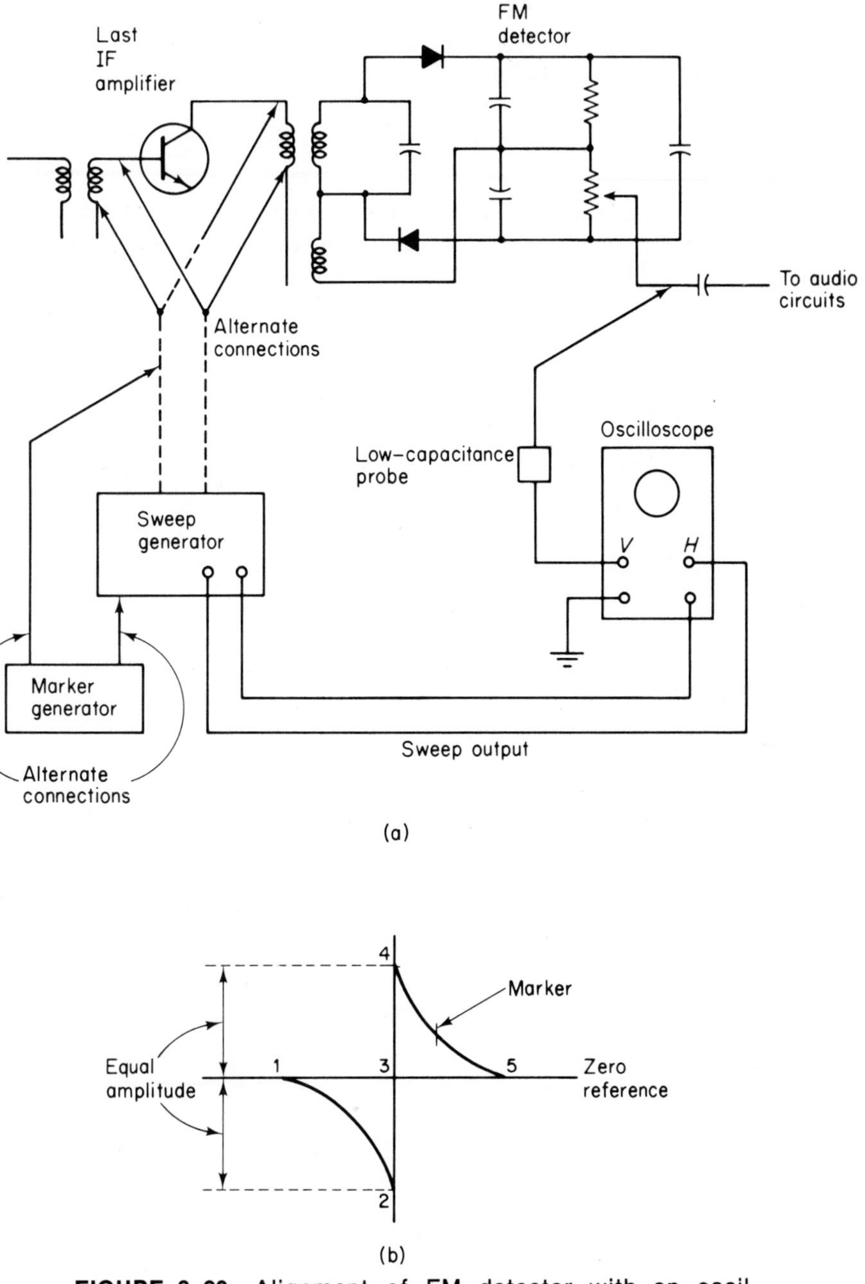

FM
detector
Last
IF
amplifier
To audio
circuits
Alternate
connections
Oscilloscope
Low–capacitance
probe
Sweep
generator
V
H
Marker
generator
Sweep output
Alternate
connections
(a)
4
Marker
Equal
amplitude
1
3
5
Zero
reference
2
(b)

FIGURE 8-20 Alignment of FM detector with an oscilloscope.

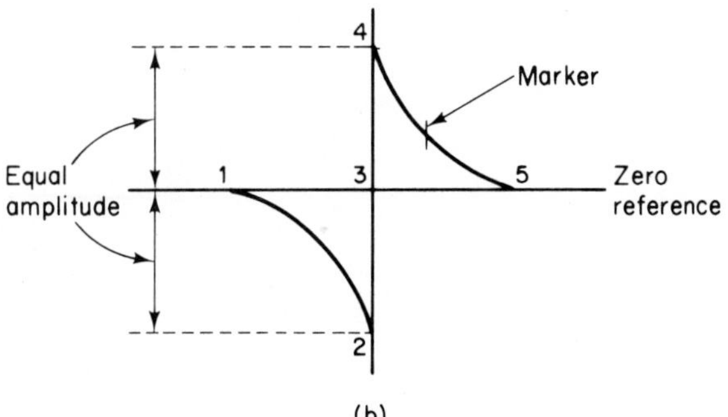

the zero line. Also, points 1, 3, and 5 of the response curve should be on the zero reference line.

6. If it is desired to determine the exact frequency at which detector response occurs, the marker generator can be adjusted until the marker pip is aligned at the point of interest. The frequency, or band of frequencies, can then be read from the marker generator frequency dial.
7. The amplitude of any point on the response curve can be measured directly on the oscilloscope (assuming that the vertical display is voltage calibrated).

9

Communications Equipment Tests

This chapter is devoted entirely to test procedures for communications equipment and circuits. The test procedures used in communications work are basically the same as those used in other fields of electronics. Most communications test procedures are performed using meters, signal generators, frequency counters, oscilloscopes, power supplies, and assorted clips, patch cords, and so on. Theoretically, all communications test procedures can be performed using conventional test equipment, provided that the oscilloscopes have the necessary gain and bandpass characteristics, the frequency counters cover the necessary range, the signal generators cover the appropriate frequencies, and so on. However, there are some specialized test instruments that greatly simplify communications test and service, such as SWR meters, field strength meters, RF wattmeters, and dummy loads. Also, there are specialized versions of basic test equipment that have been specifically developed to simplify communications tests. We cover such equipment in this chapter, in addition to describing communications tests that can be performed with the most basic of equipment.

9-1 USING OSCILLOSCOPES IN COMMUNICATIONS EQUIPMENT TESTS

There are two uses for oscilloscopes in communications work: signal tracing and modulation measurement.

9-1.1 Signal Tracing with an Oscilloscope

The signals in all circuits of a typical communications equipment (such as a CB set) may be traced with an oscilloscope, provided that the scope is equipped with the proper probes. You can check amplitude, frequency, and waveforms of signals with a scope. However, many communications service technicians do not use scopes extensively, for the following reasons.

The oscilloscope will measure signal amplitude, but a meter is easier to read. The same applies to signal frequency. The frequency counter is easier to read, and it is far simpler to measure frequency with a counter than with a scope, particularly in the typical communications frequency range (such as the CB range of 27 MHz). The oscilloscope is a superior instrument for monitoring waveforms. However, in communications work, the signals are mostly sine waves, and waveforms are not that critical.

9-1.2 Modulation Checks with an Oscilloscope

The main use for an oscilloscope in communications work is to measure percentage of modulation and uniformity or linearity of modulation. The use of an oscilloscope for modulation checks is not new. There are many variations of the basic technique, each of which is discussed in the following paragraphs of this section.

9-1.3 Direct Measurement of the Modulation Envelope with a High-Frequency Oscilloscope

If the vertical channel response of the oscilloscope is capable of handling the transmitter output frequency, the output can be applied through the oscilloscope vertical amplifier. The basic test connections are shown in Fig. 9-1. The procedure is as follows:

1. Connect the oscilloscope to the antenna jack, or the final RF amplifier of the transmitter, as shown in Fig. 9-1. Use one of the three alternatives shown, or the modulation measurement described in the transmitter service literature.
2. Key the transmitter (press the push-to-talk switch) and adjust the oscilloscope controls to produce displays as shown. You can either speak into the microphone (for a rough check of modulation), or you can introduce an audio signal (typically at 400 or 1000 Hz) at the microphone jack input (for a precise check of modulation). Note that Fig. 9-1 provides simulations of typical oscilloscope displays during modulation tests.

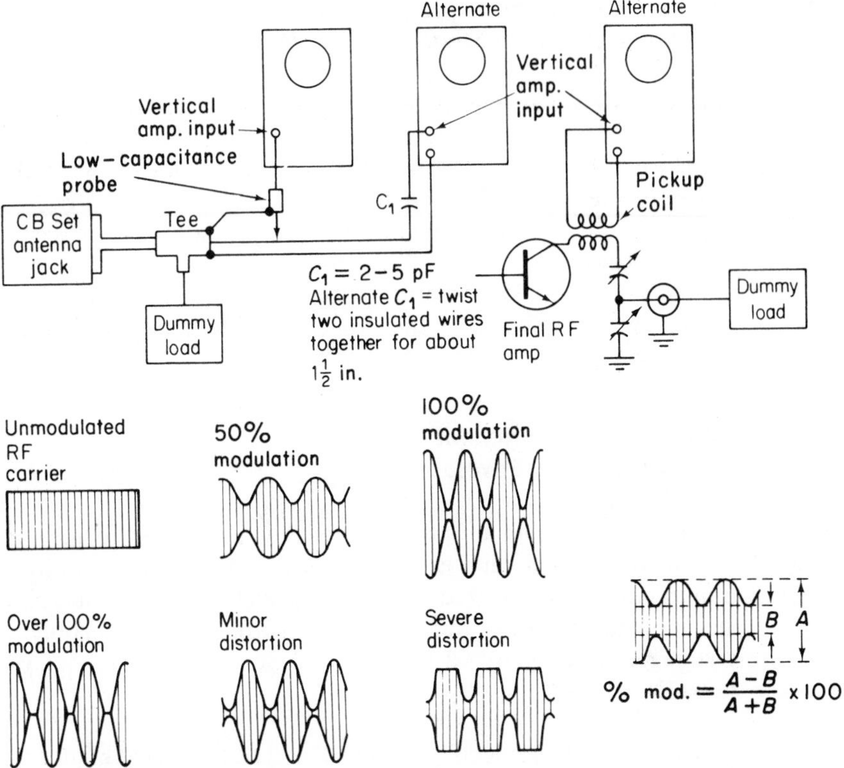

FIGURE 9-1 Direct measurement of modulation envelope with a high frequency (30 MHz or higher) oscilloscope.

3. Measure the vertical dimensions shown as A and B in Fig. 9-1 (the crest amplitude and the trough amplitude). Calculate the percentage of modulation using the equation of Fig. 9-1. For example, if the crest amplitude (A) is 63 (63 screen divisions, 6.3 V, and so on) and the trough amplitude (B) is 27, the percentage of modulation is

$$\frac{63 - 27}{63 + 27} \times 100 = 40\%$$

Make certain to use the same oscilloscope scale for both crest (A) and trough (B) measurements. Keep in mind when making modulation measurements, or any measurement that involves the transmitter, the RF output (antenna connector) must be connected to an antenna or dummy load. Dummy loads are discussed in Section 9-4. Antennas are discussed in Section 9-9.

9-1.4 Direct Measurement of the Modulation Envelope with a Low-Frequency Oscilloscope

If the oscilloscope is not capable of passing the transmitter frequency, the transmitter output can be applied directly to the vertical deflection plates of the oscilloscope cathode-ray tube. However, there are two drawbacks to this approach. First, the vertical plates may not be readily accessible. Next, the voltage output of the final RF amplifier may not produce sufficient deflection of the oscilloscope trace.

The test connections and modulation patterns are essentially the same as those shown in Fig. 9-1. Similarly, the procedures are the same as those described in Section 9-1.3.

9-1.5 Trapezoidal Measurement of the Modulation Envelope

The trapezoidal technique has an advantage in that it is easier to measure straight-line dimensions than curving dimensions. Thus, any nonlinearity in modulation may easily be checked with a trapezoid. In the trapezoidal method, the modulated carrier amplitude is plotted as a function of modulating voltage, rather than as a function of time. The basic test connections are shown in Fig. 9-2.

1. Connect the oscilloscope to the final RF amplifier and modulator. As shown in Fig. 9-2, use either the capacitor connection or the pickup coil for the RF (oscilloscope vertical input). However, for best results, connect the transmitter outputs directly to the deflection plates of the oscilloscope tube. The oscilloscope amplifiers may be nonlinear and can cause the modulation to appear distorted.
2. Key the transmitter and adjust the controls (oscilloscope controls and R_1) to produce a display as shown.
3. Measure the vertical dimensions shown as A (crest) and B (trough) on Fig. 9-2, and calculate the percentage of modulation using the equation given. For example, if the crest amplitude (A) is 80, and the trough amplitude (B) is 40, using the same scale, the percentage of modulation is

$$\frac{80 - 40}{80 + 40} \times 100 = 33\%$$

Again, make sure that the transmitter output is connected to an antenna or dummy load, before transmitting.

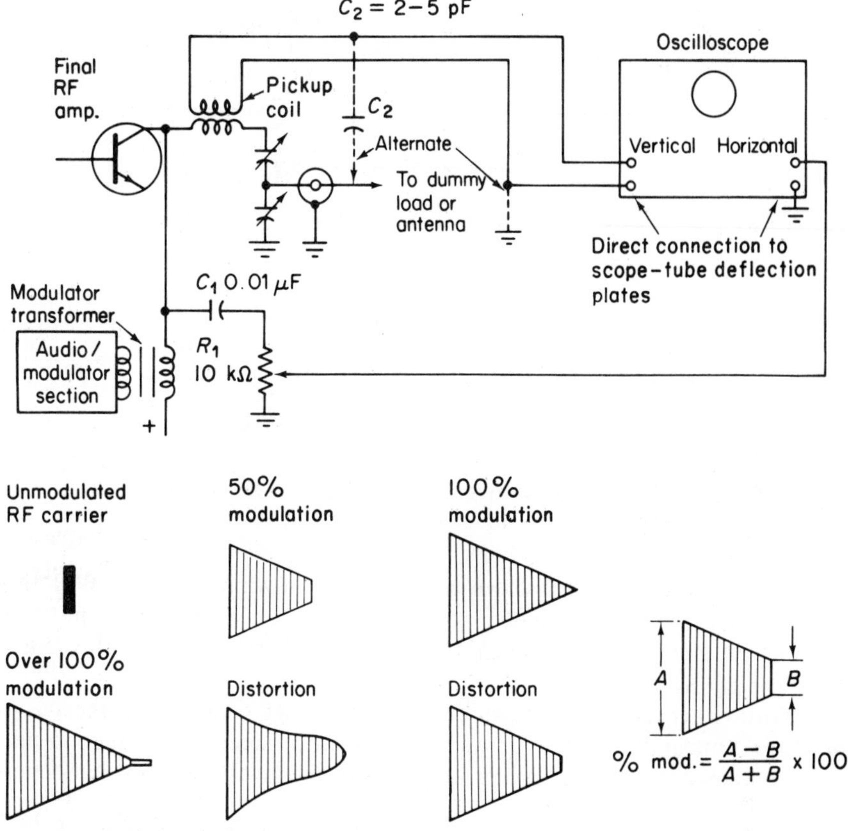

FIGURE 9-2 Trapezoidal measurement of the modulation envelope.

9-1.6 *Down-Conversion Measurement of the Modulation Envelope*

If the oscilloscope is not capable of passing the transmitter carrier signals, and the transmitter output is not sufficient to produce a good indication when connected directly to the oscilloscope tube, it is possible to use a down-converter test setup. One method requires an external RF generator and an IF transformer. The other method uses a receiver capable of monitoring the transmitter frequencies.

The RF generator method of down-conversion is shown in Fig. 9-3. In this method, the RF generator is tuned to a frequency above or below the transmitter frequency by an amount equal to the IF transmitter frequency. For example, if the IF transformer is 455 kHz, tune the RF generator to a frequency 455 kHz above (or below) the transmitter frequency.

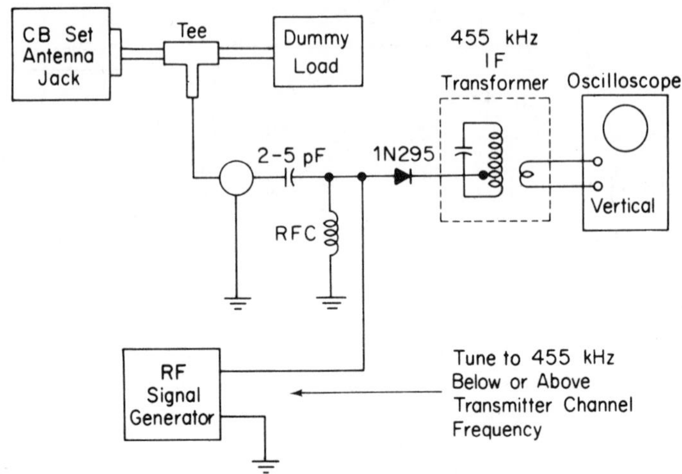

FIGURE 9-3 Down-conversion method of modulation measurement using a 455-kHz transformer.

The receiver method of down-conversion is shown in Fig. 9–4. With this method, the receiver is tuned to the transmitter frequency, and the oscilloscope input signal is taken from the last IF stage output through a 30-pF capacitor.

With either method of down-conversion, the RF generator or receiver is tuned for a maximum indication on the oscilloscope screen. Once a good pattern is obtained, the rest of the procedure is the same as described in Section 9–1.3.

The author does not generally recommend the down-conversion methods, except as a temporary measure. There are a number of relatively inexpensive oscilloscopes available that will pass signals up to and beyond the 50-MHz range.

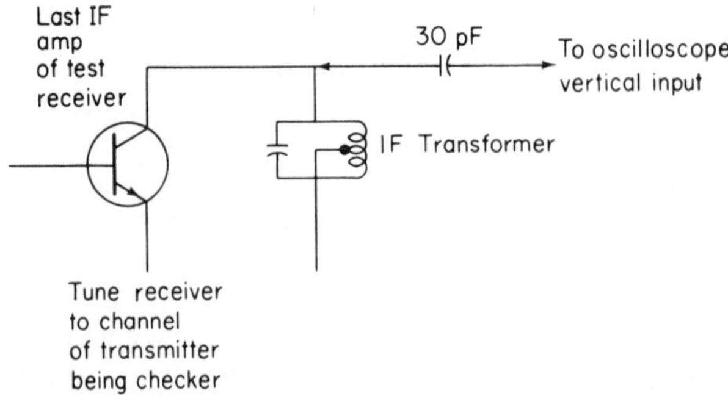

FIGURE 9-4 Down-conversion method of modulation measurement using a CB receiver.

9-1.7 *Linear Detector Measurement of the Modulation Envelope*

If you must use an oscilloscope that will not pass the carrier frequency of the transmitter, you can use a linear detector. However, the oscilloscope must have a d-c input, where the signal is fed directly to the oscilloscope vertical amplifier, not through a capacitor. Most modern oscilloscopes have both a-c (with capacitor) and d-c inputs. The basic test connections for linear detection of the modulation envelope are shown in Fig. 9-5. The basic test procedure is as follows:

1. Connect the transmitter output to the oscilloscope through the linear detector circuit as shown in Fig. 9-5. Make certain to include the dummy load (or wattmeter as shown).
2. With the transmitter not keyed, adjust the oscilloscope *position* control to place the trace on a reference line near the *bottom* of the screen, as shown in Fig. 9-5(b).
3. Key the transmitter, but do not apply modulation. Adjust the

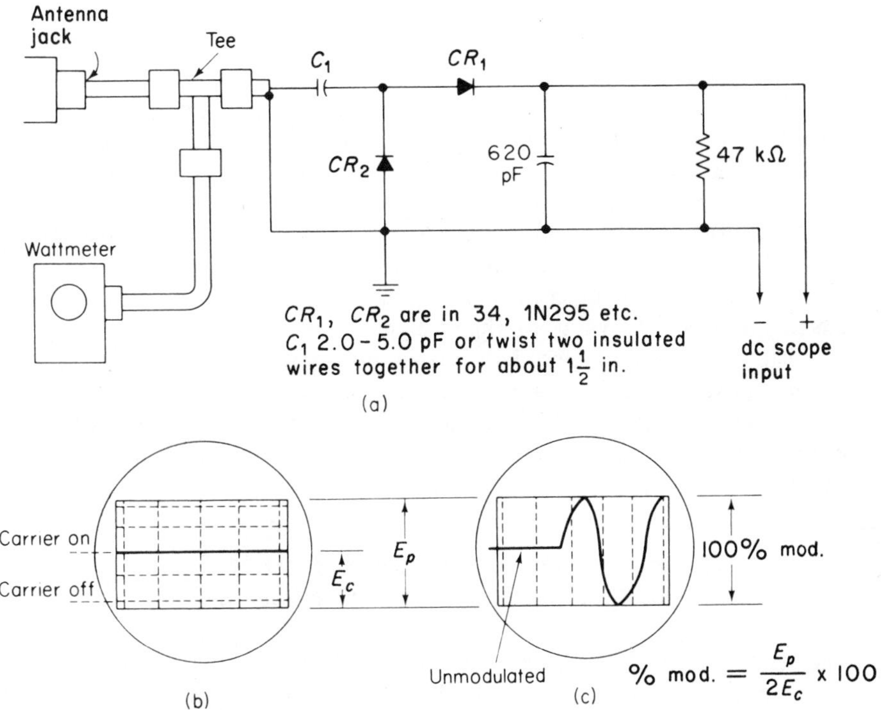

FIGURE 9-5 Pace modulation detector for measurement of the modulation envelope with a low-frequency oscilloscope.

oscilloscope gain control to place the *top* of the trace at the *center* of the screen, as shown in Fig. 9–5(b). It may be necessary to switch the transmitter off and on several times to adjust the trace properly, since the position and gain controls of most oscilloscopes interact.

4. Measure the distance (in scale divisions) of the shift between the carrier (step 3) and no-carrier (step 2) traces. For example, if the screen has a total of 10 vertical divisions, and the no-carrier trace is at the bottom or zero line, there is a shift of five scale divisions to the centerline.
5. Key the transmitter and apply modulation. Do not touch either the position or gain controls of the oscilloscope.
6. Find the percentage of modulation using the equation shown in Fig. 9–5. For example, assume that the shift between the carrier and no-carrier trace is five divisions and that the modulation produces a peak-to-peak envelope of eight divisions. The percentage of modulation is

$$\frac{8}{2 \times 5} \times 100 = 80\%$$

9–1.8 Modulation Nomogram

Figure 9–6 is a nomogram that can be used with the direct-measurement techniques (Sections 9–1.3 and 9–1.4) or the trapezoidal technique (Section 9–1.5) to find percentage of modulation. To use Fig. 9–6, measure the values of the crest (or maximum) and trough (or minimum) oscilloscope patterns.

The percentage of modulation is found by extending a straightedge from the measured value of the crest or maximum (given as *A* on Fig. 9–6) on its scale to the measured value of the trough or minimum (given as *B*) on its scale. The percentage of modulation is found where the straightedge crosses the diagonal scale. The crest and trough may be measured in any units (volts, vertical scale divisions, etc.), as long as both crest and trough are measured in the same units. The dashed line in Fig. 9–6 is used to illustrate the percentage of modulation example of Section 9–1.3.

9–1.9 Direct Measurement of the SSB Modulation Envelope with a High-Frequency Oscilloscope

If the vertical channel response of the oscilloscope is capable of handling the output frequency of a single-sideband (SSB) transmitter (including the sidebands) the output can be applied through the oscilloscope vertical

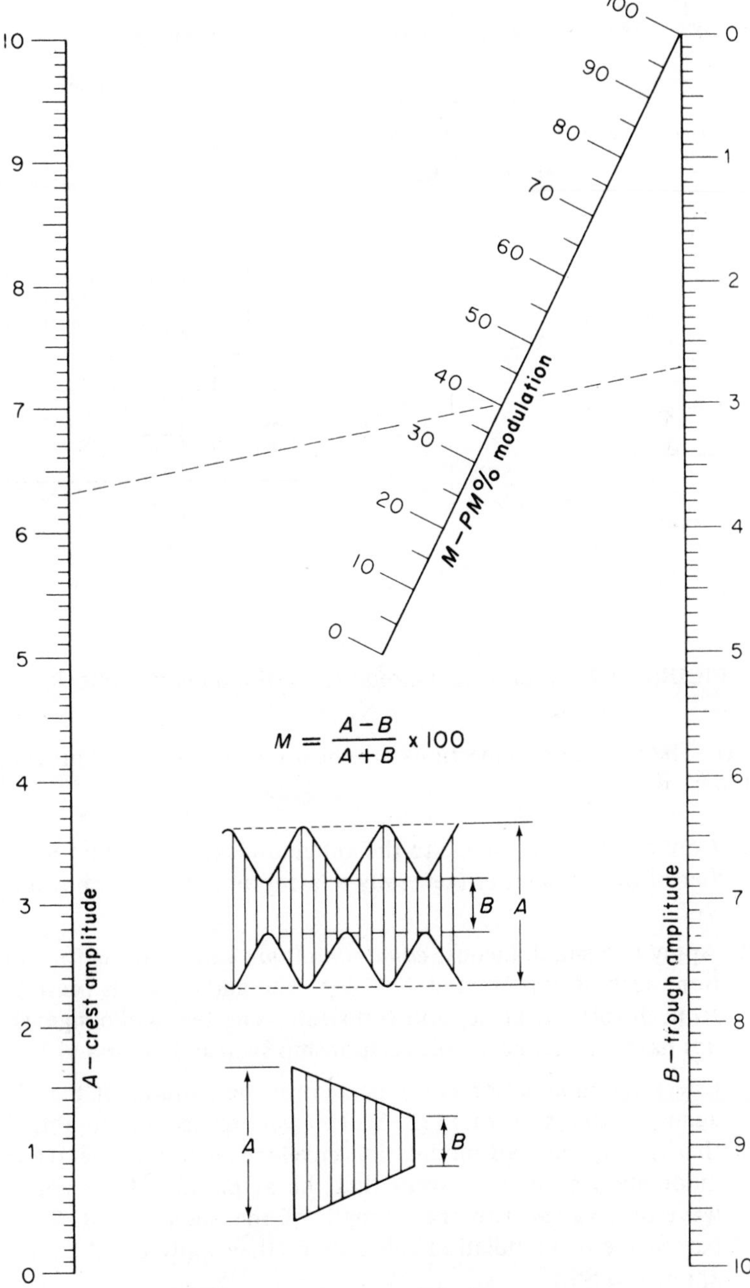

FIGURE 9-6 Pace modulation nomogram.

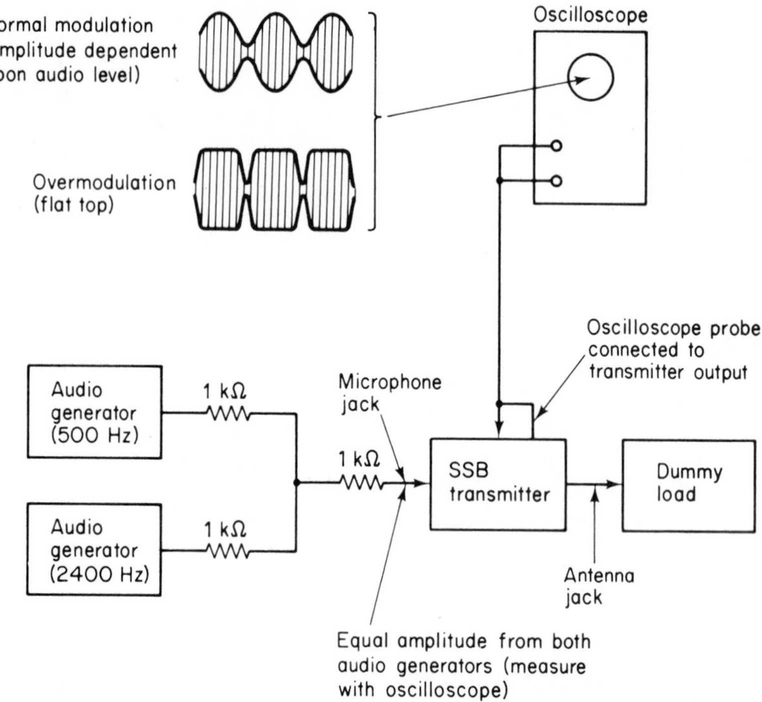

FIGURE 9-7 Typical test circuit for SSB modulation check.

amplifier. The basic test connections are shown in Fig. 9–7. The procedure is as follows:

1. Connect the oscilloscope to the antenna jack, or the final RF amplifier of the transmitter (generally a linear amplifier), as shown in Fig. 9–7.
2. Apply two simultaneous, *equal-amplitude* audio signals for modulation, such as 500 Hz and 2400 Hz. The audio signals *must be free* from distortion, noise, and transients. The two audio signals *must not* have a direct harmonic relationship such as 500 Hz and 1500 Hz.
3. Check the modulation envelope against the patterns of Fig. 9–7, or against patterns shown in the SSB transmitter service literature. Note that the typical SSB modulation envelope resembles the 100% AM modulation envelope, except that the amplitude of the entire SSB waveform varies with the strength of the audio signal. Thus, the percentage of modulation calculations that apply to AM cannot be applied to SSB.
4. Increase the amplitude of both audio modulating signals, making

certain to maintain both signals at equal amplitudes. When peak SSB power output is reached, the modulation envelope will "flat top" as shown in Fig. 9–7. That is, the instantaneous RF peaks of the SSB signal reach saturation, even with less than peak audio signal applied. This overmodulated condition results in distortion.

9–1.10 Direct Measurement of the SSB Modulation Envelope with a Low-Frequency Oscilloscope

If the oscilloscope is not capable of passing the SSB transmitter frequency, the transmitter output can be applied directly to the vertical deflection plates of the oscilloscope cathode-ray tube. However, as in the case of AM modulation test, there are two drawbacks to this approach for SSB. First, the vertical plates may not be readily accessible. Next, the voltage output of the final RF amplifier may not produce sufficient deflection of the oscilloscope trace.

The test connections and modulation patterns are essentially the same as those shown in Fig. 9–7. Similarly, the procedures are the same as those described in Section 9–1.9.

9–2 USING PROBES IN COMMUNICATIONS EQUIPMENT TESTS

In practical communications work, all meters and oscilloscopes operate with some type of probe. In addition to providing for electrical contact to the circuit being tested, probes serve to modify the voltage being measured to a condition suitable for display on an oscilloscope or readout on a meter.

9–2.1 Basic Probe

In its simplest form, the basic probe is a *test prod.* In physical appearance, the probe is a thin metal rod connected to the meter or oscilloscope input through an insulated flexible lead. The entire rod, except for the tip, is covered with an insulated handle so that the probe can be connected to any point of the circuit without touching nearby circuit parts. Sometimes, the probe tip is provided with an alligator clip so that it is not necessary to hold the probe at the circuit point.

Such probes work well on communications circuits carrying dc and audio signals. However, if the alternating current is at a high frequency, or if the gain of the meter (such as an electronic meter) or oscilloscope amplifier is high, it may be necessary to use a special *low-capacitance* probe. Hand capacitance in a simple probe or test prod can cause hum pickup, particularly if amplifier gain is high. This condition may be offset by shielding in low-capacitance probes. More important, however, is the fact that the input impedance of the meter or oscilloscope is connected directly to the circuit being tested when a simple probe is used. Such impedance may disturb circuit conditions.

9-2.2 Low-Capacitance Probes

The basic circuit of a low-capacitance probe is shown in Fig. 9-8. The series resistance R_1 and capacitance C_1, as well as the parallel or shunt R_2, are surrounded by a shielded handle. The values of R_1 and C_1 are preset at the factory and should not be disturbed unless recalibration is required.

In many low-capacitance probes, the values of R_1 and R_2 are selected to form a 10:1 voltage divider between the circuit being tested and the meter or oscilloscope input. Thus, the probes serve the dual purpose of capacitance reduction and voltage reduction. Remember that voltage indications will be one-tenth (or whatever value of attenuation is used) of the actual value when such probes are used. The capacitance value of C_1 in combination with the values of R_1 and R_2 also provide a capacitance reduction, usually in the range 3:1 to 11:1.

There are probes that combine the features of a low-capacitance probe and a basic probe (or test prod). In such probes, a switch (shown as S_1 in Fig. 9-8) is used to short both C_1 and R_1 when a direct input (simple test prod) is required. With S_1 open, both C_1 and R_1 are connected in series with the input, and the probe provides the low-capacitance and voltage-division features.

9-2.3 High-Voltage Probes

High-voltage probes are rarely, if ever, used in modern solid-state communications equipment. Most solid-state circuits operate with voltages of less than 15 V. Even vacuum-tube communications equipment uses voltage of 300 V or less, except for the very high power transmitters, such as broadcast

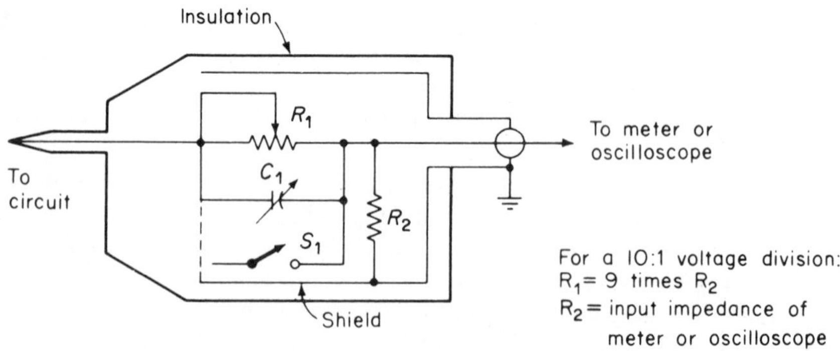

FIGURE 9-8 Typical low-capacitance probe circuit.

transmitters. However, some meters are supplied with high-voltage probes, either as accessories or built in. If you should use these probes, the voltage indications are 10:1, 100:1, or even 1000:1, depending on the attenuation factor.

9-2.4 RF Probes

As discussed in Section 8-1, when the signals to be measured are at radio frequencies and are beyond the capabilities of the meter or oscilloscope, an RF probe is required. The half-wave RF probe discussed in Section 8-1.1 is generally sufficient for most communications equipment work. A full-wave RF probe is discussed in Section 9-2.9.

9-2.5 Demodulator Probes

As discussed in Section 8-1.2, the circuit of a demodulator probe is essentially the same as that of the RF probe, but the circuit values and basic functions are somewhat different. When the high-frequency signals contain modulation (which is typical for the modulated RF carrier signals of most communications equipment), a demodulator probe is more effective for signal tracing. The demodulator probe discussed in Section 8-1.2 is generally sufficient for most communications equipment work.

9-2.6 Transistorized Signal-Tracing Probes

It is possible to increase the sensitivity of a probe with a transistor amplifier. Such an arrangement is particularly useful with a VOM for measuring small-signal voltages during test. An amplifier is usually not required for an electronic meter or oscilloscope because such instruments contain built-in amplifiers.

A transistor probe and amplifier circuit is shown in Fig. 9-9. This circuit increases the sensitivity of the probe by at least 10:1 and provides good response up to about 500 MHz. The circuit is not normally calibrated to provide a specific voltage indication. Rather, the circuit is used to increase the sensitivity of the probe for signal tracing in communications equipment circuits.

9-2.7 Probe Compensation and Calibration

Probes must be calibrated to provide a proper output to the meter or oscilloscope with which they are to be used. Probe compensation and calibration are done at the factory and require precision test equipment. The following paragraphs describe the *general procedures* for compensating and calibrating probes. Never attempt to adjust a probe unless you follow the in-

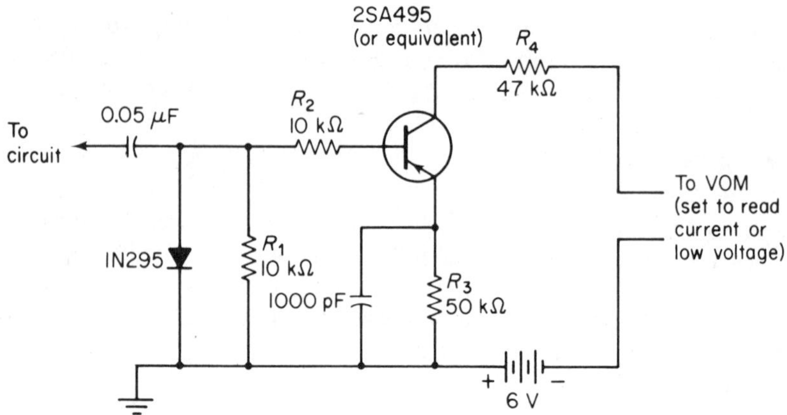

FIGURE 9-9 Typical transistorized signal-tracing probe circuit.

struction manual and have the proper test equipment. An improperly adjusted probe produces erroneous readings, and may cause undesired circuit loading.

Probe Compensation. The capacitors that compensate for excessive attenuation of high-frequency signal components (through the probe's resistance dividers) affect the entire frequency range from some midband point upward. Capacitor C_1 in Fig. 9–8 is an example of such a compensating capacitor.

Compensating capacitors must be adjusted so that the higher-frequency components are attenuated by the same amount as low frequency and direct current. It is possible to check the adjustment of the probe-compensating capacitors using a square-wave signal source. This is done by applying the square-wave signal directly to the oscilloscope input and then applying the same signals through the probe and noting any change in pattern. In a properly compensated probe, there should be no change (except for a possible reduction of the amplitude).

Figure 9–10 shows typical square-wave displays with the probe properly compensated, undercompensated (high frequencies underemphasized), and overcompensated (high frequencies overemphasized). Proper compensation of probes is often neglected, especially when probes are used interchangeably with meters or oscilloscopes having different input characteristics. It is recommended that any probe be checked with square-wave signal before it is used in a test.

Another problem related to probe compensation is that the input capacitance of the meter or oscilloscope may change with age. Also, in the case of older, vacuum-tube meters and oscilloscopes, the input capacitance may change when tubes are changed. Either way, the compensated dividers

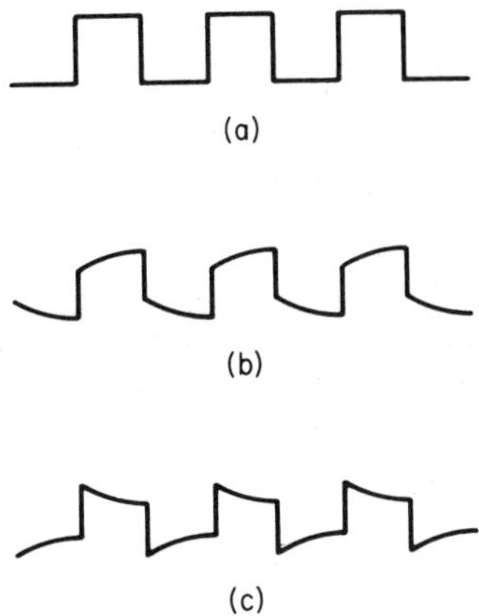

FIGURE 9-10 Typical square-wave displays showing frequency compensation of probes: (a) proper compensation; (b) under compensation of highs; (c) overcompensation of highs.

may be improperly adjusted. Readjustment of the probe will not necessarily correct for the change needed by the input circuits of the meter or oscilloscope.

Probe Calibration. The main purpose of probe calibration is to provide a specific output for a given input. For example, the value of R_1 in Fig. 8-1 is adjusted (or selected) to provide a specific amount of voltage to the meter or oscilloscope. During calibration, a voltage of known value and accuracy is applied to the input. The output is monitored, and R_1 is adjusted to produce a given value (0.707 of RF peak value, etc.).

9-2.8 Probe-Test Techniques

Although a probe is a simple instrument and does not require specific operating procedures, several points should be considered in order to use a probe effectively in test.

Circuit Loading. When a probe is used, the probe's impedance (rather than the meter's or oscilloscope's impedance) determines the amount of circuit loading. Connecting a meter or oscilloscope to a circuit may alter the signal at the point of connection. To prevent this, the impedance of the measuring device must be large in relation to that of the circuit being tested. Thus, a high-impedance probe offers less circuit loading, even though the meter or oscilloscope may have a lower impedance.

Measurement Error. The ratio of the two impedances (of the probe and the circuit being tested) represents the amount of probable error. For example, a ratio of 100:1 (perhaps a 100-MΩ probe to measure the voltage across a 1-MΩ circuit) accounts for an error of about 1%. A ratio of 10:1 produces an error of about 9%.

Effects of Frequency. The input impedance of a probe is not the same at all frequencies. Input impedance becomes smaller as frequency increases. (Capacitive reactance and impedance decrease with an increase in frequency.) All probes have some input capacitance. Even an increase at audio frequencies may produce a significant change in impedance.

Shielding Capacitance. When using a shielded cable with a probe to minimize pickup of stray signals and hum, the additional capacitance of the cable should be considered. The capacitance effects of a shielded cable can be minimized by terminating the cable at one end in its characteristic impedance. Unfortunately, this is not always possible with the input circuits of most meters and oscilloscopes.

Relationship of Loading to the Attenuation Factor. The reduction of loading (either resistive or capacitive) due to use of probes may not be the same as the attenuation factor of the probe. (Capacitive loading is almost never reduced by the same amount as the attenuation factor because of the additional capacitance of the probe cable.) For example, a typical 5:1 attenuator probe may be able to reduce capacitive loading by about 2:1. A 50:1 attenuator probe may reduce capacitive loading by about 10:1. Beyond this point, little improvement can be expected because of the stray capacitance at the probe tip.

Checking Effects of the Probe. When testing communications circuits (or any circuits for that matter), it is possible to check the effect of a probe on a circuit by making the following simple test. Attach and detach another connection of similar kind (such as another probe) and observe any difference in meter reading or oscilloscope display. If there is little or no change when the additional probe is touched to the circuit, it is safe to assume that the probe has little effect on the circuit.

Probe Length and Connections. Long probes should be restricted to the measurement of relatively slowly changing signals (direct current and low-frequency ac). The same is true for long ground leads. The ground lead should be connected where no hum or high-frequency signal components exist in the ground path between that point and the signal-pickoff point.

Measuring High Voltages. Avoid applying more than the rated voltage to a probe. Fortunately, most commercial probes will handle the highest voltages found in communications equipment, even vacuum-tube equipment.

9-2.9 *Full-Wave RF Probe for Communications Equipment Tests*

Figure 9–11 shows the circuit diagram of a probe suitable for communications tests. The probe is designed specifically for use with a VOM or digital electronic meter, and converts both audio-frequency and RF signals to direct current. Since the probe is full-wave, it produces a larger output signal (for a given input signal) than the probes described in Section 8–1. However, the probe of Fig. 9–11 is essentially a signal-tracing device and is not designed to provide accurate readings.

The probe of Fig. 9–11 will operate satisfactorily up to about 250 MHz. The meter used with the probe must be set to read direct current, because the probe output is dc. However, if the RF input signal is amplitude-modulated, the probe output will be pulsating direct current.

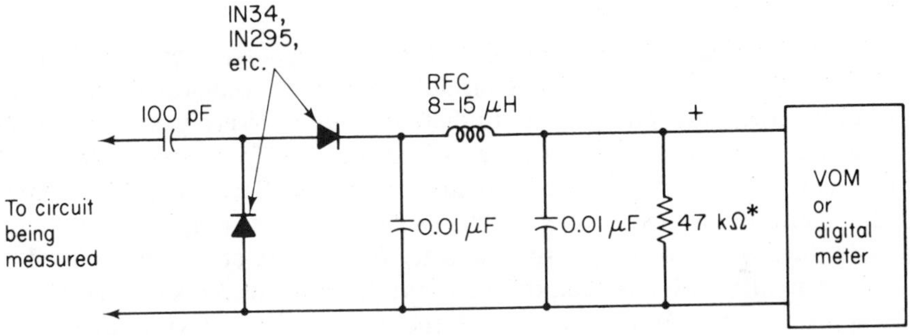

*Not used with VOM.

FIGURE 9–11 Full-wave RF probe for communications equipment test.

9-3 USING FREQUENCY METERS AND COUNTERS IN COMMUNICATIONS EQUIPMENT TESTS

There are two basic types of frequency-measuring devices for communications equipment tests: the heterodyne or zero-beat frequency meter and the digital electronic counter.

9-3.1 *Heterodyne or Zero-Beat Frequency Meter*

In the early days of radio communications, the heterodyne meter was the only practical device for frequency measurement of transmitter signals. Figure 9–12 shows the block diagram of a basic heterodyne frequency meter. The

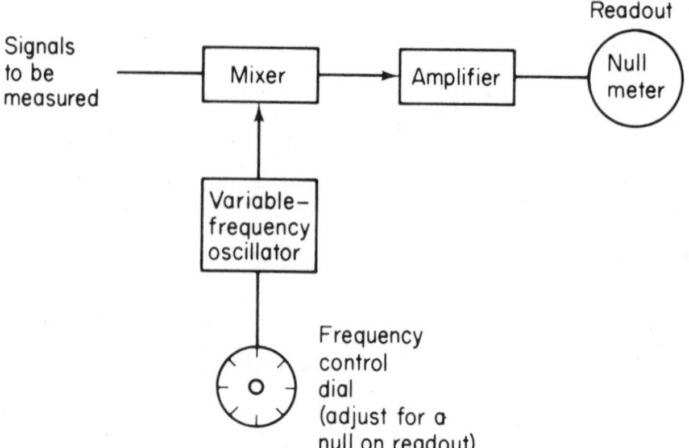

FIGURE 9-12 Basic heterodyne or zero-beat frequency meter circuit.

signals to be measured are applied to a mixer, together with the signals of known frequency (usually from a variable-frequency oscillator in the meter). The meter oscillator is adjusted until there is a null or "zero beat" on the output device, indicating that the oscillator is at the same frequency as the signals to be measured. This frequency is read from the oscillator frequency control dial. Precision frequency meters often include charts or graphs to help interpret frequency dial readings, so that exact frequencies can be pinpointed.

As an alternative system, the meter produces fixed frequency signals that are applied to the mixer, together with the signals of unknown frequency. As an example, one such frequency meter provides 40 crystal-controlled signals, one signal for each of the 40 CB (citizens'-band) channels. Both the CB transmitter and frequency meter are set to the same channel, and any deviation is read out on the frequency meter indicator.

9-3.2 Electronic Digital Counter

The electronic counter has become far more popular for communications equipment tests than the heterodyne frequency meter. One reason is that the counter is generally easier to operate and has much greater resolution or readability. Using the counter, you need only connect the test leads to the circuit or test point, select a time base and attenuator/multiplier range, and read the signal frequency on a convenient digital readout.

Digital Counter Basics. Although there are many types of digital counters, all counters have several basic functional sections in common. These sections are interconnected in a variety of ways to perform the various counter

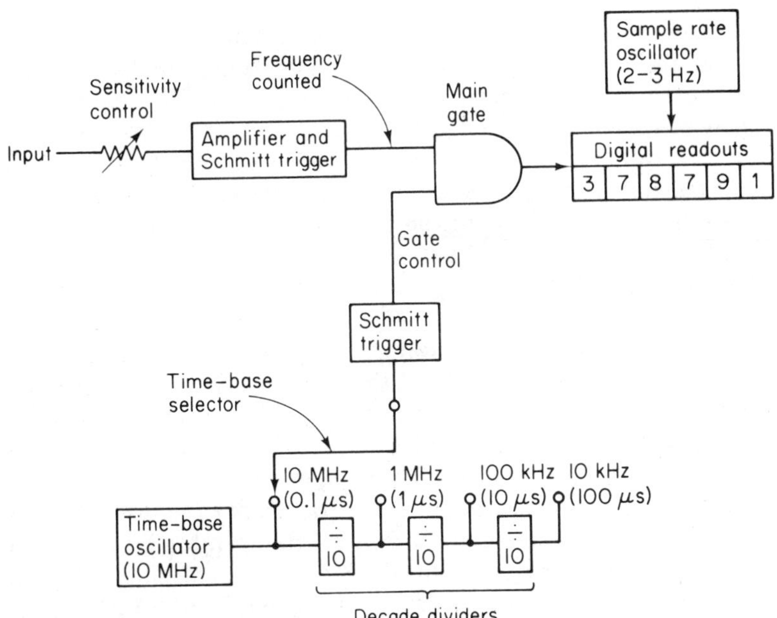

FIGURE 9–13 Basic digital counter circuit for frequency measurement operation.

functions. Figure 9–13 shows the basic counter circuit for frequency measurement operation (which is the most common of the various counter functions used in communications test).

All electronic counters have some form of *counter and readout*. Early instruments used binary counters and readout tubes that converted the binary count to a decade readout. Such instruments have (generally) been replaced by decade counters that convert the count to *binary-coded-decimal* (BCD) form, decoders that convert the BCD data to decade form (generally BCD to seven-segment decoders), and readouts that display the decade information (generally seven-segment LEDs, LCDs, etc.). One readout or display is provided for each digit. For example, eight readouts will provide for a count up to 99,999,999.

All electronic counters have some form of *main gate* that controls the count start and stop with respect to time. Usually, the main gate is some form of AND gate. All electronic counters have some form of *time base* that supplies the precise increment of time to control the gate for a frequency or pulse-train measurement. Usually, the time base is a crystal-controlled oscillator. The accuracy of the counter depends on the accuracy of the time base, plus or minus one count. For example, if the time-base accuracy is 0.005%, the overall accuracy of the counter is 0.005%, plus or minus one count. The one-count error arises because the count may start and stop in the middle of an input pulse, thus omitting the pulse from the total count. Or part of a pulse may pass through the gate before the gate closes, thus adding a pulse to the count.

Most electronic counters have *dividers* that permit variation of gate time. These dividers convert the fixed-frequency time base to several other frequencies. In addition to the four basic sections, most electronic counters have attenuator networks, amplifier and trigger circuits to shape a variety of input signals to a common form, and logic circuits to control operation of the instrument. The various modes of electronic counter operation are described in the following paragraphs.

Frequency-Measurement Operation. For frequency measurements, the digital counter circuits are arranged as shown in Fig. 9–13. As shown, the input signal is first converted to uniform pulses by the Schmitt trigger. These pulses are then routed through the main gate and into the counter/readout circuits, where the pulses are totalized. The number of pulses totalized during the "gate-open" interval is a measure of the *average input frequency* for that interval. (For example, assume that the gate is held open for 1 s and the count is 333. This indicates a frequency of 333 Hz.) The count obtained, with the correct decimal point, is then displayed and retained until a new sample is ready to be shown. The sample rate oscillator determines the time between samples (not the interval of gate opening and closing), resets the counter, and thus initiates the next measurement cycle.

The time-base selector switch selects the gating interval, thus positioning the decimal point and selecting the appropriate measurement units. The time base selector selects one of the frequencies from the time-base oscillator. If the 10-MHz signal (directly from the time base) is selected, the time interval (gate-open to gate-close) is 1/10 μs. If the 1-MHz signal (from the first decade divider) is chosen, the measurement time interval is 1 μs.

Totalizing Operation. For totaling measurements, the digital counter circuits are arranged as shown in Fig. 9–14. As shown, the main gate is controlled by a manual START/STOP switch. With the switch in START (gate open), the counter totalizes the input pulses until the main gate is closed by the switch being changed to STOP. The counter display then represents the input pulses received during the interval between manual START and manual STOP.

Period-Measurement Operation. For period measurement, the digital counter circuits are arranged as shown in Fig. 9–15. Period is the inverse of frequency (period = 1/frequency). Thus, period measurements are made with the input and time-base connections reversed from those of frequency measurement, as shown. The unknown input signal controls the main gate time, and the time-base frequency is counted and read out. For example, if the base frequency is 1 MHz, the indicated count is in microseconds (a count of 70 indicates that the gate has been held open for 70 μs). Usually, the input-

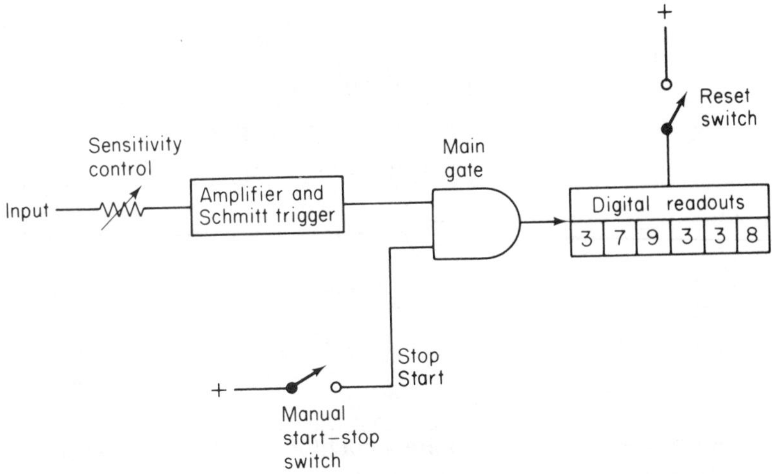

FIGURE 9-14 Basic digital counter circuit for totalizating operation.

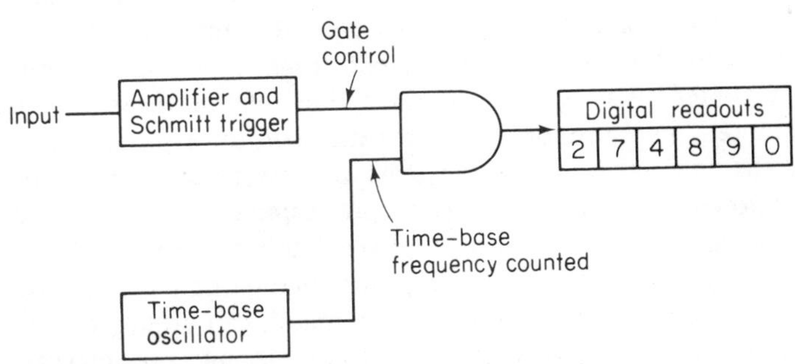

FIGURE 9-15 Basic digital counter circuit for period measurement operation.

shaping circuit selects the positive-going zero crossing of successive cycles of the unknown signal as trigger points for opening and closing of the gate.

The accuracy and resolution of period measurements can be increased by *period averaging*. The connections are shown in Fig. 9–16. These are the same connections as for regular period measurements, except that the signal to be measured is lowered in frequency by dividers, thus extending the gate-open period. For example, if the input signal is 1 kHz, the period is 1 ms with a conventional period measurement. If the time base is 1 MHz, the count is 00001000 on an eight-digit readout. If the period-average method is used, and the input frequency is reduced to 1 Hz, as shown in Fig. 9–16, the period is 1 s, and the count is 01000000 on the same eight-digit readout. Thus, the resolution is increased by 10^3.

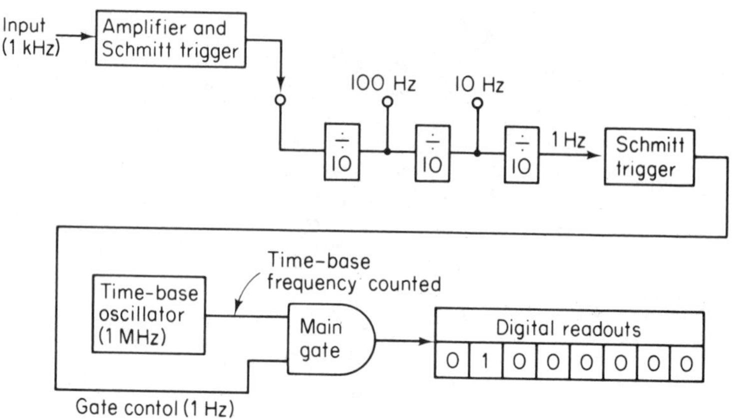

FIGURE 9-16 Basic digital counter circuit for period averaging measurement operation.

Time-Interval Operation. Time-interval measurements are essentially the same as period measurements. However, the time-interval mode is more concerned with time between two events than with the repetition rate of signals. Counters vary greatly in their time-interval-measuring capability. Some counters measure only the duration of an electrical event; others measure the interval between the start of two pulses. The most versatile models, known as universal counters, have separate inputs for the start and stop commands and have separate controls that permit setting the trigger-level amplitude, polarity, slope, and type of input coupling (ac or dc) for the start and stop channel. Since stop and start commands can originate from common or separate sources, this type of counter can measure the interval from one point on a waveform to another point on the same waveform.

The basic time-interval measurement circuit is shown in Fig. 9–17. Note that the time-base signals are counted and read out when the gate is open. Control of the gate is accomplished by two trigger circuits that receive their inputs from the signal being measured. With switch S_1 in the SEPARATE position, the two triggers receive inputs from separate lines. Assume that the START (gate-open) trigger receives its input from a signal applied to an amplifier under test, while the amplifier's output is applied to the STOP (gate-closed) trigger. Under these circumstances, the gate is open for an interval of time equal to the delay through the amplifier under test. If the time base is 1 MHz and the counter produced a readout of 33, the delay is 33 μs.

With switch S_1 in the COMMON position, the two triggers receive inputs from the same lines. With this arrangement, each trigger is adjusted so that it responds to a different portion of the same waveform. Assume that the START trigger opens the gate when the input signal rises to +10 V and the STOP trig-

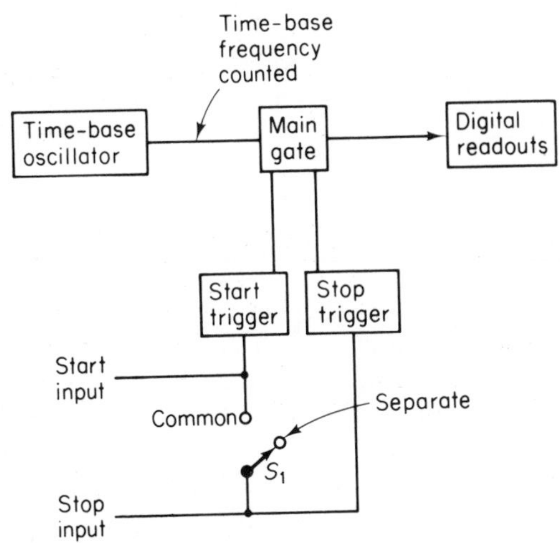

FIGURE 9–17 Basic digital counter circuit for time-interval measurement operation.

ger closes the gate when the input signal reaches +15 V. Thus, the count represents the time interval between the two points.

Counter Accuracy. As discussed, the accuracy of a frequency counter is set by the stability of the time base rather than the readout. The readout is typically accurate to within ±1 count. The time base of the Fig. 9–13 counter is 10 MHz and is stable to within ±10 ppm (parts per million), or 100 Hz. The time base of a precision laboratory counter could be in the order of 4 MHz and is stable to within ±1 ppm, or 4 Hz.

Counter Resolution. The resolution of an electronic counter is set by the number of digits in the readout. For example, assume that you must use a five-digit counter to test CB set operation. The CB operating frequencies or channels are in the 27-MHz range. Now assume that you measure a 27-MHz signal with the five-digit counter. The count could be 26.999 or 27.001, or within 1000 Hz of 27 MHz. Since the FCC requires that the operating frequency of a CB set be held within 0.005% (or about 1350 Hz in the case of a 27-MHz signal), a digital counter for CB tests must have a minimum of five digits in the readout.

Combining Accuracy and Resolution. To find out if a counter is adequate for a particular communications test, add the time-base stability (in terms of frequency) to the resolution at the operating frequency. Again using the CB set example, if the accuracy is 100 Hz, and the count can be resolved to

1000 Hz (at the measurement frequency), the maximum possible inaccuracy is 1000 + 100 Hz, or 1100 Hz. This is within the approximate 1350 Hz (0.005% of 27 MHz) required.

9-3.3 Calibration Check of Frequency Meters and Counters

The accuracy of frequency-measuring devices (both meters and counters) used for communications tests should be checked periodically, at least every 6 months. Always follow the procedures recommended in the frequency meter or counter service instructions. Generally, you can send the instrument to a calibration lab, or to the factory, or you can maintain your own frequency standard. (This latter is generally not practical for most communications service shops.)

No matter what standard is used, keep in mind that the standard must be more accurate, and have better resolution, than the frequency-measuring device, just as the meter or counter must be more accurate than the communications equipment.

9-3.4 Using WWV Signals for Frequency Calibration

In the absence of a frequency standard, or factory calibration, you can use the frequency information broadcast by U.S. government radio station WWV. These WWV signals are broadcast on 2.5, 5, 10, 15, 20, and 25 MHz continuously night and day, except for silent periods of approximately 4 minutes beginning 45 minutes after each hour. Broadcast frequencies are held accurate to within 5 parts in 10^{11}. This is far more accurate than that required for most communications equipment tests.

The hourly broadcast schedules of WWV are shown in Fig. 9-18. However, these schedules are subject to change. For full data on WWV broadcasts, refer to *NBS* (National Bureau of Standards) *Standard Frequency and Time Services* (Miscellaneous Publication 236), available from the Superintendent of Documents, U.S. Government Printing Office, Washington, D.C. 20402.

It is the continuous-wave (CW) signals broadcast by WWV that provide the most accurate means of calibrating (or checking) frequency meters and counters. It is not practical to use the signal directly, except on some special frequency meters, but the test connections for check are not complex.

Figure 9-19 shows the basic test connections for checking the accuracy of a frequency counter using WWV. Note that a receiver and signal generator are required. The accuracy of the signal generator and receiver are not critical,

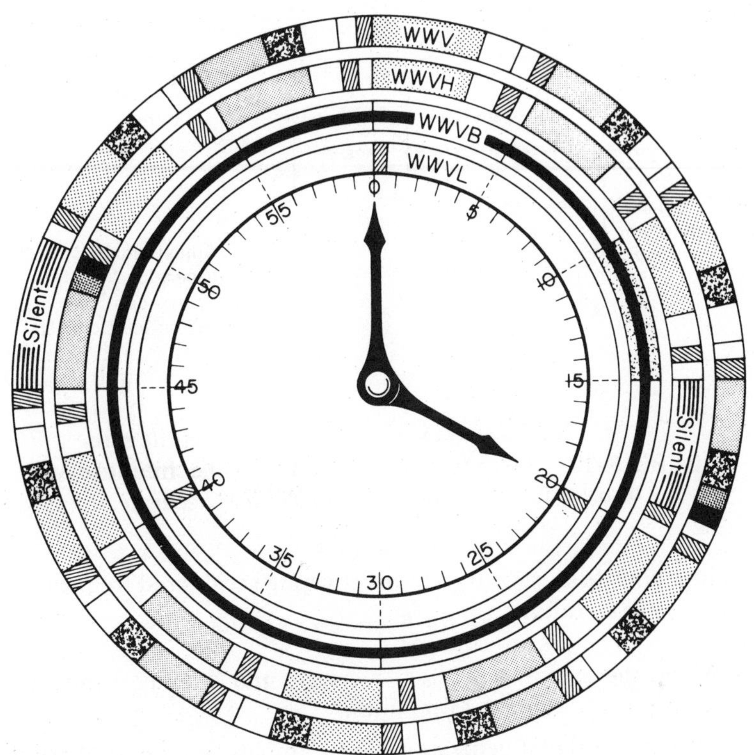

Seconds pulses – WWV, WWVH – continuous except for 59th second of each minute and during silent periods

WWVB – special time code

WWVL – none

WWV – morse code – call letters, universal time, propagation forecast

voice – mountain standard time

morse code – frequency offset (on the hour only)

WWVH – morse code – call letters, universal time,

voice – Hawaiian standard time

morse code – frequency offset (on the hour only)

WWVL – morse code – call letters, frequency offset

Station announcement

100 pps 1000 Hz modulation WWV timing code

Tone modulation 600 Hz

Tone modulation 440 Hz

Geoalerts

Identification phase shift

UT–2 time correction

Special time code

FIGURE 9-18 Hourly broadcast schedules of WWV.

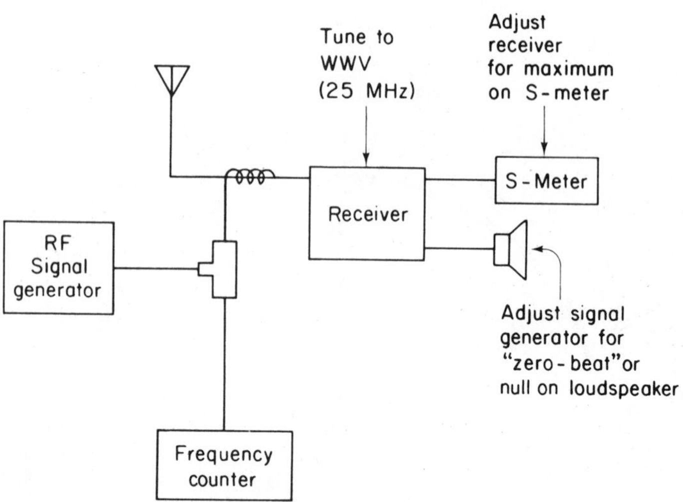

FIGURE 9-19 Basic test connections for checking the accuracy of a frequency counter using WWV.

but both instruments must be capable of covering the desired frequency range. The procedure is as follows:

1. Allow the signal generator, receiver, and counter being tested to warm up for at least 15 minutes.
2. Reduce the signal generator output amplitude to zero. Turn off the signal generator output if this is possible without turning off the entire signal generator.
3. Tune the receiver to the desired WWV frequency. It is generally best to use a WWV frequency that is near the operating frequency of the communications equipment. For example, if a 27-MHz CB set is being tested, use the 25-MHz WWV signal.
4. Operate the receiver controls until you can hear the WWV signal in the receiver loudspeaker.
5. If the receiver is of the communication type, it will have a beat-frequency oscillator (BFO) and output signal strength or S meter. Turn on the BFO, if necessary, to locate and identify the WWV signal. Then tune the receiver for maximum signal on the S meter. The receiver is now exactly on 25 MHz, or whatever WWV frequency is selected.
6. Turn on the signal generator, and tune the generator until it is at "zero beat" against the WWV signal. As the signal generator is adjusted so that its frequency is close to that of the WWV signal (so

that the difference in frequency is within the audio range), a tone, whistle, or "beat note" will be heard on the receiver. When the signal generator is adjusted to exactly the WWV frequency (25 MHz in this case), there is no "difference" signal, and the tone can no longer be heard. In effect, the tone drops to zero and the two signals (generator and WWV) are at "zero beat."

7. Read the counter. The readout should be equal to the WWV frequency. For example, with a five-digit counter at 25 MHz, the reading should be 24.999 to 25.001.
8. Repeat the procedure at other WWV broadcast frequencies.

9-4 USING A DUMMY LOAD IN COMMUNICATIONS EQUIPMENT TESTS

Never adjust a radio transmitter without an antenna or load connected to the output. This will almost certainly cause damage to the transmitter circuits. When a transmitter is connected to an antenna or load, power is transferred from the final RF stage to the antenna or load. Without an antenna or load, the final RF stage must dissipate the full power and will probably be damaged. Equally important, you should not make any major adjustments to a transmitter that is connected to a radiating antenna. You will probably cause interference.

These two problems can be overcome by means of a nonradiating load, commonly called a dummy load. There are a number of commercial dummy loads for communications equipment tests. The RF wattmeters described in Section 9-5 and the special test sets covered in Section 9-12 contain dummy loads. It is also possible to make up dummy loads suitable for most communications equipment tests. There are two generally accepted dummy loads: the fixed resistance and the lamp. Keep in mind that these loads are for routine tests; they are not a substitute for an RF wattmeter or special test set.

9-4.1 *Fixed-Resistor Dummy Load*

The simplest dummy load is a fixed resistor capable of dissipating the full power output of the transmitter. The resistor can be connected to the transmitter antenna connector by means of a plug, as shown in Fig. 9-20.

Most communications transmitters operate with a 50-Ω antenna and lead-in, and thus require a 50-Ω resistor. The nearest standard resistor is 51Ω. This 1-Ω difference is not critical. However, it is essential that the resistor be noninductive (composition or carbon), never wire-wound. Wire-wound resistors have some inductance, which changes with frequency. Thus, the load (impedance) presented by the wire-wound resistor changes with frequency.

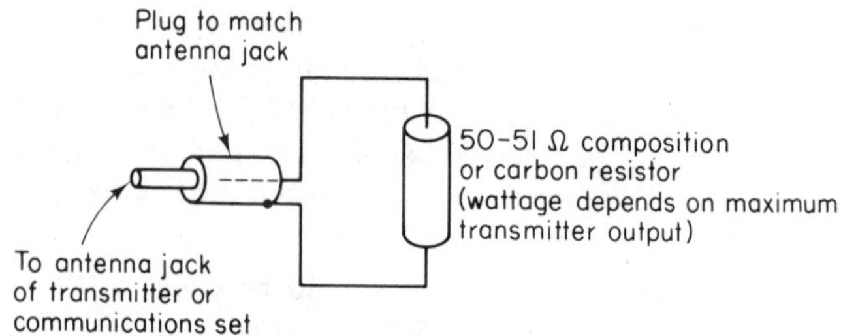

FIGURE 9-20 Fixed-resistor dummy load.

Always use a resistor with a power rating greater than the anticipated maximum output power of the transmitter. For example, an AM CB transmitter can (legally) have a 5-W input, which results in an output of about 4 W. A 7- to 10-W resistor should be used for the dummy load. An SSB CB transmitter should not produce more than 12 W of output with full modulation. Thus, a 15- to 20-W dummy-load resistor can be used.

RF Power Output Measurement with a Dummy-Load Resistor. It is possible to get an approximate measurement of RF power output from a radio transmitter with a resistor dummy load and a suitable meter. Again, these procedures are not to be considered a substitute for power measurement with an accurate RF wattmeter.

The procedure is simple. Measure the voltage across the 50-Ω dummy-load resistor and find the power with the equation

$$\text{power} = \frac{(\text{voltage})^2}{50}$$

For example, if the voltage measured is 14 V, the power output is

$$\frac{(14)^2}{50} = 3.92 \text{ W}$$

which is typical for an AM CB transmitter.

Certain precautions must be observed. First, the meter must be capable of producing accurate voltage indications at the transmitter operating frequency. This usually requires a meter with an RF probe (preferably a probe calibrated with the meter). An AM transmitter should be checked with an RMS voltmeter and with no modulation applied. An SSB transmitter must be checked with a peak-reading voltmeter and with modulation applied (since SSB produces no output without modulation). This usually involves connecting an audio generator to the microphone input of the SSB transmitter circuits.

Always follow the service literature recommendations for all RF power output measurements (frequency, channels, operating voltages, modulation, etc.). However, as guidelines for CB set operation, an AM set should produce no more than 4 W output, with or without modulation. SSB sets should not produce more than 12 W of output with full modulation (speaking loudly into the microphone).

9-4.2 Lamp Dummy Load

Lamps have been the traditional dummy loads for communications equipment tests. For example, the No. 47 lamp (often found as a pilot lamp in many electronic instruments) provides the approximate impedance and power dissipation required as a CB dummy load. The connections are shown in Fig. 9-21.

You cannot get an accurate measurement of RF power output when a lamp is used as the dummy load. However, the lamp provides an indication of relative power and shows the presence of modulation. The intensity of the light produced by the lamp varies with modulation (more modulation, brighter glow). Thus, you can tell at a glance if the transmitter is producing an RF carrier (steady glow), and if modulation is present (varying glow).

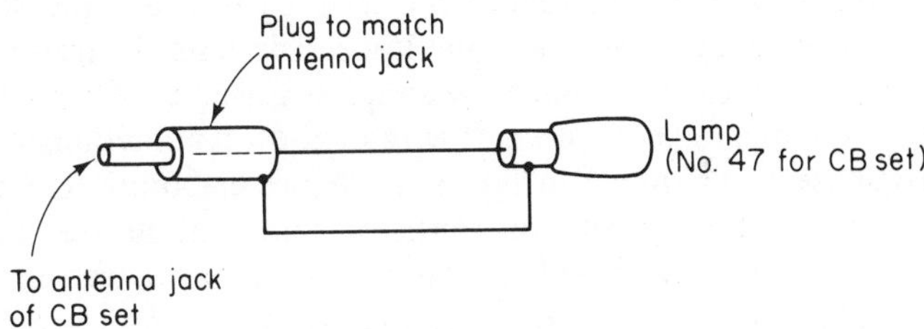

FIGURE 9-21 Lamp dummy load.

9-5 USING RF WATTMETERS IN COMMUNICATIONS EQUIPMENT TESTS

A number of commercial RF wattmeters are available for communications equipment tests. Also, the special test sets described in Section 9-12 usually include an RF wattmeter. The basic RF wattmeter consists of a dummy load (fixed resistor) and a meter that measures voltage across the load, but reads out in watts (rather than in volts), as shown in Fig. 9-22. You simply connect the RF wattmeter to the antenna connector of the set (transmitter output), key the transmitter, and read the power output on the wattmeter scale.

Although operation is simple, you must remember that SSB transmitters require a peak-reading wattmeter to indicate PEP (peak envelope power), whereas an AM set uses an RMS-reading wattmeter. Most commercial RF wattmeters are RMS-reading, unless specifically designed for SSB.

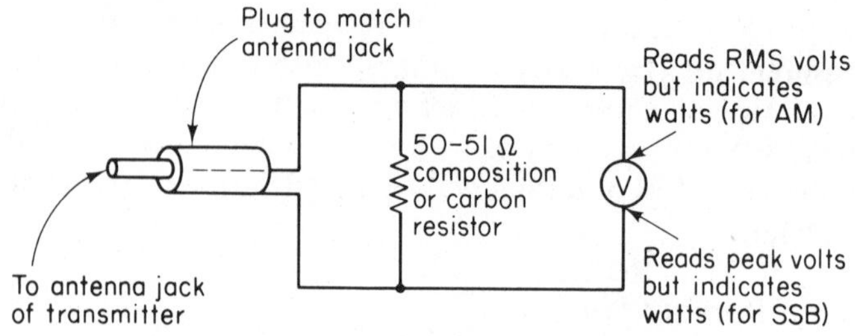

FIGURE 9-22 Basic RF wattmeter circuit.

9-6 USING FIELD-STRENGTH METERS IN COMMUNICATIONS EQUIPMENT TESTS

There are two basic types of field-strength meters: the simple relative field-strength (RFS) meter and the precision laboratory or broadcast-type instrument. Most communications equipment tests can be carried out with simple RFS instruments. An exception is where you must make precision measurements of broadcast antenna radiation patterns.

The purpose of a field-strength meter is to measure the strength of signals radiated by an antenna. This simultaneously tests the transmitter output, the antenna, and the lead-in. In the simplest form, a field-strength meter consists of an antenna (a short piece of wire or rod), a potentiometer, diodes, and a microammeter, as shown in Fig. 9–23. More elaborate RFS meters include a tuned circuit and possibly a transistor amplifier. In use, the meter is placed near the antenna at some location accessible to the transmitter or set (where you can see the meter), the transmitter is keyed, and the *relative* field

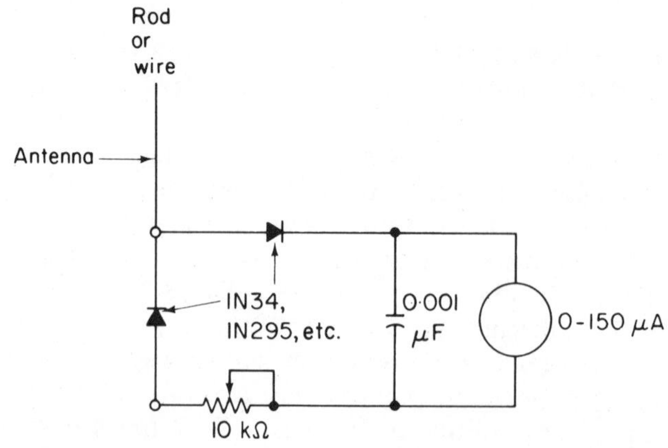

FIGURE 9-23 Basic relative-field-strength (RFS) meter circuit.

strength is indicated on the meter. Some of the special test sets described in Section 9-12 include an RFS test function.

9-7 STANDING-WAVE-RATIO MEASUREMENT

The standing-wave ratio (SWR) of an antenna is actually a measure of match or mismatch for the antenna, transmission line (lead-in), and the communications set (receiver-transmitter). When the impedances of the antenna, line, and set are perfectly matched, all the energy or signal is transferred to or from the antenna, and there is no loss. If there is a mismatch (as is the case in any practical application), some of the energy or signal is reflected back into the line. This energy cancels part of the desired signal.

If the voltage (or current) is measured along the line, there are voltage or current maximums (where the reflected signals are in-phase with the outgoing signals) and voltage or current minimums (where the reflected signal is out of phase, partially canceling the outgoing signal). The maximums and minimums are called *standing waves.* The ratio of the maximum to the minimum is the standing-wave ratio (SWR). The ratio may be related to either voltage or current. Since voltage is easier to measure, it is usually used, resulting in the common term *voltage standing-wave ratio* (VSWR). The theoretical calculations for VSWR are shown in Fig. 9-24.

An SWR of 1-to-1, expressed as 1:1, means that there are no maximums or minimums (the voltage is constant at any point along the line) and that there is a perfect match for set, line, and antenna. As a practical matter, if this 1:1 ratio should occur on one frequency, it will not occur at any other frequency, since impedance changes with frequency. It is not likely that all three elements (set, antenna, line) will change impedance by the exact same amount on all frequencies. Therefore, when checking SWR, always check on all frequencies or channels, where practical. As an alternative, check SWR at the high, low, and middle channels or frequencies.

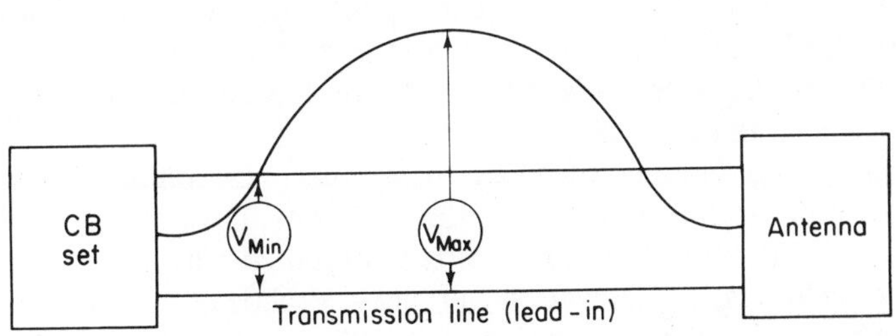

$$\text{Voltage standing wave ratio VSWR} = \frac{V_{Max}}{V_{Min}}$$

FIGURE 9-24 Calculations for voltage standing-wave ratio (VSWR).

In the case of microwave signals being measured in the laboratory, a meter is physically moved along the line to measure maximum and minimum voltages. This is not practical at most communications equipment frequencies, due to the physical length of the waves. In communications equipment, it is far more practical to measure forward and outgoing voltage and reflected voltage, and then calculate the *reflection coefficient* (reflected voltage/outgoing voltage). The relationship of reflection coefficient to SWR is as follows:

$$\text{reflection coefficient} = \frac{\text{reflected voltage}}{\text{forward voltage}}$$

For example, using a 10-V forward and a 2-V reflected voltage, the reflection coefficient is 0.2.

Reflection coefficient is converted to SWR by dividing (1 + reflection coefficient) by (1 − reflection coefficient). For example, using the 0.2 reflection coefficient, the SWR is

$$\frac{1 + 0.2}{1 - 0.2} = \frac{1.2}{0.8} = 1.5 \text{ SWR}$$

This may be expressed as 1 to 1.5, 1.5 to 1, 1.5:1, 1:1.5, or simply as 1.5, depending on the meter scale. In practical terms, an SWR of 1.5 is poor, since it means that at least 20% of the power is being reflected.

SWR can be converted to reflection coefficient by dividing (SWR − 1) by (SWR + 1). For example, using 1.5 SWR, the reflection coefficient is

$$\frac{1.5 - 1}{1.5 + 1} = \frac{0.5}{2.5} = 0.2 \text{ reflection coefficient}$$

In the commercial SWR meters used for communications work, it is not necessary to calculate either reflection coefficient or SWR. This is done automatically by the SWR meter. (The meter is actually reading reflection coefficient, but the scale reads in SWR. If you have a reflection coefficient of 0.2, the SWR reading is 1.5.)

There are a number of SWR meters used in communications work. Some communications sets, such as most CB sets, have built-in SWR meters and circuits. The SWR function is often combined with other measurement functions (field strength, power output, etc.). Practically all of the special test sets described in Section 9–12 include an SWR measurement feature, since it is so important to proper operation of communication sets.

Basic SWR meter circuits are quite simple, and it is possible to build them in the shop. However, it is not practical in most cases to do so. The basic circuit requires that a *directional coupler* and pickup wires be inserted in the transmission line. Even under good conditions, a mismatch and some power loss may result. A poorly designed pickup may result in considerable power loss, as well as inaccurate readings. Thus, it is more practical to use commercial SWR meters.

The basic SWR meter circuit is shown in Fig. 9-25. Operation of the circuit is as follows. As shown, there are two pickup wires, both parallel to the center conductor of the transmission line. Any RF voltage on either of the parallel pickups is rectified and applied to the meter through switch S_1. Each pickup wire is terminated in the impedance of the transmission line by corresponding resistors R_1 and R_2 (typically 50 to 52 Ω).

The outgoing voltage (transmitter to antenna) is absorbed by R_1. Thus, there is no outgoing voltage on the reflected pickup wire beyond point *A*. However, the outgoing voltage remains on the transmission line at the outgoing pickup wire. This voltage is rectified by CR_1, and appears as a reading on the meter, when S_1 is in the outgoing voltage position.

The opposite condition occurs for the reflected voltage (antenna to

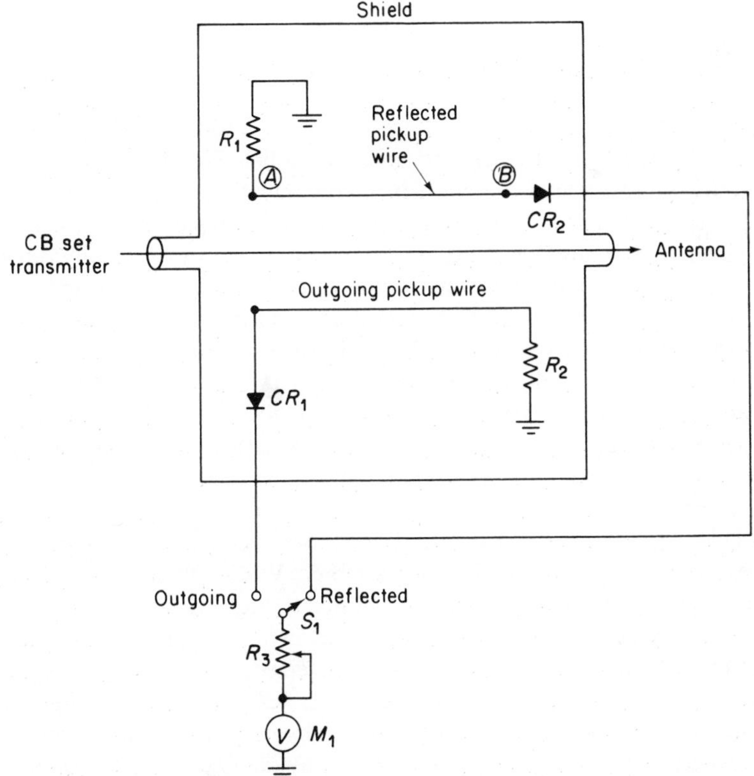

FIGURE 9-25 Basic SWR meter circuit (directional coupler).

transmitter). There is no reflected voltage on the outgoing pickup wire beyond point *B*, because the reflected voltage is absorbed by R_2. The reflected voltage does appear on the reflected pickup wire beyond this point and is rectified by CR_2. The reflected voltage appears on meter M_1 when S_1 is in the reflected voltage position.

In use, switch S_1 is set to read outgoing voltage, and resistor R_3 is adjusted until the meter needle is aligned with some "set" or "calibrate" line (near the right-hand end of the meter scale). Switch S_1 is then set to read reflected voltage, and the meter needle moves to the SWR indication.

As a practical matter, SWR meters often do not read beyond 1:3. This is because an indication above 1:3 indicates a poor match. Make certain you understand the scale used on the SWR meter. For example, a typical SWR meter is rated at 1:3, meaning that it reads SWR from 1:1 (perfect) to 1:3 (poor). However, the scale indications are 1, 1.5, 2, and 3. These scale indications mean 1:1, 1:1.5, 1:2, and 1:3, respectively. The scale indications between 1 and 1.5 are the most useful, since a good antenna system will show a typical 1.1 or 1.2. Anything between 1.2 and 1.5 is on the borderline.

9-8 DIP METERS

The dip meter, or grid-dip meter, has long been a tool in radio communications service work, particularly in amateur radio. The dip meter has many uses, but its most useful function in communications work is in presetting "cold" transmitter and receiver resonant circuits (no power applied to the set). This makes it possible to adjust the resonant circuits of a badly tuned set, or a set where new coils and transformers must be installed as a replacement.

As an example, it is possible that the replacement coil or transformer is tuned to an undesired frequency when shipped from the factory. Using a dip meter, it is possible to install the coil, tune it to the correct frequency, and then apply power to the set and adjust the circuit for "peak" as described in the service literature. (Most service literature assumes that the circuits are not badly tuned and only require "peaking.")

There are many types of dip meters and circuits. A typical dip meter is a hand-held, battery-operated device. The circuit is essentially an RF oscillator with external coil, a tuning dial, and a meter. When the coil is held near the circuit to be tested, and the oscillator is tuned to the resonant frequency of the test circuit, part of the RF energy is absorbed by the test circuit, and the meter indication "dips." The procedure can be reversed, where the dip meter is set to a desired frequency and the test circuit is tuned to produce a "dip" indication on the meter. We will not go into the many uses of the dip meter here. Instead, we shall describe how a *dip adapter* may be used to preset resonant circuits or to check the frequency of resonant circuits.

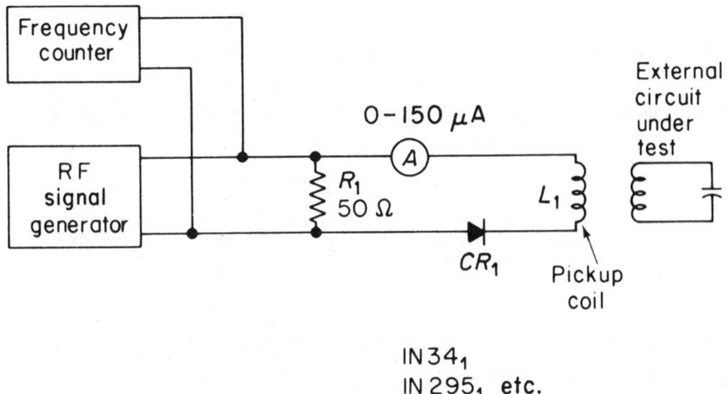

FIGURE 9-26 Basic dip-adapter circuit.

9-8.1 Basic Dip Adapter

A basic dip adapter circuit is shown in Fig. 9–26. Such a circuit may be fabricated in the shop with little difficulty. Resistor R_1 should match the impedance of the signal generator (typically 50 Ω). Both diode CR_1 and the microammeter should match the output of the signal generator. The pickup coil L_1 consists of a few turns of insulated wire. The accuracy of the dip adapter circuit depends on the counter accuracy, or on the signal generator accuracy if the counter is omitted.

9-8.2 Setting Resonant Frequency with a Dip Adapter

The frequency of a resonant circuit may be set using a dip adapter. The following procedure is applicable to both series and parallel resonant circuits.

1. Couple the dip adapter to the resonant circuit using pickup coil L_1 of Fig. 9-26. Usually, the best coupling has a few turns of L_1 passed over the coil of the resonant circuit. Make certain that the communications set is off.
2. Set the signal generator to the desired resonant frequency, as indicated by the frequency counter. Adjust the signal generator output amplitude control for a convenient reading on the adapter meter.
3. Tune the resonant circuit for a maximum dip on the adapter meter. The resonant circuits of the set may be tuned by means of adjustable slugs in the coil and/or adjustable capacitors.

4. Most resonant circuits are designed so as not to tune across both the fundamental frequency and any harmonics. However, it is possible that the circuit will tune to a harmonic and produce a dip. To check this condition, tune the resonant circuit for maximum dip, and set the signal generator to the first harmonic (twice the desired resonant frequency) and to the first subharmonic (one-half the resonant frequency). Note the amount of dip at both harmonics. The harmonics should produce substantially smaller dips than the fundamental resonant frequency.
5. For maximum accuracy, check the dip frequency from both high and low sides of the resonant circuit tuning. A significant difference in frequency readout from either side indicates overcoupling between the dip adapter circuit and the resonant circuit under test. Move the adapter coil L_1 away from the test circuit until the dip indication is just visible. This amount of coupling should provide maximum accuracy. (If there is difficulty in finding a dip, overcouple the adapter until a dip is found, then loosen the coupling and make a final check of frequency. Generally, the dip is more pronounced when it is approached from the direction that causes the meter reading to rise.)
6. If there is doubt as to whether the adapter is measuring the resonant frequency of the desired circuit or some nearby circuit, ground the circuit under test. If there is no change in the adapter dip reading, the resonance of another circuit is being measured.
7. The area surrounding the circuit being measured should be free of wiring scraps, solder drippings, and so on, as the resonant circuit can be affected by them (especially at high frequencies), resulting in inaccurate frequency readings. Keep fingers and hands as far away as possible from the adapter coil (to avoid adding body capacitance to the circuit under test).
8. All other factors being equal, the nature of a dip indication provides an approximate indication of the test circuit's Q factor. Generally, a sharp dip indicates a high Q, whereas a broad dip shows a low Q.
9. The dip adapter may also be used to measure the frequency to which a resonant circuit is tuned. The procedure is essentially the same as that for presetting the resonant frequency (steps 1 through 8), except that the signal generator is tuned for a maximum dip (communications set still cold and the test circuit untouched). The resonant frequency to which the test circuit is tuned is then read from the counter, or the signal generator dial if the counter is omitted. When making this test, watch for harmonics, which also produce dip indications.

9-9 ANTENNA AND TRANSMISSION-LINE MEASUREMENTS

In general, antennas and transmission lines (lead-ins) used with radio communications sets are best tested using commercial SWR meters, field-strength meters, and the special test sets described in Section 9–12. However, it is possible to make a number of significant tests using basic meters (voltmeter, ohmmeter, ammeter). The following paragraphs describe these test procedures.

9-9.1 Antenna Length and Resonance Measurements

Most antennas are cut to a length related to the wavelength of the signals being transmitted or received. Generally, antennas are cut to one-half wavelength (or one-quarter wavelength) of the center operating frequency. The electrical length of an antenna is always greater than the physical length, owing to capacitance and end effects. Therefore, two sets of calculations are required: one for electrical length and one for physical length. The calculations for antenna length and resonant frequency are shown in Fig. 9–27.

9-9.2 Practical Resonance Measurements for Antennas

With a short antenna it is possible to measure the exact physical length and find the electrical length (and hence the resonant frequency) using the equations of Fig. 9–27. Obviously, this is not practical for long antennas. Also, the exact resonant frequency (electrical length) may still be in doubt for short antennas due to the uncertain K factor of Fig. 9–27. Therefore, for practical purposes the electrical length and resonant frequency of an antenna should be determined electrically.

There are three practical methods for determining antenna resonant frequency: dip adapter circuit, antenna ammeter, and wavemeter.

Dip-Adapter Measurement of the Antenna Resonant Frequency. The dip adapter (Section 9–8) can be used to measure resonant frequency of both grounded and ungrounded antennas. The basic technique is to couple the adapter to the antenna as if the antenna were a resonant circuit, tune for a dip, and read the resonant frequency. However, there are certain precautions to be observed.

The measurement can be made as a conventional resonant circuit provided that the antenna is accessible, allowing the dip adapter to be coupled directly to the antenna elements. If the antenna is a simple grounded element

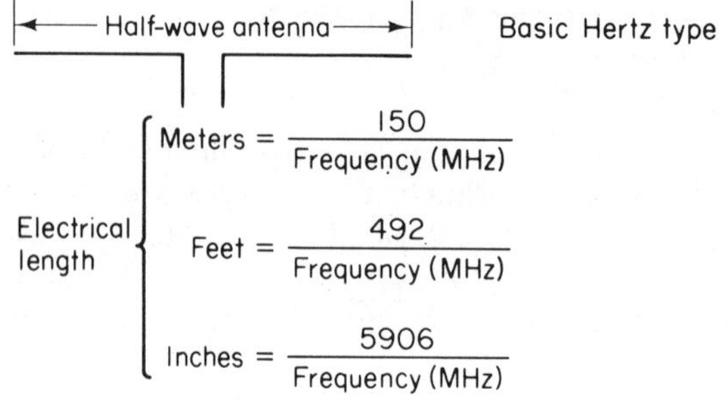

Quarter-wave antenna

Basic Marconi type

Electrical length

$$\text{Meters} = \frac{75}{\text{Frequency (MHz)}}$$

$$\text{Feet} = \frac{246}{\text{Frequency (MHz)}}$$

$$\text{Inches} = \frac{2953}{\text{Frequency (MHz)}}$$

Physical length (approx.) = electrical length x K factor

K factor = 0.96 for frequencies below 3 MHz
= 0.95 for frequencies between 3 and 30 MHz
= 0.94 for frequencies above 30 MHz

FIGURE 9-27 Calculations for antenna length.

(no matching problems between antenna and transmission line), the adapter can be coupled to the transmission line. However, if the antenna is fed by a coaxial line or any system in which the line is matched to the antenna, the adapter coil must be coupled directly to the antenna elements.

No matter how carefully the antenna and lead-in are matched, there is some mismatch, at least over a range of frequencies. This means that there are two reactances (or impedances) that interact to produce extra resonances. If resonance is measured under such conditions, a dip is produced at the correct antenna frequency (and at harmonics), and another dip (plus harmonics) at the extra frequency. It can be very confusing to tell them apart. Also, antenna resonance measurements should be made with the antenna in the actual operating position. The nearness of directive or reflective elements, as well as the height, affects antenna characteristics and possibly changes resonant frequency.

The dip-adapter procedure for grounded antennas is as follows:

1. Couple the dip adapter to the antenna tuner if the antenna is used for the transmission line and the feed line is tuned.
2. If the antenna is untuned or it is not practical to couple to the antenna tuner, disconnect the antenna lead-in and couple the lead-in to the adapter through a pickup coil, as shown in Fig. 9–28.
3. Set the generator to its lowest frequency. Adjust the signal generator output for a convenient reading on the adapter meter.
4. Slowly increase the generator frequency, observing the meter for a dip indication. Tune for the bottom of the dip.
5. Note the frequency at which the first (lowest frequency) dip occurs. This should be the primary resonant frequency of the antenna. As the signal generator frequency is increased, additional dips should be noted. These are harmonics and should be exact multiples of the primary resonant frequency. Check two or three of these frequencies to be sure that they are harmonics. Then go back to the lowest frequency dip to ensure that the lowest frequency is the primary resonant frequency.

The dip-adapter procedure for ungrounded antennas is as follows:

1. Disconnect the antenna lead-in or feed line from the antenna.
2. If the antenna is center-fed, short across the feed point with a piece of wire.
3. Couple the dip-adapter coil directly to the antenna. Usually, the best results are obtained by coupling at a maximum current (low-impedance) point. For example, the maximum current point occurs at the center of a half-wave antenna.
4. Starting at the lowest signal generator frequency and working upward, tune the signal generator for a dip on the meter as described

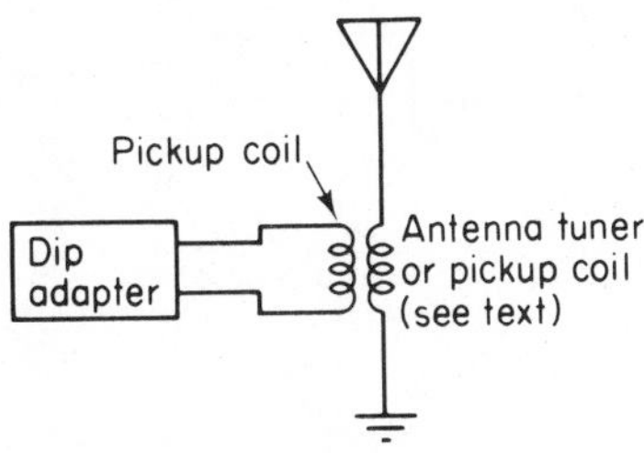

FIGURE 9–28 Dip-adapter measurement of antenna resonant frequency.

for grounded antennas. The lowest dip is the primary resonant frequency.

Series Ammeter Measurement of the Antenna Resonant Frequency. A series ammeter can be used to find the resonant frequency of an antenna. The basic circuit is shown in Fig. 9-29, and the procedure is quite simple. The signal generator is tuned for a maximum reading on the ammeter, indicating a maximum transfer of energy from the generator into the antenna (as a result of both being at the same frequency). The antenna frequency is then read from the generator dial or frequency counter (if used). The series ammeter method has the advantage of measuring the combined resonant frequency of both the antenna and transmission line. This is most practical, since in normal operation the antenna is operated with the transmission line.

A version of the series ammeter method is often used in transmitters as an indicator for antenna tuning. Most transmitters are crystal-controlled and operated at a specific frequency with the *antenna tuned to that frequency.* The electrical length (and consequently the resonant frequency) of the antenna is varied by a reactance in series with the lead-in, as shown in Fig. 9-30. The reactance can be a variable capacitor or variable inductance. With such an arrangement, the transmitter is tuned to its operating frequency, and then the antenna is tuned to that frequency as indicated by a maximum reading on the series ammeter.

The series ammeter method has certain drawbacks, one being the operating frequency limit of the series ammeter. Another is the fact that the series ammeter consumes some power in operation. However, the series ammeter has an advantage in that true antenna power can be calculated (as discussed in Section 9-9.3).

Wavemeter Measurement of the Antenna Resonant Frequency. A wavemeter can be used to find the resonant frequency of an antenna. The

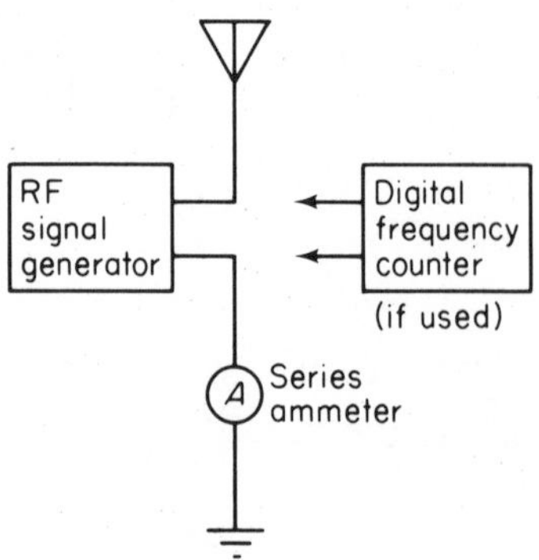

FIGURE 9-29 Series ammeter measurement of antenna resonant frequency.

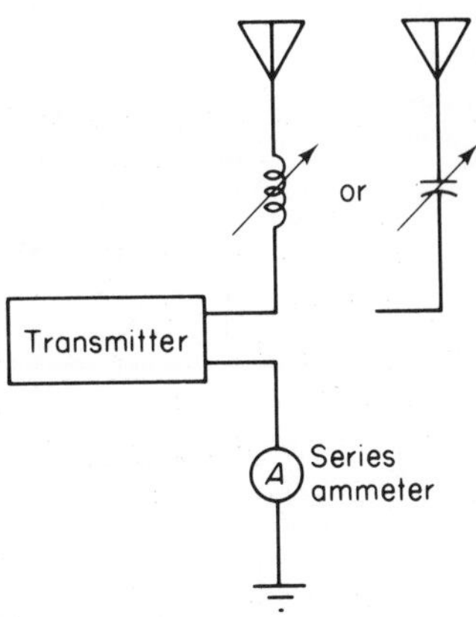

FIGURE 9-30 Tuning an antenna to resonant frequency of associated transmitter using the series ammeter method.

basic wavemeter circuit, shown in Fig. 9-31, is essentially a tuned resonant circuit, detector, and indicator. A commercial wavemeter has a precision-calibrated tuning dial so that exact frequency can be measured, and an amplifier circuit so that weak signals can be measured. When used to measure antenna resonant frequency, the wavemeter is tuned to the approximate resonant frequency of the antenna, and then the signal generator is tuned for a maximum reading on the wavemeter. In this case, the wavemeter resonant circuit is broadly tuned. When used to tune an antenna, the wavemeter is tuned to the transmitter operating frequency, and then the antenna is tuned for a maximum reading on the wavemeter. Not all wavemeters are provided with precision tuning. Such wavemeters serve only as a maximum (or peak) readout device, similar to the field strength meter (Section 9-6).

9-9.3 Antenna Impedance and Radiated Power Measurements

The impedance of an antenna is not constant along the entire length of the antenna. In a typical half-wave antenna as shown in Fig. 9-32, the impedance is minimum at the center and maximum at the ends. In theory, the impedance is zero at the center. Since the antenna is fed at some point away from the exact center, there is some impedance for any antenna. A typical antenna used in CB transmitters has an impedance of 50 to 72 Ω, whereas a half-wave TV antenna has 300-Ω impedance.

Antenna impedance is determined using the basic Ohm's law equation $Z = E/I$, with voltage and current being measured at the *antenna feed point.* However, such measurements are not usually made in practical applications.

Radiation resistance is a more meaningful term. When the d-c resistance of the antenna is disregarded (antenna d-c resistance is usually a few ohms or

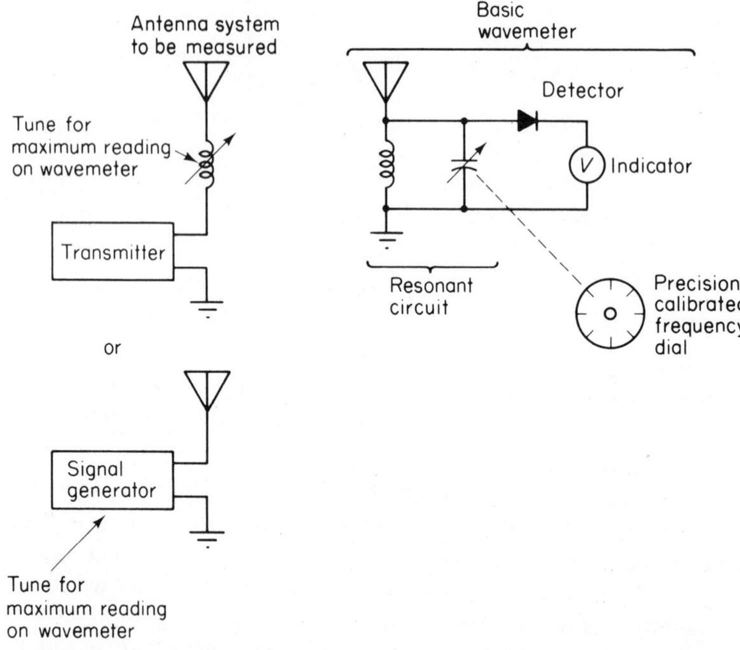

FIGURE 9-31 Wavemeter measurement of antenna resonant frequency.

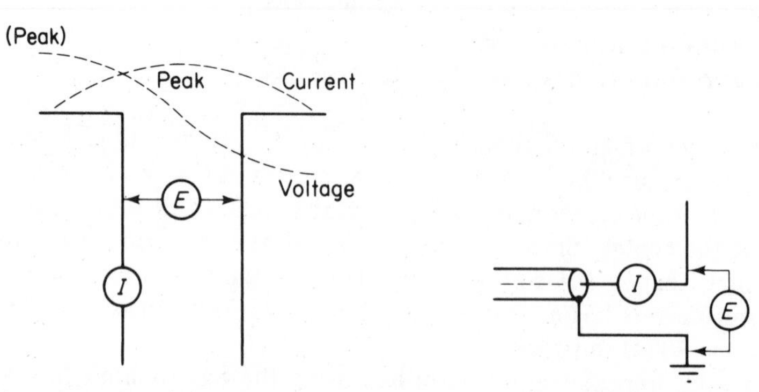

FIGURE 9-32 Theoretical calculations for antenna impedance and radiated power: (a) half-wave; (b) quarter-wave.

less, except in low-frequency long-wire antennas), the antenna impedance can be considered as the radiation resistance. Radiated power can then be determined using the basic Ohms's law equation $P = I^2R$.

Practical Impedance and Radiated Power Measurement for Antennas. On those antennas designed to be used with coaxial or twin-lead transmission lines (such as CB and TV antennas, respectively), the antenna and transmission-line impedance must be matched. In this case, the impedance match between transmission line and antenna is of greater importance than actual impedance value (both antenna and transmission line must be 50 Ω, 72 Ω, 300 Ω, etc.). The condition of match (or mismatch) between antenna and transmission line can best be tested by the standing-wave ratio (SWR), as discussed in Section 9–7.

On those antennas where the lead-in and antenna are considered as one piece with no match or mismatch problems, it is necessary to measure the actual antenna impedance (or radiation resistance) in order to calculate radiated power.

The following procedure can be used to find the impedance and radiated power of any antenna system. However, it should be noted that the impedance and power obtained are for the *complete antenna system* (antenna and transmission line), as seen from the measurement end.

1. Connect the equipment as shown in Fig. 9–33.
2. Set switch S_1 to position 1. Adjust the signal generator to the center frequency at which the antenna is used (or for any desired operating frequency to which the antenna can be tuned). If the antenna is to be used with a transmitter, the transmitter may be substituted for the signal generator.
3. Tune the antenna to the operating frequency by adjusting L_1 for a maximum indication on the ammeter. Record the indicated current.
4. If a digital frequency counter or precision wavemeter is available, verify that the signal generator (or transmitter) and antenna are tuned to the correct frequency.
5. Set switch S_1 to position 2. Adjust capacitor C_1 for a maximum indication on the ammeter. If the frequency counter or wavemeter is available, verify that the operating frequency has not been changed.

   NOTE: If a capacitor has been used to tune the antenna instead of L_1, a precision inductor must be used with resistor R_1.

6. Adjust load resistor R_1 until the indicated current is the same as the antenna current recorded in step 3.

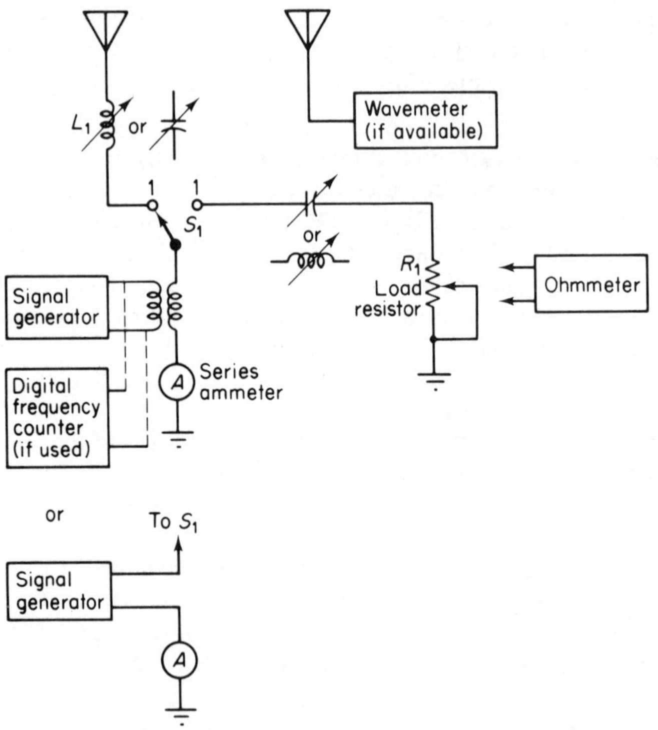

FIGURE 9-33 Measuring antenna impedance and radiated power using resistance substitution (ammeter method).

7. Remove power from the circuit. Measure the d-c resistance of R_1 with an ohmmeter. This resistance is equal to the antenna system impedance (or radiation resistance), at the frequency of measurement.
8. Calculate the actual power delivered to the antenna (or radiated power) using:

$$\text{radiated power} = I^2 \times R \text{ (or } Z\text{)}$$

where I = indicated current (amperes)
R = radiation resistance (or antenna system impedance)

An alternative method must be used when the operating frequency is beyond the range of the available ammeter, when no ammeter is available, or when the ammeter presents an excessive load. A precision 1-Ω, noninductive resistor and voltmeter can be used in place of the ammeter, as shown in Fig. 9-34. With a 1-Ω resistor, the indicated voltage is equal to the current passing

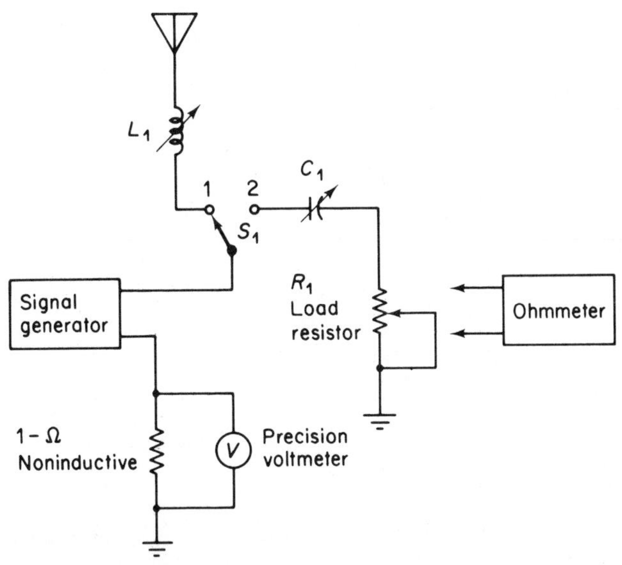

FIGURE 9-34 Measuring antenna impedance and radiated power using resistance substitution (voltmeter method).

through the resistor (and antenna system). Except for the connections, the procedure is identical to the one that requires an ammeter.

9-9.4 Antenna Gain

Antenna gain is a term usually applied to receiving antennas. Such gain is measured by comparing the voltage produced at the terminals of the antenna with that of a thin-wire dipole of the same size operating at the same frequency and location. Antenna gain is normally expressed in decibels, since gain is essentially a ratio between two voltages. Antenna gain is often shown in a gain curve or gain chart. Usually, the 0 = dB line or reference indicates the gain of the thin-wire dipole to which the antenna is being compared. Typical gain curves and calculations for antenna gain are shown in Fig. 9-35.

There are several practical methods for measuring antenna gain, as shown in Fig. 9-36. The simplest is to measure the antenna voltage with a voltmeter and RF probe [Fig. 9-36(a)]. This is usually satisfactory for laboratory work in which a transmitted signal can be directed toward the antenna. In cases where the transmitted signal is very weak, or it is desired to measure antenna gain at some specific frequency, it is necessary to use a tuning circuit, detector, and amplifier, as shown in Fig. 9-36(b). Some commercial wavemeters incorporate such circuits.

Receivers can be used to measure antenna gain. However, the automatic

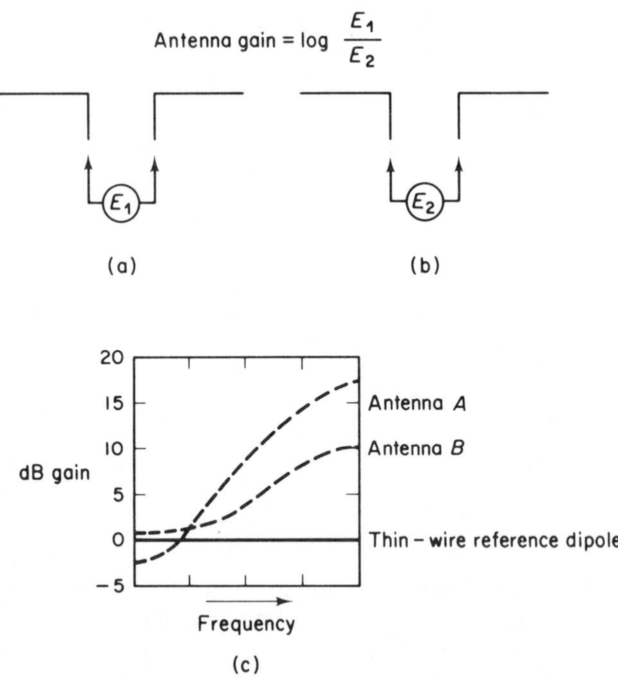

FIGURE 9-35 Calculations for antenna voltage gain and typical gain curves. (a) Antenna being measured. (b) Thin-wire reference dipole. (c) Typical antenna gain curve.

gain control (AGC) or automatic volume control (AVC) circuits of the receiver must be disabled. Also, the receiver must be provided with an output meter that reads in volts or decibels. The important consideration in making antenna gain measurements is that both the thin-wire dipole and antenna must be tested under *identical conditions* (same frequency, physical location, and test instruments). In some cases, the change of a few inches in physical location completely changes the antenna gain pattern.

9-9.5 Transmission-Line Measurements

Transmission lines are devices used to transfer energy from one unit to another. The most common type of transmission lines found in practical applications are those used to transfer signals to and from antennas. There are two basic types: parallel-wire (such as used in many TV applications) and concentric or coaxial (such as used in most CB and mobile radio applications). In either case, there is capacitance between the lines, as well as inductance set up around the lines. Because of this capacitance and inductance (and the resistance of the wire), a transmission line has characteristic or surge impedance. Such impedance depends on the dimensions of the wire or conduc-

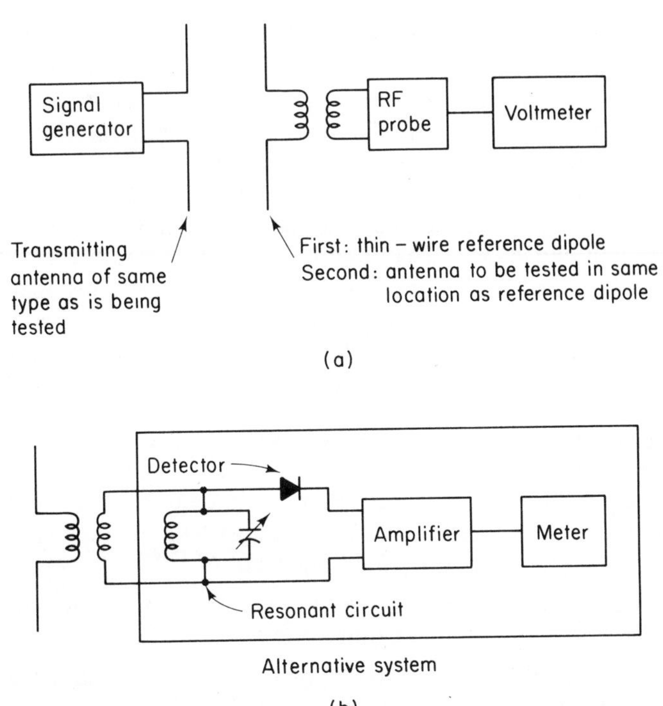

FIGURE 9-36 Basic antenna gain measurement techniques.

tor, spacing between the conductors, and dielectric constant of the material between the conductors.

As in the case of antennas, transmission lines can be cut to wavelengths of signals passing through them. The same basic equations (Section 9-9.1 and Fig. 9-27) can be used to determine electrical length. The physical length of the transmission line will be shorter, however, since radio waves travel faster in free space than along a conductor. The ratio of actual velocity along a transmission versus velocity of the same signals in free space is known as the *velocity factor*. This factor is always less than 1 and usually varies between about 0.6 and 0.97 for typical transmission lines. The velocity factors for a number of transmission lines are as follows:

Twisted-pair line (rubber dielectric)	0.6
Two-wire line (plastic dielectric)	0.75
Coaxial line (solid plastic dielectric)	0.67
Coaxial line (air dielectric)	0.85
Parallel tubing (air dielectric)	0.95
Two-wire open line (air dielectric)	0.97

(Free-space velocity is 300×10^6 meters/s, or 984×10^6 ft/s.)

Practical Transmission-Line Measurements. In practical applications, it may be necessary to find transmission-line impedance and length. In the case of coaxial lines, the impedances are listed in manufacturers' data (and in various tables) for coaxial types. The impedance should be the same for all coaxial cables of the same type without regard to the manufacturer. However, the impedance data may not be readily available. Also, transmission-line impedance can change due to deterioration. In addition to measuring transmission-line impedance, it is often convenient to measure transmission-line length by electrical means. An example is where a line is installed at an inaccessible location or when a large length of transmission line is coiled. A dip adapter, such as described in Section 9–8, can be used to measure both impedance and length of transmission lines.

Measuring Transmission-Line Impedance with a Dip Adapter

1. Select a variable resistance whose range covers the supposed impedance of the line. The resistance must be noninductive (such as a carbon or composition potentiometer).
2. Connect the equipment as shown in Fig. 9–37, but do not connect the variable resistance at this time.
3. Set the signal generator to its lowest frequency. Adjust the signal generator output for a convenient reading of the dip-adapter meter.
4. Slowly increase the signal generator frequency, observing the meter for a dip indication. Tune for the bottom of a dip. This is the resonant frequency of the transmission line. Additional harmonic dip indications should be found at multiples of this frequency. The lowest or primary resonant frequency is the point at which the transmission line is a *quarter wavelength.*
5. With the signal generator tuned to the bottom of a dip at the primary resonant frequency, connect the variable resistance to the open end of the line.
6. Adjust the variable resistance to the supposed impedance of the line. This should remove the dip.

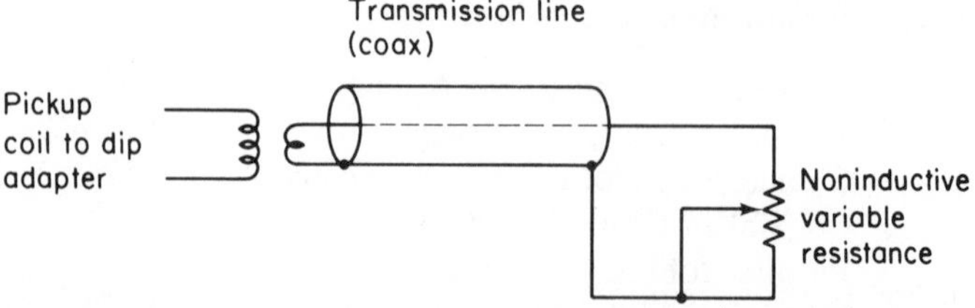

FIGURE 9–37 Measuring transmission-line characteristics with dip-adapter circuit.

7. If the dip is still present, continue adjustment of the resistance near the supposed impedance until the dip is completely eliminated. Check the settings by removing the resistance. The dip should reappear.
8. Once you have determined that the line is terminated in its characteristic impedance, measure the resistance of the potentiometer with an ohmmeter (taking care not to disturb the setting to the disconnected potentiometer). This resistance is equal to the transmission-line impedance.

Measuring Transmission-Line Length with a Dip Adapter

1. Couple the dip adapter to the accessible end of the line as shown in Fig. 9-37. Do not connect a terminating resistor.
2. Set the signal generator to its lowest frequency. Adjust the signal generator output for a convenient reading of the adapter meter.
3. Slowly increase the generator frequency, observing the meter for a dip indication. Tune for the bottom of a dip. Continue to increase generator frequency, watching for additional dips.
4. Loosen the coupling until the dips become small and then record the frequency readings of any two adjacent dips.
5. Using the two dip-indication frequency readings, calculate the line length using the equation

$$D = \frac{984 \times K}{2\,(F_1 - F_2)}$$

where D = distance from measurement end to open or short (feet)
K = velocity factor in percentage of the free-space propagation velocity of radio waves
F_1 = frequency of the first dip (MHz)
F_2 = frequency of the second (adjacent) dip (MHz)

It should be noted that the distance calculated is from the coupling loop end to the first open or short in the line. For example, if a 100-ft line is measured, and this line has a short at 70 ft, the distance measured is 70 ft, not 100 ft.

9-10 OSCILLATOR TESTS

Most communications equipment uses some form of oscillator for signal generation and frequency control. For example, the frequency-control circuits of a CB set can be quite simple or fairly complex, depending on the set. In the simplest form, such as a hand-held CB, there are two oscillators, one for transmitter frequency control and one for the receiver local oscillator. In a "typical" communications set there are two or three oscillators, combined with one or more mixers, to form a *frequency synthesizer* that produces transmitter and receiver local oscillator signals. A PLL (phase-locked loop) set contains a standard oscillator (probably within the PLL IC, but having an external crystal) and one or two other oscillators to form the complete frequency synthesizer.

No matter how complex the circuits appear, they are essentially oscillators, and can be treated as such from a practical test standpoint. That is, each circuit contains oscillators, which produce signals (probably crystal-controlled). These signals must have a given amplitude and must be at a given frequency (or capable of tuning across a given frequency range) for the set to operate properly. Thus, if you measure the signals and find them to be of the correct frequency and amplitude, the oscillators are good from a test standpoint.

9-10.1 Oscillator-Test Procedures

The first step in testing any oscillator circuit is to measure both the amplitude and frequency of the output signal. Many oscillators have a built-in test point. If not, the signal may be monitored at the collector or emitter (plate or cathode for a vacuum-tube oscillator), as shown in Fig. 9-38. Signal amplitude is monitored with a meter or oscilloscope using an RF probe (Chapter 8 and Section 9-2). The simplest way to measure oscillator signal frequency is with a frequency counter (Section 9-3).

Oscillator Frequency Problems. When you measure the oscillator signal, the frequency is (1) right on, (2) slightly off, or (3) way off.

If the frequency is slightly off, it is possible to correct the problem with adjustment. Most oscillators are adjustable, even those that are crystal-controlled. Usually, the RF coil or transformer is slug-tuned. The most precise adjustment is obtained by monitoring the oscillator signal with a frequency counter and adjusting the circuit for exact frequency. However, it is also possible to adjust an oscillator using a meter or oscilloscope.

When the circuit is adjusted for *maximum signal amplitude,* the oscillator is at the crystal frequency. However, it is possible (but not likely) that the oscillator is being tuned to a harmonic (multiple or submultiple) of the crystal frequency. The frequency counter will show this, whereas the meter or oscilloscope will not.

If the oscillator frequency is way off, look for a defect rather than improper adjustment. For example, the coil or transformer may have shorted turns, the transistor or capacitor may be leaking badly, or the wrong crystal is installed in the right socket (this does happen).

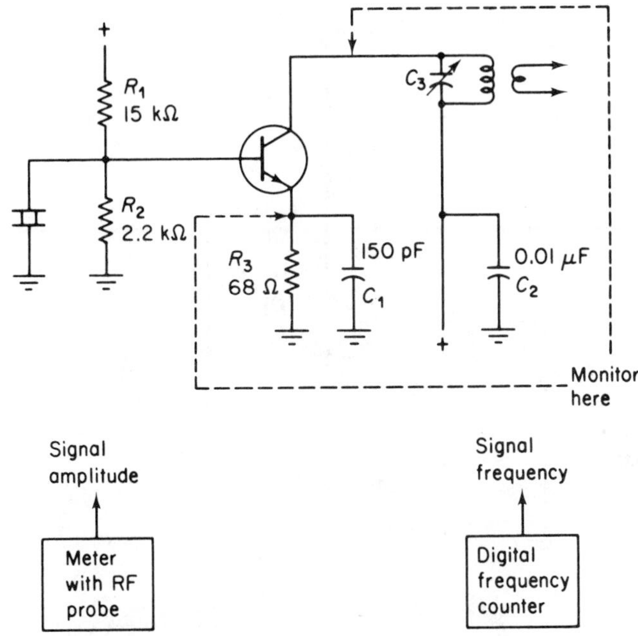

FIGURE 9-38 Oscillator signal test points.

Oscillator Signal Amplitude Problems. When you measure the oscillator signal, the amplitude is (1) right on, (2) slightly low, or (3) very low.

If the amplitude is slightly low, it is possible to correct the problem with adjustment. Monitor the signal with a meter or oscilloscope, and adjust the oscillator for maximum signal amplitude. This also locks the oscillator on the correct frequency.

If the amplitude is very low, look for defects such as low power-supply voltages, leaking transistor and/or capacitors, and shorted coil or transformer turns. Usually, when signal output is very low, there are some other indications, such as abnormal voltage and resistance values.

Oscillator Bias Problems. One of the problems in testing solid-state oscillator circuits is the bias arrangement. RF oscillators are generally reverse-biased, so that they conduct on half-cycles. However, the transistor is initially forward-biased by d-c voltages (through the bias networks, as shown in Fig. 9-39). This turns the transistor on so that the collector circuit starts to conduct. Feedback occurs, and the transistor is driven into heavy conduction.

During the time of heavy conduction, a capacitor connected to the transistor base is charged in the forward-bias direction. When saturation is reached, there is no further feedback, and the capacitor discharges. This reverse-biases the transistor and maintains the reverse bias until the capacitor

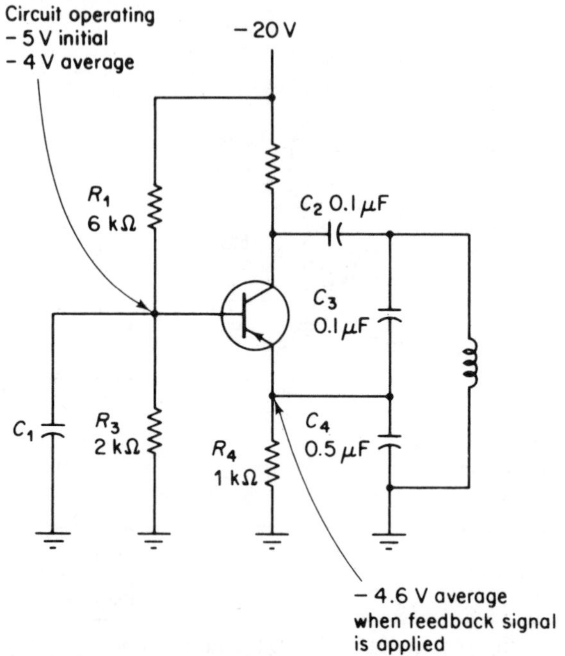

FIGURE 9-39 Class C RF oscillator (reverse-biased or zero-biased with circuit operating).

has discharged to a point where the fixed forward bias again causes conduction.

This condition presents a problem in the operation of class C solid-state RF oscillators. If the capacitor is too large, it may not discharge in time for the next half-cycle. In that case, the class C oscillator acts as a blocking oscillator, controlling the frequency by the capacitance and resistance of the circuit. If the capacitor is too small, the class C oscillator may not start at all. This same condition is true if the capacitor is leaking badly. From a practical test standpoint, the measured condition of bias on a solid-state oscillator can provide a good clue to operation, if you know how the oscillator is supposed to operate.

The oscillator in Fig. 9–39 is initially forward-biased through R_1 and R_3. As Q_1 starts to conduct and in-phase feedback is applied to the emitter (to sustain oscillation), capacitor C_1 starts to charge. When saturation is reached (or approached) and the feedback stops; capacitor C_1 then discharges in the opposite polarity, reverse-biasing Q_1. The value of C_1 is selected so that C_1 discharges to a voltage less than the fixed forward bias before the next half-cycle. Thus, transistor Q_1 conducts on slightly less than the full half-cycle. Typically, a class C RF oscillator such as the one shown in Fig. 9–39 conducts on about 140° of the 180° half-cycle.

Exploring the subject of bias further, it is commonly assumed that transistor junctions (and diodes) start to conduct as soon as forward voltage is applied; this is not true. Figure 9–40 shows characteristic curves for three different types of transistor junctions. All three junctions are silicon, but the same condition exists for germanium junctions. None of the junctions conduct noticeably at 0.6 V, but current starts to rise at that point. At 0.8 V, one junction draws almost 80 mA. At 1 V, the d-c resistance is on the order of 2 or 3 Ω, and the transistor draws almost 1 A. In a germanium transistor, noticeable current flow occurs at about 0.3 V.

For test purposes, bias measurements provide a clue to the performance of solid-state oscillators, although such measurements do not provide positive proof. The one sure test of an oscillator is to measure output signal amplitude and frequency.

Oscillator Quick-Test. It is possible to check whether an oscillator circuit is oscillating by using a voltmeter and large-value capacitor (typically 0.01 μF or larger). Measure either the collector or emitter voltage with the oscillator operating normally, and then connect the capacitor from base to ground as shown in Fig. 9–41. This should stop oscillation, and the emitter or collector voltage should change. When the capacitor is removed, the voltage will return to normal. If there is no change when the capacitor is connected, the oscillator is probably not oscillating. In some oscillators, you will get

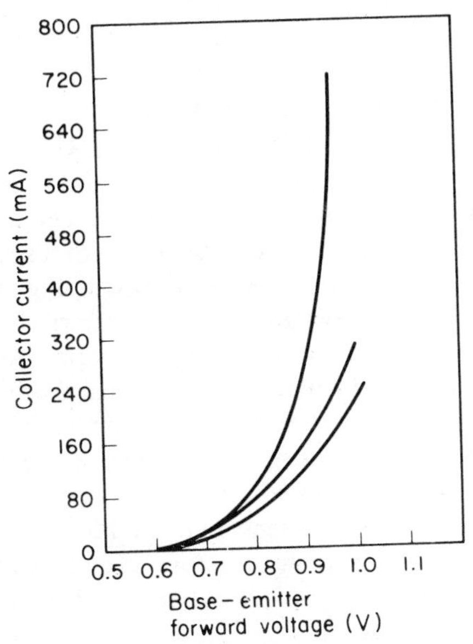

FIGURE 9–40 Characteristic curves for silicon transistor junctions.

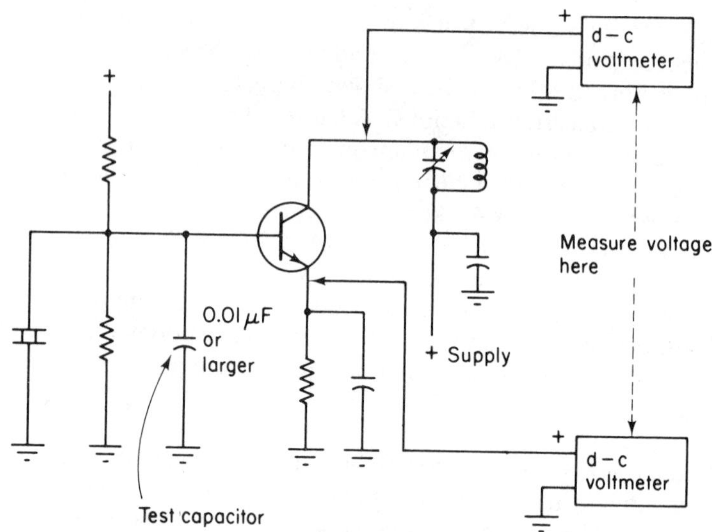

FIGURE 9-41 Test connections for oscillator quick-test.

better results by connecting the capacitor from the collector to ground. Also, do not expect the voltage to change on an element without a load. For example, if the collector is connected directly to the supply voltage, or through a few turns of wire as shown in Fig. 9-41, this voltage does not change, with or without oscillation.

9-11 USING SPECTRUM ANALYZERS IN COMMUNICATIONS EQUIPMENT TESTS

Communications waveforms (particularly those resulting from frequency modulation, or FM) are often measured by means of a spectrum analyzer. The basic circuit of a spectrum analyzer is shown in Fig. 9-42. The spectrum analyzer is essentially a narrowband receiver, electrically tuned over a given frequency range, combined with an oscilloscope. As shown, the local oscillator is swept over a given range of frequencies by a sweep generator (Chapter 8). Since the IF amplifier passband remains fixed, the input circuits and mixer are swept over a corresponding range of frequencies.

For example, if the intermediate frequency is 10 kHz and the local oscillator sweeps a band of frequencies from 100 to 200 kHz, the input is capable of receiving waveforms in the range 110 to 210 kHz. The output of the IF amplifier is further amplified and supplied to the vertical deflection plates of a cathode-ray tube.

The cathode-ray tube horizontal plates obtain their signal from the same sweep generator used to tune the local oscillator. Thus, the length of the

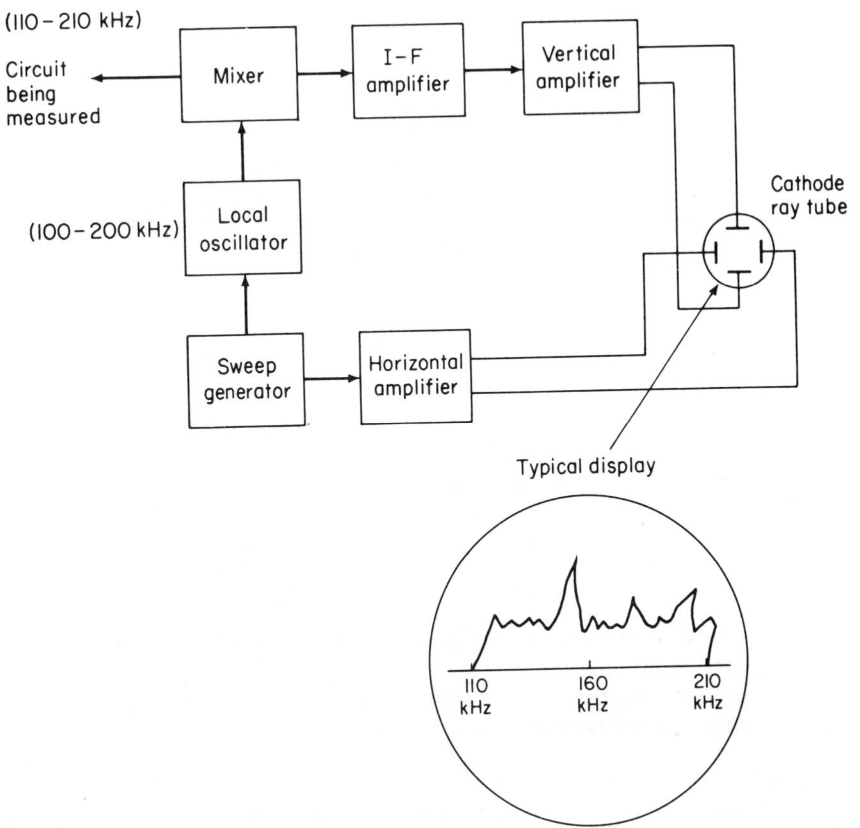

FIGURE 9-42 Basic spectrum analyzer circuit.

horizontal sweep represents the total sweep spectrum. For example, if the sweep is from 110 to 210 kHz, the left-hand end of the horizontal trace represents 110 kHz and the right-hand end represents 210 kHz. Any point along the horizontal trace represents a corresponding frequency. For example, the midpoint on the trace represents 160 kHz.

9-11.1 *Time-Amplitude Displays versus Frequency-Amplitude Displays*

To gain a better understanding of the usefulness and application of a spectrum analyzer in communications equipment tests, it is important to understand what the spectral display is and how to interpret it.

Figure 9-43 shows the relationship of time-amplitude and frequency-amplitude displays. The conventional oscilloscope produces a time-amplitude display. For example, pulse rise time and width are read directly on the *X* axis

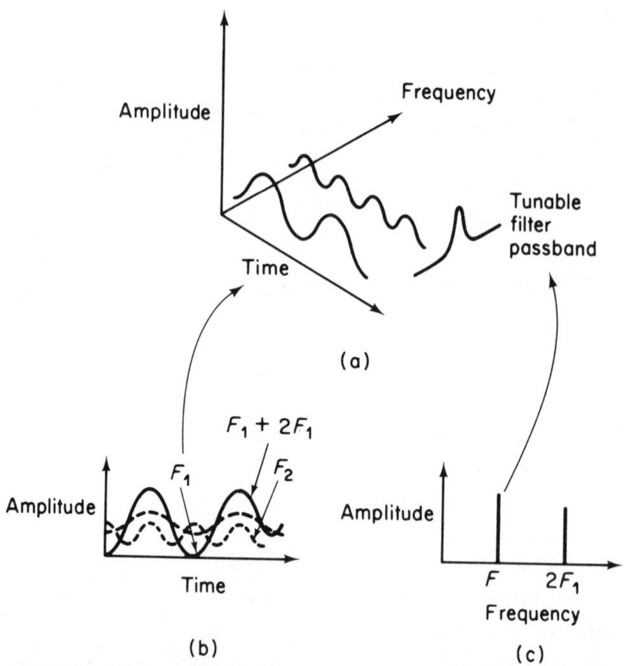

FIGURE 9-43 Relationship of time-amplitude and frequency-amplitude displays: (a) combined time-frequency-amplitude; (b) time-amplitude (oscilloscope-type display); (c) frequency-amplitude (spectrum analyzer-type display).

of the cathode-ray tube. A spectrum analyzer produces a frequency–amplitude display where signals (unmodulated, AM, FM, or pulse) are broken down into their individual frequency components, and displayed on the cathode-ray-tube X axis.

In Fig. 9–43(a), both the time–amplitude and frequency–amplitude coordinates are shown together. The example given is that showing the addition of a fundamental frequency and its second harmonic.

In Fig. 9–43(b), only the time–amplitude coordinates are shown. The solid line (which is the composite of fundamental F_1 and second harmonic $2F_1$) is the only display that appears on a conventional oscilloscope.

In Fig. 9–43(c), only the frequency–amplitude coordinates are shown. Note how the components (F_1 and $2F_1$) of the composite signal are clearly seen here.

9-11.2 *Practical Spectrum Analysis*

Spectrum analyzers are often used in conjunction with Fourier analysis and transform analysis. Both of these techniques are quite complex and beyond the scope of this book. Instead, we concentrate on the practical aspects of spectrum analysis during communications equipment tests. That is, we discuss what display results from a given input signal, and how the display can be interpreted.

Unmodulated Signal Displays. If the spectrum analyzer's local oscillator sweeps through an unmodulated or continuous-wave (CW) signal slowly, the resulting response on the analyzer screen is simply a plot of the analyzer's IF amplifier passband. A pure CW signal has, by definition, energy at only one frequency, and should therefore appear as a single spike on the analyzer screen [Fig. 9–44(a)]. This occurs provided that the total sweep width or so-called "spectrum width" is wide enough compared to the IF bandwidth in the analyzer. As spectrum width is reduced, the spike response begins to spread out until the IF bandpass characteristics begin to appear, as shown in Fig. 9–44(b).

Amplitude-Modulated Signal Displays. A pure sine wave represents a single frequency. The spectrum of a pure sine wave is shown in Fig. 9–45, and is the same as the unmodulated signal display of Fig. 9–44 (a single vertical line). The height of line F_0 represents the power contained in the single frequency. Figure 9–45(b) shows the spectrum for a single sine-wave frequency F_0, amplitude-modulated by a second sine wave, F_1. In this case, two sidebands are formed, one higher than and one lower than the frequency F_0. These sidebands correspond to the sum and difference frequencies, as shown. If more than one modulating frequency is used (as is the case with most practical amplitude-modulated signals), two sidebands are added for each frequency.

Note that if the frequency, spectrum width, and vertical response of the analyzer are calibrated (as they are with any modern laboratory instrument) it is possible to find (1) the carrier frequency, (2) the modulation frequency, (3) the modulation percentage, and (4) the nonlinear modulation (if any) and incidental FM (if any).

An amplitude-modulated spectrum display can be interpreted as follows:

The *carrier frequency* is determined by the position of the center vertical line F_0 on the X axis. For example, if the total spectrum is from 100 to 200

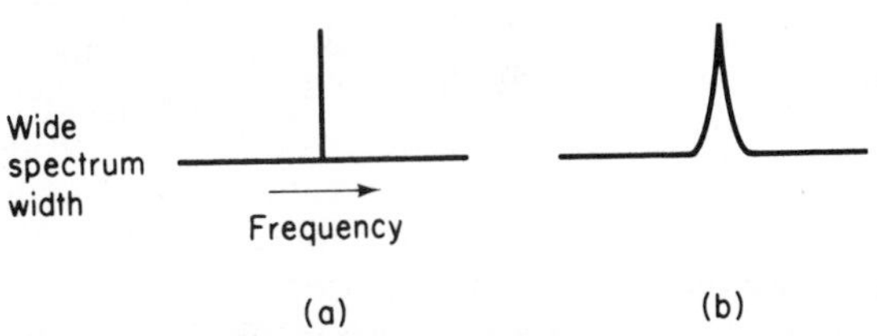

FIGURE 9–44 Unmodulation (CW) spectrum analyzer display. (a) Single spike at unmodulated signal frequency. (b) Spreading due to analyzer's IF bandpass characteristics.

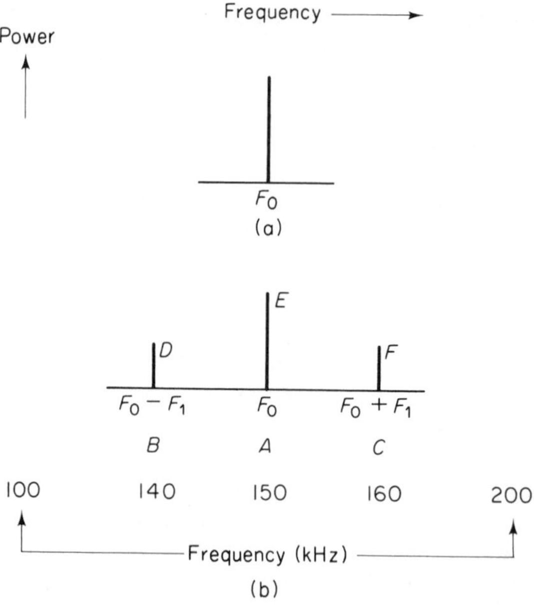

Position of A = carrier frequency (150 kHz)
Distance between B and A, or A and C = frequency modulation frequency (10 kHz)
Ratio of D to E, or F to E = one-half percentage of modulation

FIGURE 9-45 Frequency spectrum for single-tone amplitude-modulated carrier, and rules for interpretation: (a) unmodulated; (b) amplitude modulated with single frequency.

kHz and F_0 is in the center as shown in Fig. 9–45(b), the carrier frequency is 150 kHz.

The *modulation frequency* is determined by the position of the sideband lines $F_0 - F_1$ or $F_0 + F_1$ on the X axis. For example, if sideband $F_0 - F_1$ is at 140 kHz, and F_0 is at 150 kHz as shown, the modulating frequency is 10 kHz. Under these conditions, the upper sideband $F_0 + F_1$ should be 160 kHz. The distance between the carrier line F_0 and either sideband is sometimes known as the *frequency dispersion* and is equal to the modulation frequency.

The *modulation percentage* is determined by the ratio of the sideband amplitude to the carrier amplitude. The amplitude of either sideband with respect to the carrier voltage is one-half of the percentage of modulation. For example, if the carrier amplitude is 100 mV and either sideband is 50 mV, this indicates 100% modulation. If the carrier amplitude is 100 mV and either sideband is 33 mV, this indicates 66% modulation.

Nonlinear modulation is indicated when the sidebands are of unequal amplitude or are not equally spaced on both sides of the carrier frequency. Unequal amplitude indicates nonlinear modulation that results from a form of undesired frequency modulation combined with the amplitude modulation.

Incidental FM is indicated by a shift in the vertical signals along the X axis. For example, any horizontal "jitter" of the signals indicates rapid frequency modulation of the carrier.

The rules for interpreting amplitude-modulated spectrum analyzer displays are summarized in Fig. 9–45.

In practical tests, carrier signals are often amplitude-modulated at many frequencies simultaneously. This results in many sidebands (two for each modulating frequency) on the display. To resolve this complex spectrum, the operator should make sure that the analyzer bandwidth is less than the lowest modulating frequency or less than the difference between any two modulating frequencies, whichever is the smaller.

Overmodulation also produces extra sideband frequencies. The spectrum for overmodulation is very similar to multifrequency modulation. However, overmodulation is usually distinguished from multifrequency modulation by the facts that (1) the spacing between overmodulated sidebands is equal, while multifrequency sidebands may be arbitrarily spaced (unless the modulating frequencies are harmonically related); and (2) the amplitude of the overmodulated sidebands decreases progressively out from the carrier, but the amplitude of the multifrequency modulated signals is determined by the modulation percentage of each frequency and can be arbitrary.

Frequency-Modulated Signal Displays. The mathematical expression for a frequency-modulated waveform is long and complex, involving a special mathematical operator known as the *Bessel function*. However, the spectrum representation of the FM waveform is straightforward. Such a representation is shown in Fig. 9–46, which illustrates the frequency spectrum of a carrier that has been frequency-modulated by a single (1-kHz) sine wave. Figure 9-46(a) shows the unmodulated carrier spectrum waveform. Figure 9–46(b) shows the relative amplitudes of the waveform when the carrier is frequency-modulated with a deviation of 1 kHz (modulation index of 1.0). Figure 9-46(c) shows the relative amplitudes of the waveform when the carrier is frequency-modulated with a deviation of 5 kHz (modulation index of 5.0). Note that the modulation index is given by:

$$\text{modulation index} = \frac{\text{maximum frequency deviation}}{\text{modulating frequency}}$$

The term *maximum frequency deviation* is theoretical. If a CW signal F_C is frequency-modulated at a rate F_R, an infinite number of sidebands result. These sidebands are located at intervals of $F_C \pm N_F$, where $N = 1, 2, 3$, and so on.

However, as a practical matter, only the sidebands containing significant power are usually considered. For a quick approximation of the bandwidth

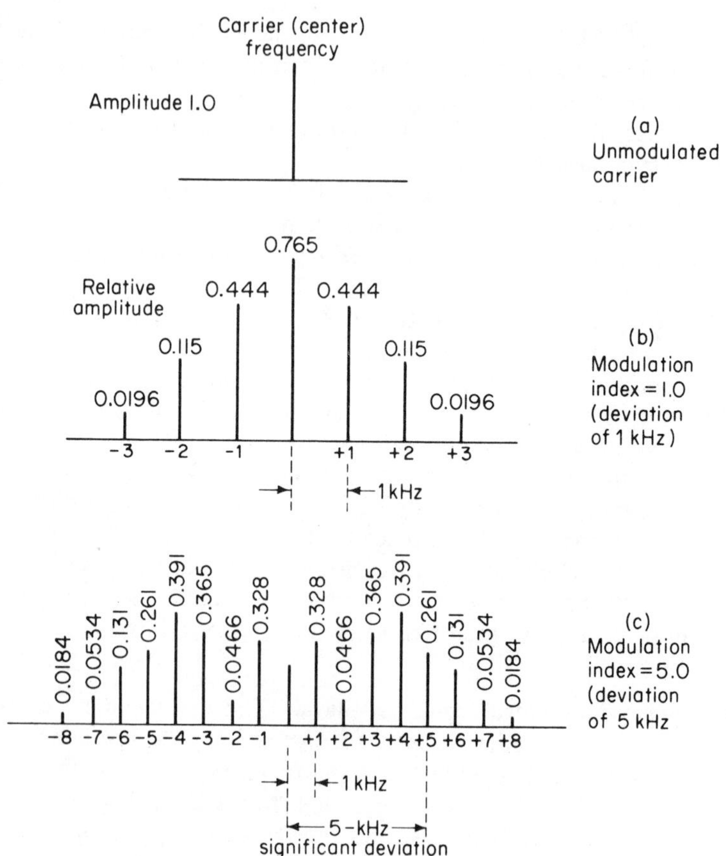

FIGURE 9-46 Frequency spectrum for single-tone (1-kHz) frequency-modulated carrier.

occupied by the significant sidebands, multiply the sum of the carrier deviation and the modulating frequency by 2:

$$\text{bandwidth} = 2\,(\text{carrier deviation} + \text{modulating frequency})$$

As a guideline, when using a spectrum analyzer to find the maximum deviation of an FM signal, locate the sideband where the amplitude begins to drop and continues to drop as the frequency moves from the center. For example, in Fig. 9-46(c), sidebands 1, 2, 3, and 4 rise and fall, but sideband 5 falls, and all sidebands after 5 continue to fall. Since each sideband is 1 kHz from the center, this indicates a practical or significant deviation of 5 kHz. (It also indicates a modulation index of 5.0.)

As in the case of amplitude modulation, the center frequency and modulating frequency can be determined by the spectrum analyzer display.

The *carrier frequency* is determined by the position of the center vertical line on the X axis. (The centerline is not always the highest amplitude, as shown in Fig. 9–46.)

The *modulating frequency* is determined by the position of the sidebands in relation to the center line or the distance between sidebands (frequency dispersion).

9-12 SPECIAL TEST SETS FOR COMMUNICATIONS EQUIPMENT

There are a number of test sets designed specifically for communications equipment service. Some of the sets are for field use, whereas others are for the bench or shop. Still other sets may be used in either the shop or field. The following paragraphs describe some of these sets. Keep in mind that these are not the only test sets available, now and in the future, but represent a cross section. Also, the most important required functions of a test set used in communications work are described. Thus, the information may be used as a guide in selecting the right test set, or combination of sets, for your particular communications equipment service needs.

9-12.1 *Power/SWR Meter*

Figure 9–47 is the schematic of a combined power/SWR test set (the Archer 21-520 by Radio Shack) suitable for both amateur radio and CB. The test set measures RF power up to 1000 W, SWR up to 1:3, and has an impedance of 50 Ω at frequencies between 2 and 175 MHz. In use, the test set is connected between the communications equipment and the antenna to monitor power output and SWR continually.

The test set is calibrated for SWR when the FWD/REF switch S_1 is set to FWD, the transmitter is keyed, and the CAL potentiometer VR_1 is adjusted until the SWR meter M_1 is at full scale. The POWER meter M_2 has one scale that indicates 10, 100, or 1000 W (full scale), depending on the position of the selector switch.

As shown in Fig. 9–47, a *directional coupler* is used for SWR measurements, as discussed in Section 9–7. For power measurement, part of the RF signal in the directional coupler is picked off by C_3, rectified by CR_3, and appears on power meter M_2. Variable resistors R_5, R_6, and R_7 permit each of the POWER meter scales to be calibrated separately. Normally, only the 10-W scale is used for CB, since an AM CB set produces only 4 W output (legally). The test will operate with SSB, but does not provide accurate power indications, since the meter is not normally adjusted to read PEP (as discussed in Section 9–5).

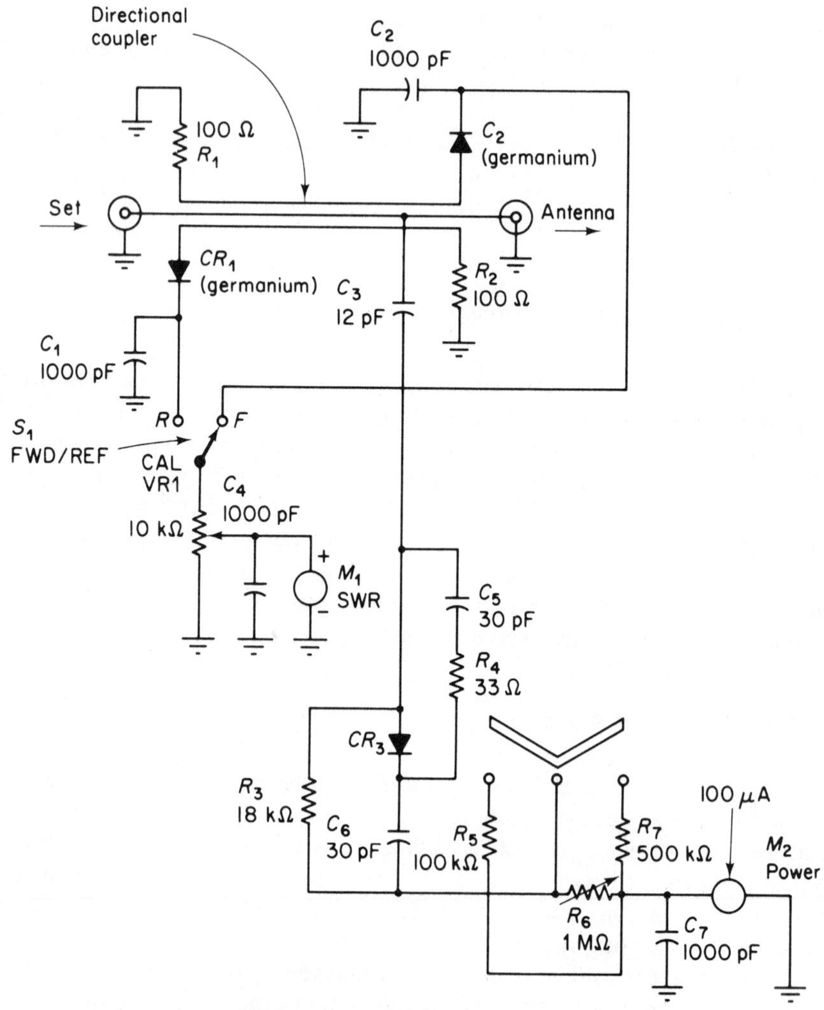

FIGURE 9-47 Power/SWR meter circuit (simplified).

9-12.2 SWR Meter/Antenna Switch

Figure 9-48 is the schematic of a combined SWR meter and antenna switch (the Archer 21-521 by Radio Shack) suitable for both amateur radio and CB. The test set operates at power outputs up to 1000 W, measures SWR up to 1:3, and has an impedance of 50 Ω at frequencies between 2 and 30 MHz. In use, the set is connected between the communications equipment and two antennas. Operation with either antenna (but not both simultaneously) is selected by the ANT 1/ANT 2 switch. The test set continuously monitors SWR of the selected antenna.

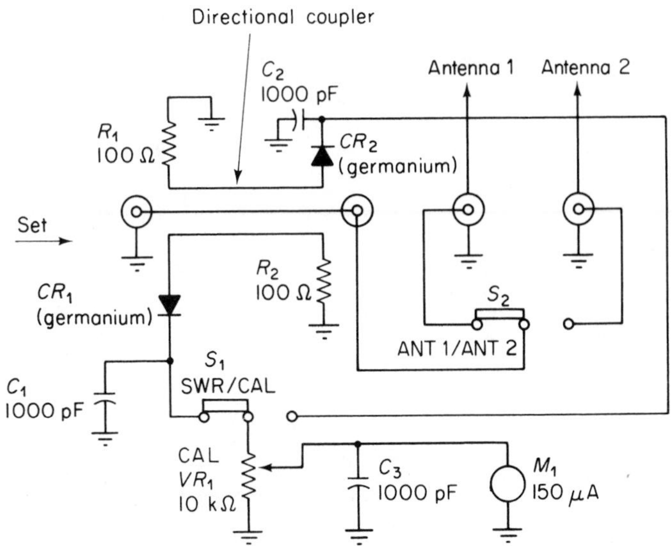

FIGURE 9-48 SWR meter/antenna switch circuit (simplified).

The set is calibrated for SWR when the SWR/CAL switch S_1 is set to CAL, the transmitter is keyed, and the CAL potentiometer VR_1 is adjusted until the SWR meter M_1 is at full scale. As shown in Fig. 9-48, a *directional coupler* is used for SWR measurements, as discussed in Section 9-7.

The desired antenna is selected by ANT 1/ANT 2 switch S_2. If one antenna is not connected, even temporarily, the manufacturer recommends that a dummy load (Section 9-4) be connected at the open antenna connection. This prevents the transmitter from operating without a load should the ANT switch be inadvertently set to the unused or open antenna connector.

9-12.3 Field Strength/SWR Meter

Figure 9-49 is the schematic of a combined field-strength and SWR meter (the Archer 21-525 by Radio Shack) suitable for both amateur radio and CB. The test set operates at power outputs up to 1000 W, measures SWR up to 1:3, has an impedance of 50 Ω at frequencies between 2 and 175 MHz, and includes a removable rod-type antenna for field-strength measurement. For SWR measurement, the test set is connected between the communications equipment and antenna, and provides continual SWR readings. For field-strength operation, the removable rod antenna is connected to the test set FS ANT jack, and the communications set is operated with its own antenna.

The set is calibrated for SWR when the FWD/REF switch S_1 is set to FWD,

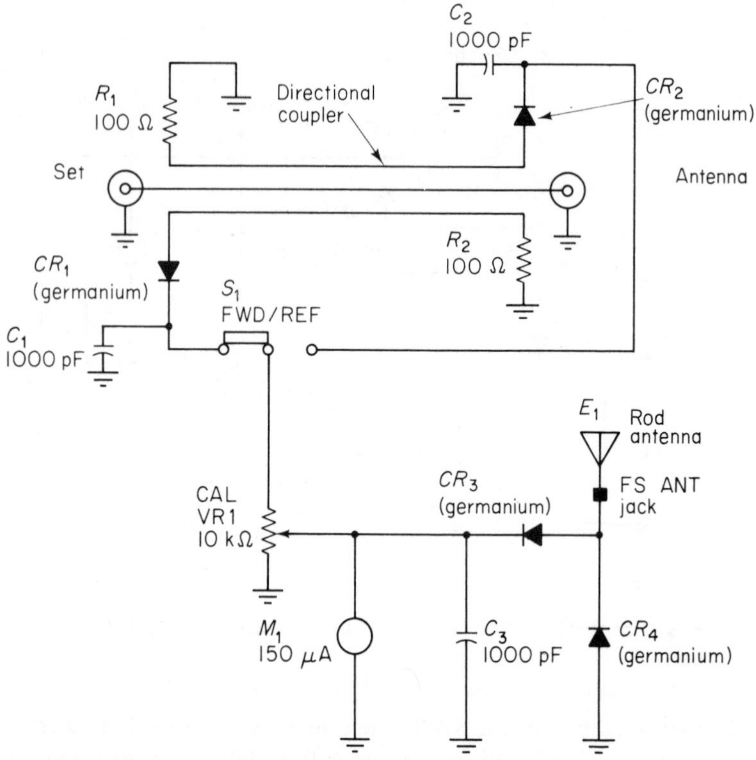

FIGURE 9-49 Field-strength/SWR meter circuit (simplified).

the transmitter is keyed, and the CAL potentiometer VR_1 is adjusted until the SWR meter M_1 is at full scale. As shown in Fig. 9-49, a *directional coupler* is used for SWR measurements, as discussed in Section 9-7.

For field-strength measurement, the transmitted signal is picked up by rod antenna E_1, rectified by CR_3 and CR_4, and appears on the meter M_1 (the same meter that is used for SWR). Keep in mind that the readout is *relative field strength* (as discussed in Section 9-6), with the meter marked off in arbitrary RFS divisions of 1 through 5.

9-12.4 CB Tester

Figure 9-50 is the schematic of a multipurpose tester (the Archer 21-526) suitable for both amateur radio and CB. The set measures power outputs up to 10 W, SWR up to 1:3, percentage of modulation between 10 and 100%, and has an impedance of 50 Ω at frequencies between 2 and 175 MHz. In use, the test set is connected between the communications equipment and antenna to continually monitor power output, SWR, and percentage of modulation, as selected by a front-panel switch.

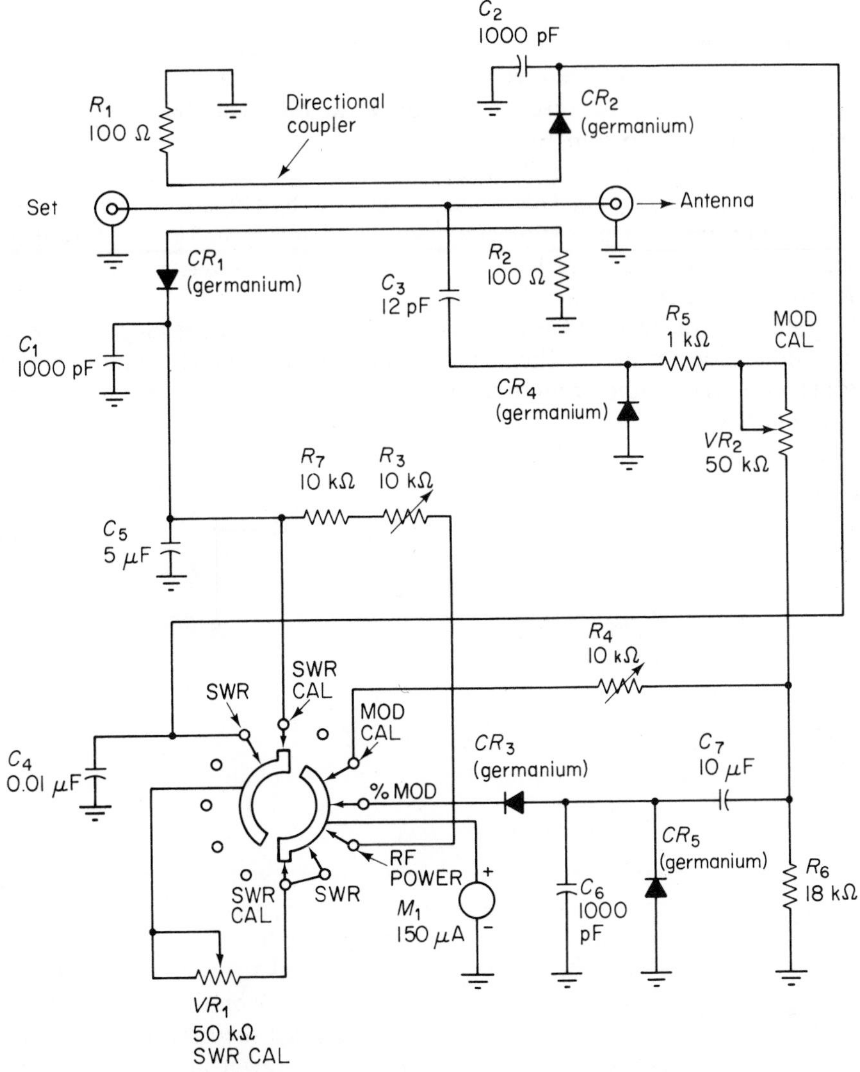

FIGURE 9-50 CB test circuit (simplified).

The same meter is used for all three functions. Operation of the meter is controlled by the selector switch as follows:

In SWR CAL, the meter M_1 is connected to read the forward voltage developed across R_2 in the directional coupler. This voltage is rectified by CR_1 and applied to meter M_1 through SWR CAL potentiometer VR_1, which is adjusted until M_1 is at full scale.

In SWR, the meter M_1 is connected to read the reflected or return voltage developed across R_1 in the directional coupler. This voltage is rectified by CR_2 and applied to meter M_1 through VR_1.

In RF POWER, the meter M_1 is connected to read the forward voltage developed across R_2 in the directional coupler. This voltage is rectified by CR_1 and applied to meter M_1 through resistors R_3 and R_7. Variable resistor R_3 permits the meter M_1 to read 10 W on full scale.

In MOD CAL, the meter M_1 is connected to read voltage tapped from the directional coupler by C_3. This voltage is rectified by CR_4 and applied to meter M_1 through MOD CAL potentiometer VR_2, which is adjusted until M_1 is at full scale with an unmodulated RF output from the transmitter. Variable resistor R_4 provides calibration for M_1 in the MOD CAL position.

In MOD, the meter M_1 is again connected to read voltage tapped from the directional coupler by C_3. The meter M_1 reads modulated RF output from the transmitter. The audio or modulation portion of the RF carrier is rectified by CR_3 and CR_4.

Index

A

Admittance:
  FET, 53
  Transistor, 29
Alpha, 8
Amplification factor, FET, 53
Antenna:
  measurements, 265
  switcher, 290
Audio circuit, 159

B

Background noise, audio, 172
Bandwidth, audio, 173
Bessel function, 287
Beta, 8
  AC with curve tracer, 21
  DC with curve tracer, 21
Blocking voltage, SCR, 145
  with curve tracer, 155
Breakdown:
  FET, 42
  FET (dual gate), 47
  JFET, 44
  MOSFET, 44
  transistor, 5
  transistor with curve tracer, 27

C

Capacitance, FET elements, 54
CB tester, 292
Channel resistance, FET, 58
Coil (RF) inductance, 206
Communications:
  equipment, 229
  test sets, 289
Conduction angle, SCR, 139

Continuity, diode, 102
Control rectifier, 125
  with curver tracer, 154
Counter, digital, 245
Cross modulation, FET, 65
Current, dual-gate FET, 48
Curve tracer tests:
  diode, 119–124
  FET, 67–73
  SCR, 154
  transistor, 18–33
  tunnel diode, 122
  UJT, 98–100
  Zener, 122

D

Demodulator probes, 202, 241
Diac, 131
  with curve tracer, 157
Digital counter, 245
Diode, 101
  with curve tracer, 119–124
Dip meters, 262
Distortion:
  audio, 166
  harmonic, 168
  intermodulation, 170
  transistor, 24
Distributed capacitance (of coil), 208
Dual-gate FET, 46
Dummy loads, 255
Dynamic tests:
  diode, 105
  FET, 48

F

Feedback, audio, 173
FET, 34
  control voltage, 39
  curve tracer, 67–73
  dynamic, 48
  operating current, 41
  operating modes, 34
  operating voltage, 40
Field strength meters, 259, 291
FM modulation tests, 287
Forward transadmittance, 49
Forward voltage:
  diode drop, 105
  SCR, 153
  SCR with curve tracer, 156
Frequency meters, 245
Frequency response, 159

G

Gain:
  audio, 163
  FET, 60
  FET with curve tracer, 70
  transistor, 6
  transistor, RF, 11
  transistor with curve tracer, 21
  transistor with ohmmeter, 10

H

Harmonic distortion, audio, 168
Heterodyne frequency meter, 245
High voltage probes, 240
Holding characteristics, SCR, 151
  with curve tracer, 156
Hybrid (h) transistor characteristics, 7

I

Impedance
  audio, 165

resonant circuits, 212
transistor, 29
In-circuit tests, transistor, 17
Inductance, coil, 206
Intermodulation, FET, 67
Intermodulation distortion, audio, 170
Internal resistance, 191

J

JFET breakdown, 44

L

Latching, SCR, 151
LC circuits, resonance, 203
Leakage:
FET, 45
SCR current, 145
transistor, 2
Linearity, transistor, 24
Load sensitivity, audio, 164
Low capacitance probe, 240

M

Modulation, 230–236
MOSFET:
breakdown, 44
handling, 36
protection, 37

N

Noise:
audio, 172
figure, 63

O

Oscillator circuits, 277
Oscilloscopes in communications tests, 229
Output, power supply, 189

P

Pinch off, FET, 70
PNPN switch, 128
Power, audio, 163
Power supply circuit, 189
Power/SWR meters, 289
Probes:
demodulator, 202, 241
high voltage, 240
RF, 201
Pulse definitions, 12
PUT, 74

Q

Q of resonant circuits, 209

R

Rate of rise, SCR, 140
Receiver circuits, 215–228
Rectifier tests, 106
Regulation, power supply, 189
Resonant circuit Q, 209
Resonant LC circuit, 203
Reverse leakage, diode, 104
Reverse transadmittance, FET, 52
RF circuits, 200
gain in transistors, 11
probes, 241
Ripple, power supply, 192

S

Saturation, transistor, 28
SBS, 132
SCR, 125
SCR with curve tracer, 154
SCS, 128
Self resonance, coil, 208
Signal tracing:
  audio, 166
  oscilloscope, 230
  probes, 241
Sinewave analysis, audio, 166
Small signal diode, 108
Spectrum analyzer, 282
Squarewave definitions, 12
SSB modulation, 236
SUS, 132
Switching:
  diode, 109
  FET, 59
  transistor, 12
SWR:
  measurement, 259
  meters, 289

T

Temperature:
  transistor, 31
Tester, transistor, 9
Thyristor, 125
  with curve tracer, 154
Transadmittance, FET, 52
Transconductance, FET, 49
  with curve tracer, 70
Transformer characteristics, 194
Transistors:
  curve tracer tests, 18–33
  gain, 6
  gain with ohmmeter, 10
  hybrid (h) characteristics, 7
  in-circuit tests, 16
  ohmmeter tests, 4
  switching tests, 12
Transmission line measurements, 267
Transmitter circuits, 214
Triac, 131
  with curve tracer, 157
Trigger, SCR, 147
  with curve tracer, 156
Tunnel diode, 115
  with curve tracer, 122
Turn-on/Turn-off, SCR, 142
Two-junction transistor tests, 1

U

UJT, 14
  tester, 91
  with curve tracer, 98

V

Voltage, RF, 201
Voltage gain, 163

W

Wattmeters, RF, 257
WWV, 252
y-parameters, 49

Z

Zener, 111
  with curve tracer, 122
Zero beat frequency meter, 245
Zero temperature coefficient, FET, 71